2011 中国环境统计年报

ANNUAL STATISTIC REPORT ON ENVIRONMENT IN CHINA

中国环境统计年报

中华人民共和国环境保护部 编

MINISTRY OF ENVIRONMENTAL PROTECTION
OF THE PEOPLE'S REPUBLIC OF CHINA

U0351555

中国环境出版社·北京

图书在版编目（CIP）数据

中国环境统计年报．2011/中国环境监测总站编
．—北京：中国环境科学出版社，2012.12
ISBN 978-7-5111-1253-8

Ⅰ．①中… Ⅱ．①中… Ⅲ．①环境统计—统计资料—
中国—2011—年报 Ⅳ．①X508.2-54

中国版本图书馆 CIP 数据核字（2012）第 311316 号

责任编辑 殷玉婷
责任校对 扣志红
封面设计 玄石至上

出版发行 中国环境科学出版社
　　　　　（100062 北京市东城区广渠门内大街 16 号）
　　　　　网　　址：http://www.cesp.com.cn
　　　　　联系电话：010-67112765（编辑管理部）
　　　　　　　　　　010-67187041（学术著作图书出版中心）
　　　　　发行热线：010-67125803，010-67113405（传真）
印　　刷 北京中科印刷有限公司
经　　销 各地新华书店
版　　次 2012 年 12 月第 1 版
印　　次 2012 年 12 月第 1 次印刷
开　　本 889×1194　1/16
印　　张 19.75
字　　数 450 千字
定　　价 120.00 元

《中国环境统计年报·2011》编委会

组织编写　环境保护部污染物排放总量控制司

　　　　　中　国　环　境　监　测　总　站

资料提供　各省、自治区、直辖市环境保护厅（局）

批　　准　中　华　人　民　共　和　国　环　境　保　护　部

目录

2 各地区环境统计　　　　　　　　　　　39 ~ 90

3 重点城市环境统计　　　　　　　　　91 ~ 188

4 各工业行业环境统计 189 ~ 204

5 流域及入海陆源废水排放统计 205 ~ 256

全国环境统计概要
QUANGUO HUANJING TONGJI GAIYAO

ANNUAL STATISTIC REPORT ON ENVIRONMENT IN CHINA

2011

综述

2011年，国务院印发了《关于加强环境保护重点工作的意见》和《国家环境保护"十二五"规划》，召开了第七次全国环境保护大会，进一步明确了"十二五"环境保护目标任务、重点工作及政策措施。全国环保系统坚决贯彻中央关于环境保护的决策部署，以建设生态文明、推进环境保护历史性转变、积极探索环境保护新道路为指导思想，着力解决影响科学发展和损害群众健康的突出环境问题，推动我国生态环境保护领域从认识到实践发生重要变化，环境保护各项工作取得积极进展，重点流域区域污染防治不断深化，主要污染物总量减排扎实推进。与2010年相比，化学需氧量排放量下降2.04%，氨氮排放量下降1.55%，二氧化硫排放量下降2.20%，氮氧化物由于增量大、减排工程多处于启动建设阶段、完成项目少、效益尚未显现，排放量比2010年上升5.74%。

2011年全国废水排放总量659.2亿吨。其中，工业废水排放量230.9亿吨，城镇生活污水排放量427.9亿吨。废水中化学需氧量（COD）排放量2 499.9万吨，其中工业源化学需氧量排放量为354.8万吨，农业源化学需氧量排放量为1 186.1万吨，城镇生活化学需氧量排放量为938.8万吨。废水中氨氮排放量260.4万吨，其中工业源氨氮排放量为28.1万吨，农业源氨氮排放量为82.7万吨，城镇生活氨氮排放量为147.7万吨。

全国废气中二氧化硫（SO_2）排放量2 217.9万吨。其中，工业二氧化硫排放量为2 017.2万吨，城镇生活二氧化硫排放量为200.4万吨。全国废气中氮氧化物（NO_x）排放量2 404.3万吨。其中，工业氮氧化物排放量为1 729.7万吨，城镇生活氮氧化物排放量为36.6万吨，机动车氮氧化物排放量为637.6万吨。全国废气中烟（粉）尘排放量1 278.8万吨。其中，工业烟（粉）尘排放量为1 100.9万吨，城镇生活烟（粉）尘排放量为114.8万吨，机动车烟（粉）尘排放量为62.9万吨。

全国一般工业固体废物产生量32.3亿吨，综合利用量19.5亿吨，贮存量6.0亿吨，处置量7.0亿吨，倾倒丢弃量433万吨，全国一般工业固体废物综合利用率为59.9%。全国工业危险废物产生量3 431.2万吨，综合利用量1 773.1万吨，贮存量823.5万吨，处置量916.5万吨，倾倒丢弃量0.01万吨，全国工业危险废物综合利用处置率为76.5%。

全国共调查统计了城镇污水处理厂3 974座，设计处理能力达到1.4亿吨/日；生活垃圾处理厂（场）2 039座，填埋设计容量达35.9亿米^3，堆肥设计处理能力达到2.1万吨/日，焚烧设计处理能力达到2.2万吨/日；危险废物集中处理（置）厂（场）644座，医疗废物集中处理（置）厂（场）260座，危险废物设计处置能力达到35.3万吨/日。

1.1 统计调查企业基本情况

1.1.1 工业企业基本情况

（1）工业企业污染物治理情况

2011年，发表调查了153 027家工业企业，对其他工业企业的污染排放量按比率进行估算。

发表调查的企业中，共有91 506套废水治理设施，形成了31 406万吨/日的废水处理能力，投入运行费732.1亿元。共处理580.6亿吨工业废水，去除化学需氧量3 092.7万吨，氨氮145.3万吨，石油类35.5万吨，挥发酚6.8万吨，氰化物0.6万吨。

企业在用的工业锅炉和炉窑数分别为10.4万台和10.2万台，共安装21.6万套废气治理设施（其中，脱硫设施20 539套，脱硝设施497套，除尘设施169 788套），形成了156.9亿米^3/时的废气处理能力，投入运行费1 579.5亿元。共去除二氧化硫3 967.5万吨，氮氧化物86.3万吨，烟（粉）尘80 790.1万吨。

（2）工业企业污染治理投资情况

2011年，调查工业企业中，共实施工业污染治理项目9 257个，其中废水治理项目3 738个，废气治理项目3 478个，工业固体废物治理项目577个。当年竣工的项目数7 005个，其中废水治理项目2 633个，废气治理项目2 787个，工业固体废物治理项目420个。

工业污染治理项目本年完成投资总额为444.4亿元，其中，工业废水治理项目完成投资157.7亿元，新增设计处理能力1 494万吨/日。工业废气治理项目完成投资211.7亿元，新增设计处理能力52 513万米^3（标态）/时。工业固体废物治理项目完成投资31.4亿元，新增设计处理能力496万吨/日。

1.1.2 农业源调查基本情况

2011年，发表调查了133 277家规模化畜禽养殖场，9 749家规模化畜禽养殖小区，对种植业、水产养殖业和其他养殖专业户按产排污强度等进行了核算。

1.2 废水

1.2.1 废水及主要污染物排放情况

（1）废水排放情况

2011年，全国废水排放量659.2亿吨。其中，工业废水排放量230.9亿吨，占废水排放总量的35.0%。生活污水排放量427.9亿吨，占废水排放总量的64.9%。集中式污染治理设施废水（不含城镇污水处理厂，下同）排放量0.4亿吨，占废水排放总量的0.1%。

表1　全国废水及其主要污染物排放情况

排放量 \ 排放源	合计	工业源	农业源	城镇生活源	集中式污染治理设施
废水/亿吨	659.2	230.9	—	427.9	0.4
化学需氧量/万吨	2499.9	354.8	1 186.1	938.8	20.1
氨氮/万吨	260.4	28.1	82.7	147.7	2.0

注：1. 自2011年起环境统计中增加农业源的污染排放统计。农业源包括种植业、水产养殖业和畜禽养殖业排放的污染物。
　　2. 集中式污染治理设施排放量指生活垃圾处理厂（场）和危险废物（医疗废物）集中处理（置）厂垃圾渗滤液/废水及其污染物的排放量。

（2）化学需氧量排放情况

2011年，全国废水中化学需氧量排放量2 499.9万吨。其中工业废水中化学需氧量排放量354.8万吨，占化学需氧量排放总量的14.2%。农业源排放化学需氧量1 186.1万吨，占化学需氧量排放总量的47.4%。城镇生活污水中化学需氧量排放量938.8万吨，占化学需氧量排放总量的37.6%。集中式污染治理设施废水中化学需氧量排放量20.1万吨，占化学需氧量排放总量的0.8%。

（3）氨氮排放情况

2011年，全国废水中氨氮排放量260.4万吨。其中工业废水氨氮排放量28.1万吨，占氨氮排放总量的10.8%。农业源氨氮排放量82.7万吨，占氨氮排放总量的31.8%。城镇生活污水中氨氮排放量147.7万吨，占氨氮排放总量的56.7%。集中式污染治理设施废水中氨氮排放量2.0万吨，占氨氮排放总量的0.8%。

（4）废水中其他主要污染物排放情况

2011年，全国废水中石油类排放量2.1万吨，挥发酚排放量2 430.6吨，氰化物排放量 217.9 吨。废水中重金属铅、镉、汞、总铬及砷排放量分别为 155.2 吨、35.9 吨、1.4 吨、293.2 吨和 146.6 吨。

表 2　全国废水中重金属及其他污染物排放量　　　　　　　　　单位：吨

排放源＼污染物	石油类	挥发酚	氰化物	砷	铅	镉	汞	总铬
工业源	20 589.1	2 410.5	215.4	145.2	150.8	35.1	1.2	290.3
集中式	423.0	20.0	2.5	1.4	4.4	0.8	0.2	2.9

1.2.2 各地区废水及主要污染物排放情况

（1）各地区废水排放情况

2011 年，废水排放量大于 30 亿吨的省份依次为广东、江苏、山东、浙江、河南、福建，6 个省份废水排放总量为 293.7 亿吨，占全国废水排放量的 44.6%。工业废水排放量最大的是江苏，城镇生活污水排放量最大的是广东，集中式污染治理设施渗滤液/废水排放量最大的是浙江。

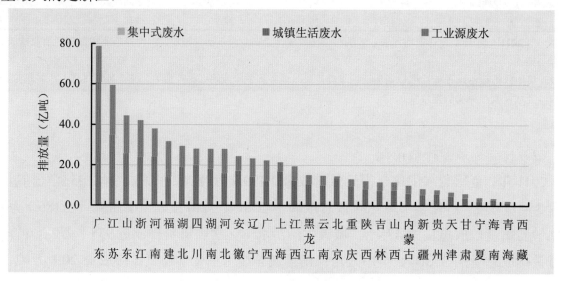

图 1　各地区废水排放情况

（2）各地区化学需氧量排放情况

化学需氧量排放量大于 100 万吨的省份依次为山东、广东、黑龙江、河南、河北、辽宁、湖南、四川、江苏和湖北，10 个省份的化学需氧量排放量为 1 457.1 万吨，占全国化学需氧量排放量的 58.3%。工业化学需氧量排放量最大的是广东，农业化学需氧量排放量最大的是山东，城镇生活化学需氧量排放量最大的是广东，集中式污染治理设施化学需氧量排放量最大的是湖北。

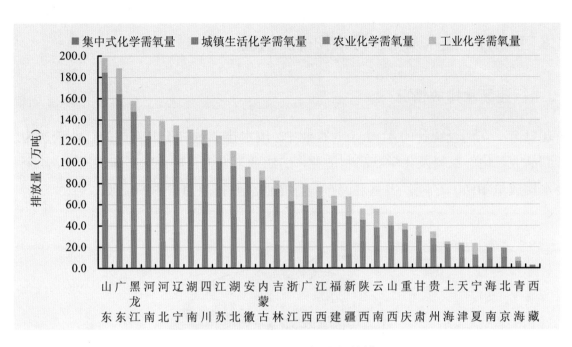

图2　各地区化学需氧量排放情况

（3）各地区氨氮排放情况

氨氮排放量大于10万吨的省份依次为广东、山东、湖南、江苏、河南、四川、湖北、浙江、河北、辽宁和安徽，11个省份的氨氮排放量为160.5万吨，占全国氨氮排放量的61.6%。工业氨氮排放量最大的是湖南，农业氨氮排放量最大的是山东，城镇生活氨氮排放量最大的是广东，集中式污染治理设施化学需氧量排放量最大的是湖北。

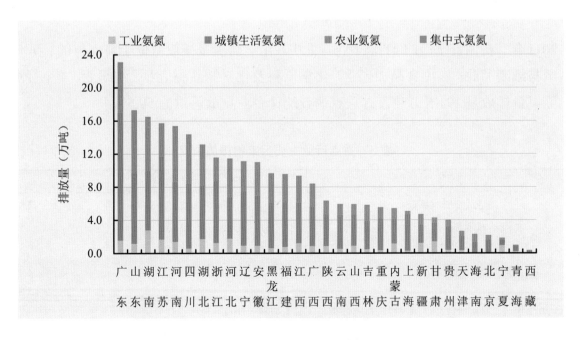

图3　各地区氨氮排放情况

1.2.3 工业行业废水及主要污染物排放情况

1.2.3.1 行业废水排放情况

2011 年，在调查统计的 41 个工业行业中，废水排放量位于前 4 位的行业依次为造纸与纸制品业，化学原料及化学制品制造业，纺织业，电力、热力生产和供应业，4 个行业的废水排放量 107.0 亿吨，占重点调查统计企业废水排放总量的 50.3%。

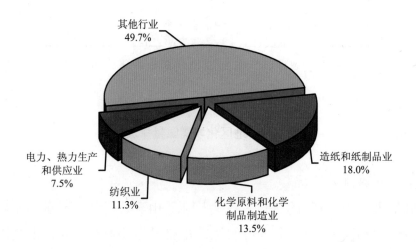

图 4　重点行业废水排放情况

1.2.3.2 行业化学需氧量排放情况

2011 年，在调查统计的 41 个工业行业中，化学需氧量排放量位于前 4 位的行业依次为造纸与纸制品业，农副食品加工业，化学原料及化学制品制造业，纺织业，4 个行业的化学需氧量排放量 191.5 万吨，占重点调查统计企业排放总量的 59.5%。

表 3　重点行业主要污染物排放情况　　　　　　　　　　　　　　　　单位：万吨

行业	化学需氧量	氨氮
造纸及纸制品业	74.2	2.5
农副食品加工业	55.3	2.1
化学原料及化学制品制造业	32.8	9.3
纺织业	29.2	2.0
累计	191.5	15.9

注：自 2011 年起，环境统计按《国民经济行业分类》（GB/T 4754—2011）标准执行分类统计。

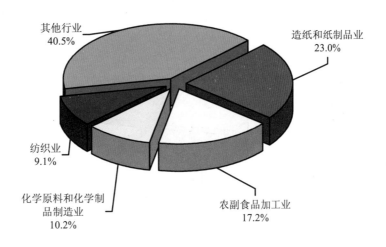

图 5　重点行业化学需氧量排放情况

1.2.3.3 行业氨氮排放情况

2011 年，在调查统计的 41 个工业行业中，氨氮排放量位于前 4 位的行业依次为化学原料及化学制品制造业，造纸与纸制品业，农副食品加工业，纺织业，4 个行业的氨氮排放量 15.9 万吨，占重点调查统计企业排放总量的 60.6%。

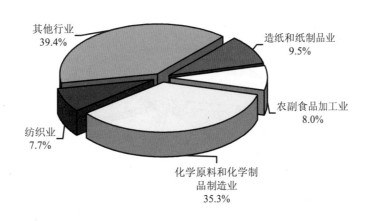

图 6　工业行业氨氮排放情况

1.2.3.4 行业重金属等污染物排放情况

2011 年，重金属（铅、镉、汞、总铬、砷）排放量位于前 4 位的行业依次为有色金属冶炼和压延加工业，皮革、毛皮、羽毛及其制品，金属制品业和制鞋业。4 个行业重金属排放量为 489.6 吨，占重点调查统计企业排放量的 78.6%。

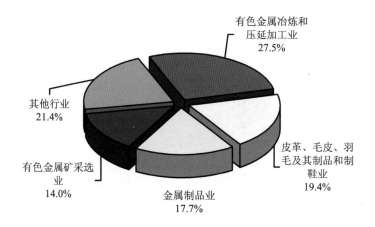

图 7　工业行业重金属排放情况

1.2.3.5 行业石油类排放情况

2011 年，石油类排放量位于前 4 位的行业依次为煤炭开采和洗选业，黑色金属冶炼和压延加工业，石油加工、炼焦和核燃料加工业，化学原料和化学制品制造业，4 个行业石油类排放量为 11 774.2 吨，占重点调查统计企业石油类排放量的 57.2%。

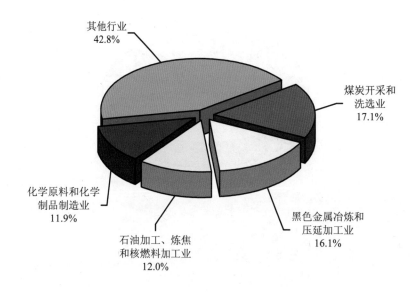

图 8　工业行业石油类污染物排放情况

1.2.4 十大流域废水及主要污染物排放与治理情况

1.2.4.1 废水及主要污染物排放情况

（1）废水

2011 年，松花江、辽河、海河、黄河、淮河、长江、珠江、东南诸河、西南诸河和西北诸河十大流域废水排放量分别为 25.9 亿吨、27.9 亿吨、67.1 亿吨、40.9 亿吨、90.3 亿吨、213.9 亿吨、110.9 亿吨、64.1 亿吨、6.1 亿吨和 12.2 亿吨。

其中，工业废水排放量、城镇生活污水量、集中式污染治理设施渗滤液/废水排放量居第一位的均为长江流域，分别占十大流域片工业废水排放量、城镇生活污水量、集中式污染治理设施渗滤液/废水排放量的 30.7%、33.4% 和 45.5%。

表4　十大流域废水排放情况　　　　　　　　　　　　单位：亿吨

排放源＼流域	松花江	辽河	海河	黄河	淮河	长江	珠江	东南诸河	西南诸河	西北诸河
工业	8.1	10.8	22.8	14.8	33.1	70.8	31.1	31.6	2.8	5.0
生活	17.8	17.1	44.3	26.1	57.2	142.9	79.7	32.4	3.2	7.2
集中式	0.007	0.011	0.035	0.011	0.033	0.183	0.055	0.059	0.004	0.004
总计	25.9	27.9	67.1	40.9	90.3	213.9	110.9	64.1	6.1	12.2

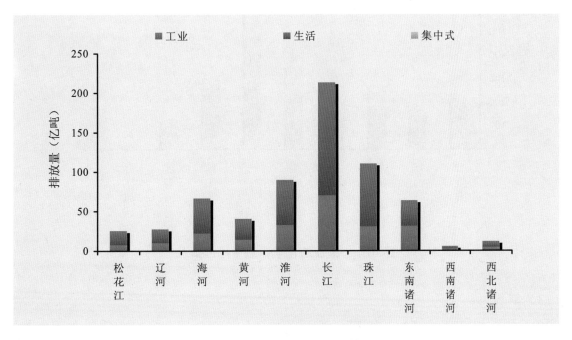

图9　十大流域废水排放情况

（2）化学需氧量

2011 年，松花江、辽河、海河、黄河、淮河、长江、珠江、东南诸河、西南诸河和西北诸河十大流域废水化学需氧量排放量分别为 242.6 万吨、176.8 万吨、274.9 万吨、197.8 万吨、358.1 万吨、673.9 万吨、317.2 万吨、130.2 万吨、33.1 万吨和 95.2 万吨。

其中，长江流域工业废水化学需氧量、农业源化学需氧量、城镇生活污水化学需氧量、集中式污染治理设施渗滤液/废水化学需氧量排放量均居十大流域第一位，分别占十大流域工业废水化学需氧量、农业源化学需氧量、城镇生活污水化学需氧量、集中式污染治理设施渗滤液/废水化学需氧量排放总量的 26.4%、21.0%、34.4%和 41.3%。

表5　十大流域化学需氧量排放情况　　　　　　　单位：万吨

排放源＼流域	松花江	辽河	海河	黄河	淮河	长江	珠江	东南诸河	西南诸河	西北诸河
工业	17.9	15.9	33.4	40.2	36.1	93.7	51.9	25.1	13.4	27.3
农业	167.7	118.1	175.2	88.7	198.6	249	102.4	34.3	3.2	48.8
生活	55.7	41.8	64.5	67.8	121.4	322.9	160.7	69.8	15.8	18.5
集中式	1.27	1.09	1.78	1.07	1.97	8.3	2.24	1.08	0.69	0.62
总计	242.6	176.8	274.9	197.8	358.1	673.9	317.2	130.2	33.1	95.2

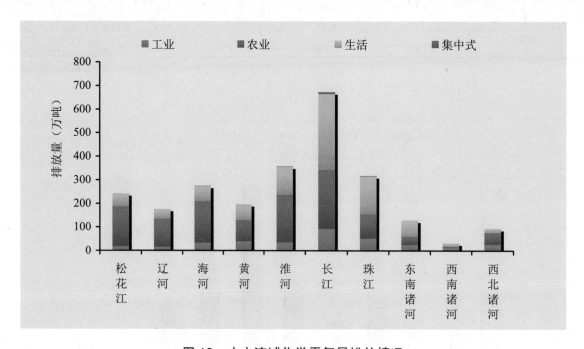

图10　十大流域化学需氧量排放情况

（3）氨氮

2011 年，松花江、辽河、海河、黄河、淮河、长江、珠江、东南诸河、西南诸河和西北诸河十大流域氨氮排放量分别为 15.1 万吨、14.0 万吨、24.2 万吨、19.3 万吨、37.6

万吨、85.0 万吨、37.6 万吨、18.1 万吨、2.7 万吨和 6.8 万吨。

其中，长江流域工业废水、农业源、城镇生活污水、集中式污染治理设施渗滤液/废水氨氮排放量均居十大流域第一位，分别占十大流域工业废水、农业源、城镇生活污水、集中式污染治理设施渗滤液/废水氨氮排放总量的 31.0%、32.9%、32.7%和 39.0%。

表 6　十大流域氨氮排放情况　　　　　　　　　　单位：万吨

排放源 \ 流域	松花江	辽河	海河	黄河	淮河	长江	珠江	东南诸河	西南诸河	西北诸河
工业	1.1	1.3	3.0	3.7	3.1	8.7	3.1	1.7	0.2	2.2
农业	5.2	4.5	9.2	4.0	14.8	27.2	10.7	5.0	0.5	1.5
生活	8.6	8.1	11.9	11.5	19.4	48.3	23.6	11.3	2.0	3.0
集中式	0.14	0.14	0.15	0.13	0.21	0.78	0.23	0.09	0.09	0.04
总计	15.1	14.0	24.2	19.3	37.6	85.0	37.6	18.1	2.7	6.8

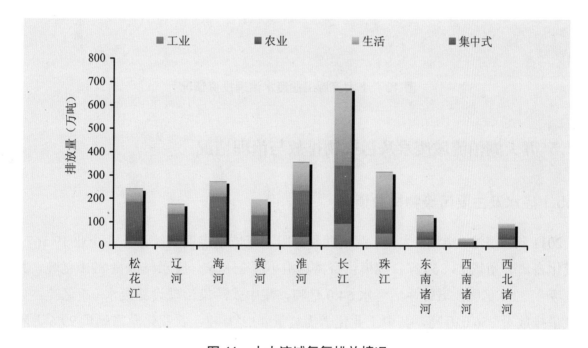

图 11　十大流域氨氮排放情况

1.2.4.2 废水及主要污染物治理与投资情况

2011 年，十大流域共有 91 506 套废水治理设施，形成了 31 406 万吨/日的废水处理能力，投入运行费 732.1 亿元。

十大流域共实施工业废水治理项目 3 738 个，竣工 2 633 个，工业废水治理项目完成投资 157.7 亿元，工业废水治理竣工项目新增设计处理能力 1 493.8 万吨/日。

十大流域纳入统计的污水处理厂共 3 974 座，共形成 13 990.9 万吨/日的处理能力，处理生活污水 349.7 亿吨。共去除化学需氧量 1 014.9 万吨，氨氮 85.8 万吨，油类 5.5 万吨，总氮 67.4 万吨，总磷 15.0 万吨。

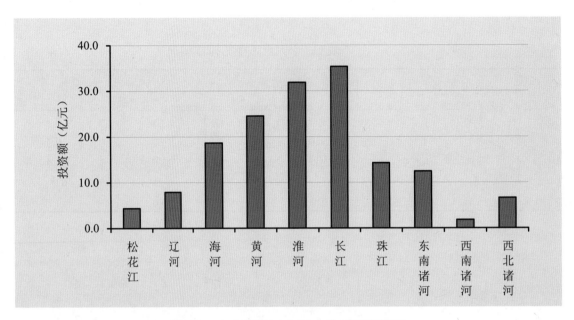

图 12　十大流域工业废水治理投资情况

1.2.5　五大湖泊流域废水及污染物排放与治理情况

1.2.5.1　废水及主要污染物排放情况

2011 年，滇池、巢湖、太湖、洞庭湖和鄱阳湖流域重点调查统计工业企业 19 567 家，规模化畜禽养殖场 23 865 家，规模化畜禽养殖小区 258 家。废水排放量 87.1 亿吨，其中工业废水 33.0 亿吨，城镇生活污水 54.0 亿吨，集中式污染治理设施废水 0.1 亿吨。化学需氧量排放量 256.0 万吨，其中工业化学需氧量 40.3 万吨，农业化学需氧量 99.6 万吨，城镇生活化学需氧量 113.0 万吨，集中式污染治理设施化学需氧量 3.1 万吨。氨氮排放量 32.9 万吨，其中工业氨氮 4.7 万吨，农业氨氮 11.3 万吨，城镇生活氨氮 16.6 万吨，集中式污染治理设施氨氮 0.3 万吨。

表 7　五大湖泊流域废水及主要污染物排放情况

污染物	废水/亿吨			化学需氧量/万吨				氨氮/万吨			
	工业	生活	集中式	工业	农业	生活	集中式	工业	农业	生活	集中式
排放量	33.1	54.0	0.1	40.3	99.6	113.0	3.1	4.7	11.3	16.6	0.3

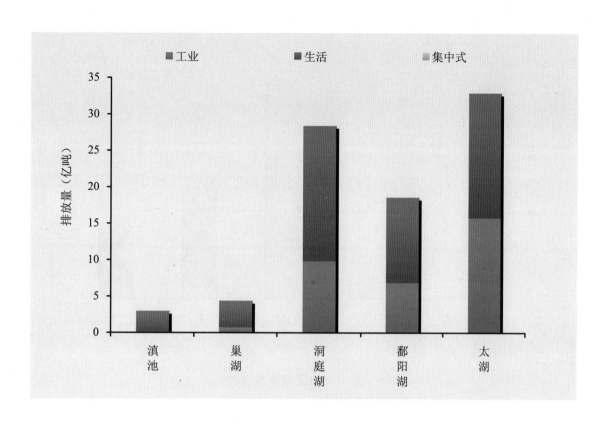

图 13　五大湖泊流域废水排放情况

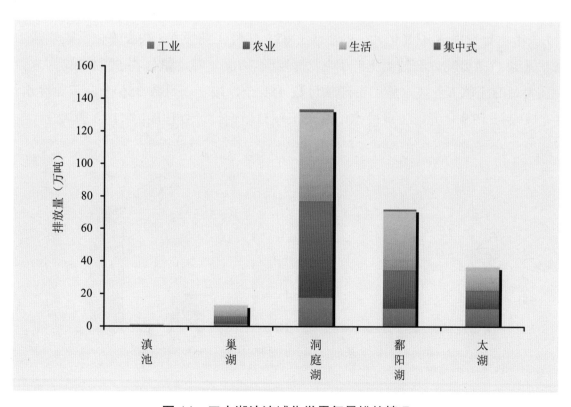

图 14　五大湖泊流域化学需氧量排放情况

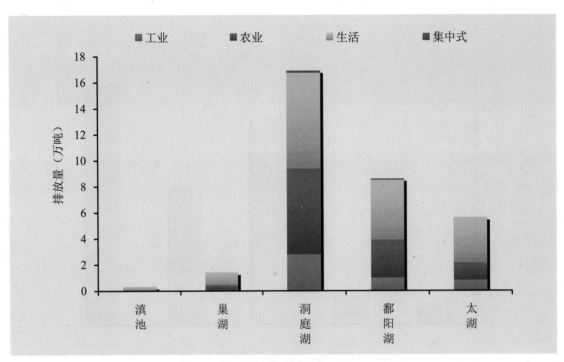

图 15   五大湖泊流域氨氮排放情况

1.2.5.2 废水及主要污染物治理与投资情况

2011 年，五大湖泊流域共有废水治理设施 15 682 套，形成了 3 666 万吨/日的废水处理能力，年运行费用达 93.9 亿元，共处理了 62.7 亿吨工业废水，共去除化学需氧量 198.4 万吨、氨氮 11.7 万吨、石油类 1.5 万吨、挥发酚 3 386.1 吨、氰化物 509.3 吨。

五大湖泊流域工业废水施工治理项目数 451 个，竣工项目数 356 个，工业废水治理项目完成投资 15.9 亿元，工业废水治理竣工项目新增设计处理能力 114.1 万吨/日。

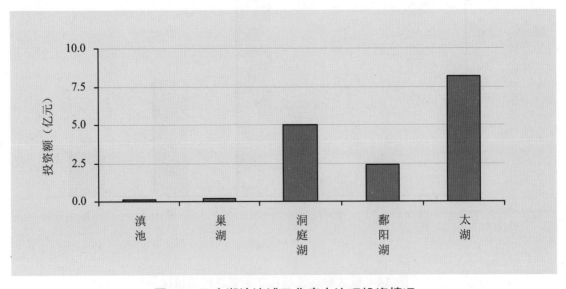

图 16   五大湖泊流域工业废水治理投资情况

五大湖泊流域纳入统计的污水处理厂共 580 座，共形成 1 745 万吨/日的处理能力，处理生活污水 42.4 亿吨。共去除化学需氧量 114.4 万吨、氨氮 9.0 万吨、油类 0.5 万吨、总氮 6.4 万吨、总磷 1.2 万吨。

1.2.6 三峡库区及其上游流域废水和主要污染物排放情况

（1）废水及污染物排放情况

2011 年，重点调查了三峡库区及其上游流域（含库区、影响区及上游区共 319 个区县）工业企业 14 502 家，规模化畜禽养殖场 10 541 家，规模化畜禽养殖小区 848 家。

三峡库区及其上游流域共排放废水 52.8 亿吨。其中，工业废水 14.4 亿吨，城镇生活污水 38.4 亿吨，集中式污染治理设施废水 0.04 亿吨。

三峡库区及其上游流域化学需氧量排放量为 208.7 万吨。其中，工业化学需氧量为 25.0 万吨，农业化学需氧量 75.0 万吨，城镇生活化学需氧量 106.3 万吨，集中式污染治理设施化学需氧量 2.3 万吨。

三峡库区及其上游流域氨氮排放量为 24.6 万吨。其中，工业氨氮为 1.2 万吨，农业氨氮 8.3 万吨，城镇生活氨氮 14.9 万吨，集中式污染治理设施氨氮 0.2 万吨。

表 8　三峡库区及其上游流域主要污染物排放情况

污染物	废水/亿吨			化学需氧量/万吨				氨氮/万吨			
	工业	生活	集中式	工业	农业	生活	集中式	工业	农业	生活	集中式
排放量	14.4	38.4	0.04	25.0	75.0	106.3	2.3	1.2	8.3	14.9	0.2

三峡库区及其上游流域排放废水量、化学需氧量排放量、氨氮排放量最大的均是四川。

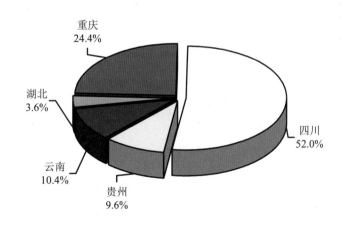

图 17　三峡库区及其上游流域废水排放区域构成

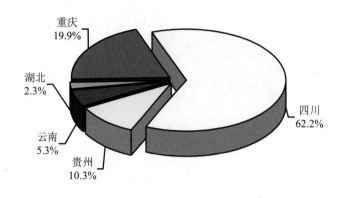

图 18 三峡库区及其上游流域化学需氧量排放区域构成

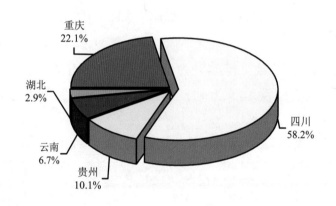

图 19 三峡库区及其上游流域氨氮排放区域构成

（2）废水及污染物治理与投资情况

2011 年，三峡库区及其上游流域共有废水治理设施 7 297 套，形成了 1 882 万吨/日的废水处理能力，年运行费用达 76.9 亿元，共处理了 40.2 亿吨工业废水，共去除化学需氧量 111.0 万吨、氨氮 14.3 万吨、石油类 1.0 万吨、挥发酚 3 999.0 吨、氰化物 270.0 吨。

2011 年，三峡库区及其上游流域工业废水治理施工项目数 401 个，竣工项目数 290个，工业废水治理项目完成投资 10.9 亿元，工业废水治理竣工项目新增设计处理能力 67.7万吨/日。

三峡库区及其上游流域纳入统计的污水处理厂 442 座，共形成 908 万吨/日的处理能力，处理生活污水 25.5 亿吨/年，共去除化学需氧量 56.0 万吨、氨氮 5.0 万吨、油类 0.3万吨、总氮 4.7 万吨、总磷 0.7 万吨。

1.2.7 入海陆源废水及主要污染物排放情况

2011 年，入海陆源的统计范围为我国沿海 12 个地区的 235 个县（区、市）。我国沿海地区重点调查的入海陆源工业企业数为 23 402 家，规模化畜禽养殖场 13 864 家，规模化畜禽养殖小区 500 家。

表 9　近岸海域废水及主要污染物接纳情况

污染物	废水/亿吨			化学需氧量/万吨				氨氮/万吨			
	工业	生活	集中式	工业	农业	生活	集中式	工业	农业	生活	集中式
排放量	45.8	59.8	0.1	42.5	110.6	109.7	2.1	3.2	9.6	18.5	0.2

沿海地区入海陆源的废水排放总量为 105.7 亿吨。其中，工业废水排放量为 45.8 亿吨，占入海陆源废水排放总量的 43.4%。城镇生活污水排放量为 59.8 亿吨，占入海陆源废水排放总量的 56.5%。集中式污染治理设施废水排放量为 0.1 亿吨，占入海陆源废水排放总量的 0.1%。其中，东海沿海地区入海陆源工业废水排放量、生活污水排放量均为四大海域最大。

沿海地区入海陆源排放的化学需氧量 264.9 万吨。其中，工业废水化学需氧量为 42.5 万吨，农业源化学需氧量 110.6 万吨，城镇生活污水化学需氧量 109.7 万吨，集中式污染治理设施化学需氧量 2.1 万吨。四大海域沿海地区中，东海沿海地区工业化学需氧量排放量最大，渤海农业源化学需氧量排放量最大，东海生活污水化学需氧量排放量最大。

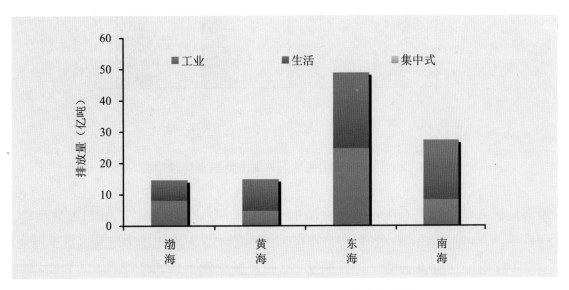

图 20　四大海域沿海地区入海陆源废水排放情况

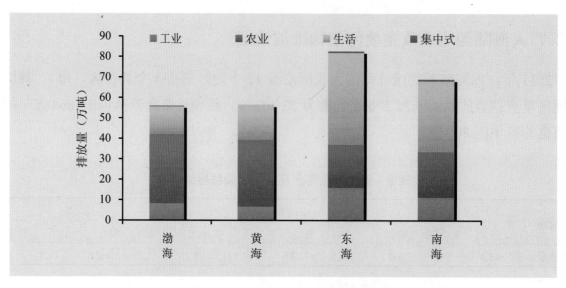

图 21　四大海域沿海地区入海陆源化学需氧量排放情况

沿海地区入海陆源氨氮排放量为 31.5 万吨。其中，工业废水氨氮为 3.2 万吨，农业源氨氮为 9.6 万吨，城镇生活污水氨氮为 18.5 万吨，集中式污染治理设施废水氨氮为 0.2 万吨。四大海域沿海地区中，工业废水、农业源、城镇生活污水氨氮排放量最大的均为东海沿海，集中式污染治理设施废水氨氮排放量最大的是南海沿海地区。

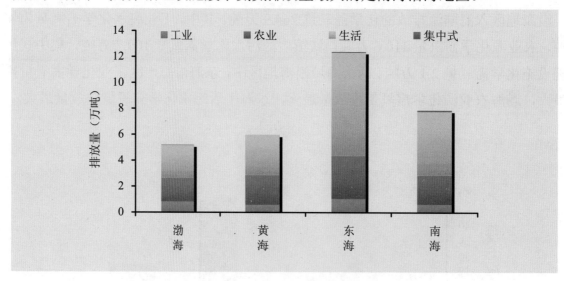

图 22　四大海域沿海地区入海陆源氨氮排放情况

沿海地区入海陆源共有废水治理设施 14 315 套，形成了 8 381 万吨/日的处理能力，年运行费用达 135.3 亿元，共处理了 77.0 亿吨工业废水，去除化学需氧量 348.5 万吨，氨氮 12.1 万吨，石油类 6.0 万吨，挥发酚 2 972 吨，氰化物 1 197 吨。

沿海地区入海陆源工业废水治理施工项目数 666 个，竣工项目数 425 个，工业废水治理项目完成投资 28.0 亿元，工业废水治理竣工项目新增设计处理能力 278.8 万吨/日。

沿海地区入海陆源纳入统计的集中式污水处理厂585座，共形成2 755.0万吨/日的处理能力，处理生活污水65.8亿吨。共去除化学需氧量227.8万吨、氨氮16.5万吨、油类1.3万吨、总氮14.0万吨、总磷2.4万吨。

1.3 废气

1.3.1 废气及废气中主要污染物排放情况

（1）二氧化硫排放情况

2011年，全国工业废气排放量674 509.3亿米^3（标态）。

全国二氧化硫排放量2 217.9万吨。其中，工业二氧化硫排放量2 017.2万吨，占全国二氧化硫排放总量的91.0%；生活二氧化硫排放量200.4万吨，占全国二氧化硫排放总量的9.0%；集中式污染治理设施二氧化硫排放量0.3万吨。

表10　全国二氧化硫排放量

单位：万吨

排放源	合计	工业	生活	集中式
排放量	2 217.9	2 017.2	200.4	0.3

注：集中式污染治理设施包括生活垃圾处理厂（场）和危险废物（医疗废物）集中处理（置）厂焚烧废气中排放的污染物。（下同）

（2）氮氧化物排放情况

2011年，全国氮氧化物排放量2 404.3万吨。其中，工业氮氧化物排放量1 729.7万吨，占全国氮氧化物排放总量的71.9%；生活氮氧化物排放量36.6万吨，占全国氮氧化物排放总量的1.5%；机动车氮氧化物排放量637.6万吨，占全国氮氧化物排放总量的26.5%；集中式污染治理设施氮氧化物排放量0.3万吨。

表11　全国氮氧化物排放量

单位：万吨

排放源	合计	工业	生活	机动车	集中式
排放量	2 404.3	1 729.7	36.6	637.6	0.3

注：自2011年起机动车排气污染物排放情况与生活源分开单独统计。

（3）烟（粉）尘排放情况

2011年，全国烟（粉）尘排放量1 278.8万吨。其中，工业烟（粉）尘排放量1 100.9

万吨，占全国烟（粉）尘排放总量的86.1%；生活烟尘排放量114.8万吨，占全国烟（粉）尘排放总量的9.0%；机动车颗粒物排放量62.9万吨，占全国烟（粉）尘排放总量的4.9%；集中式污染治理设施烟（粉）尘排放量0.2万吨。

表12 全国烟（粉）尘排放量   单位：万吨

排放源	合计	工业	生活	机动车	集中式
排放量	1 278.8	1 100.9	114.8	62.9	0.2

注：1. 自2011年起不再单独统计烟尘和粉尘，统一以烟（粉）尘进行统计；2. 机动车的烟（粉）尘排放量指机动车的颗粒物排放量。

1.3.2 各地区废气中主要污染物排放情况

（1）二氧化硫排放情况

2011年，二氧化硫排放量超过100万吨的省份依次为山东、河北、内蒙古、山西、河南、辽宁、贵州和江苏，8个省份的二氧化硫排放量占全国排放量的48.3%。各地区中，山东工业二氧化硫排放量最大，贵州生活二氧化硫排放量最大，浙江集中式污染治理设施二氧化硫排放量最大。

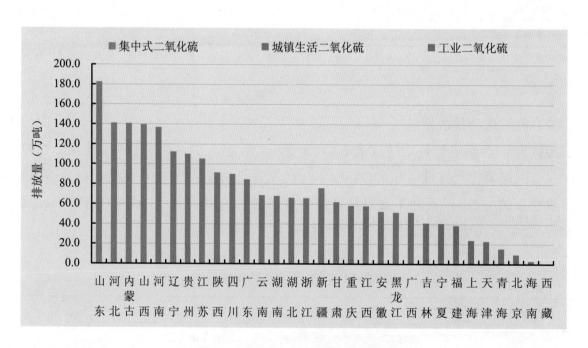

图23 各地区二氧化硫排放情况

（2）氮氧化物排放情况

氮氧化物排放量超过100万吨的省份依次为河北、山东、河南、江苏、内蒙古、广

22

东、山西和辽宁，8 个省份氮氧化物排放量占全国氮氧化物排放量的 49.7%。工业氮氧化物排放量最大的是山东，生活氮氧化物排放量最大的是黑龙江，机动车氮氧化物排放量最大的是河北，集中式污染治理设施氮氧化物排放量最大的是江苏。

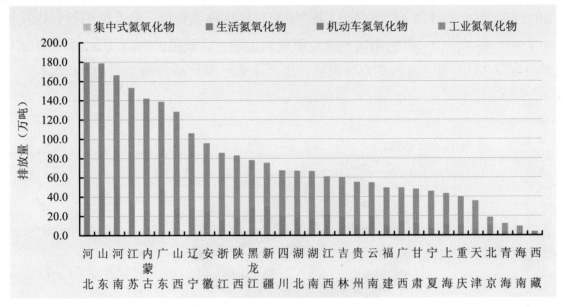

图 24　各地区氮氧化物排放情况

（3）烟（粉）尘排放情况

烟（粉）尘排放量超过 50 万吨的省份依次为河北、山西、山东、内蒙古、辽宁、河南、黑龙江、新疆和江苏，9 个省份烟尘排放量占全国烟（粉）尘排放量的 55.1%。各地区中，河北工业烟（粉）尘排放量最大，黑龙江生活烟（粉）尘排放量最大，河北机动车颗粒物排放量最大，安徽集中式污染治理设施烟（粉）尘排放量最大。

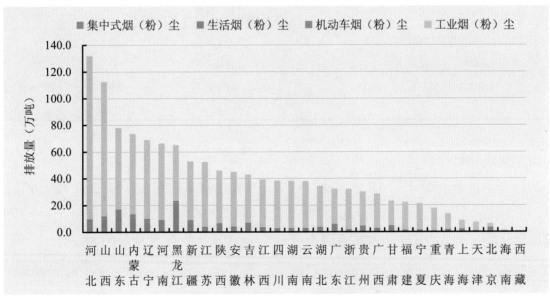

图 25　各地区烟（粉）尘排放情况

1.3.3 工业行业废气中主要污染物排放情况

（1）二氧化硫排放情况

2011 年，调查统计的 41 个工业行业中，二氧化硫排放量位于前 3 位的行业依次为电力、热力生产和供应业、黑色金属冶炼及压延加工业、非金属矿物制品业，3 个行业共排放二氧化硫 1 354.3 万吨，占重点调查统计工业企业二氧化硫排放总量的 71.4%。

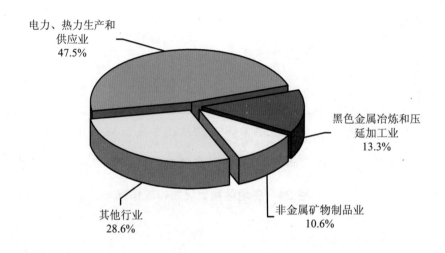

图 26　工业行业二氧化硫排放情况

表 13　重点行业废气中主要污染物排放情况　　　　　　　单位：万吨

行业	二氧化硫	氮氧化物	烟（粉）尘
电力、热力生产和供应业	901.2	1 106.8	215.6
黑色金属冶炼及压延加工业	251.4	95.1	206.2
非金属矿物制品业	201.7	269.4	279.1
合计	1 354.3	1 471.3	700.9

（2）氮氧化物排放情况

2011 年，在调查统计的 41 个工业行业中，氮氧化物排放量位于前 3 位的行业依次为电力、热力生产和供应业、非金属矿物制品业、黑色金属冶炼及压延加工业，3 个行业共排放氮氧化物 1 471.3 万吨，占重点调查统计企业氮氧化物排放总量的 88.6%。

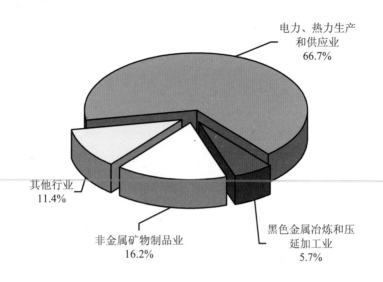

图 27　工业行业氮氧化物排放情况

（3）烟（粉）尘排放情况

2011 年，在调查统计的 41 个工业行业中，烟（粉）尘排放量位于前 3 位的行业依次为非金属矿物制品业、电力、热力生产和供应业、黑色金属冶炼及压延加工业，3 个行业共排放烟（粉）尘 700.9 万吨，占重点调查统计企业烟（粉）尘排放量的 68.2%。

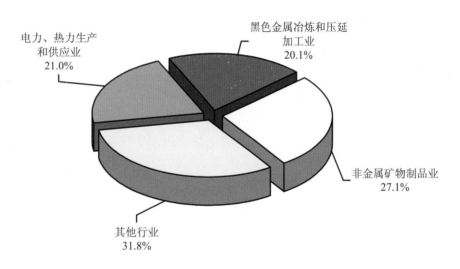

图 28　工业行业烟（粉）尘排放情况

1.3.4 火电行业主要污染物排放及处理情况

2011 年，纳入重点调查统计范围的电力企业 3 311 家。其中，独立火电厂 1 828 家，自备电厂 1 483 家。

独立火电厂二氧化硫排放量为 819 万吨，占全国工业二氧化硫排放量的 40.6%。独立火电厂二氧化硫排放量大于 50 万吨的省份依次为山东、山西、内蒙古和贵州，占全国独立火电厂排放量的 33.5%。安装 3 379 套脱硫设施，共去除二氧化硫 2 394 万吨，二氧化硫去除率达到 74.5%。

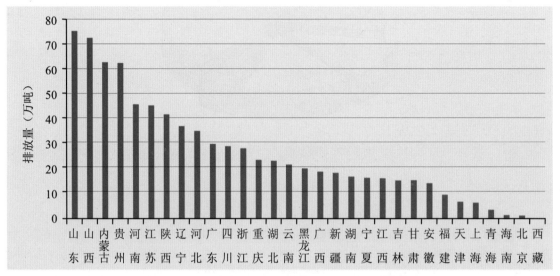

图 29　各地区独立火电厂二氧化硫排放情况

独立火电厂氮氧化物排放量为 1 073 万吨，占全国工业氮氧化物排放量的 62.0%。氮氧化物排放量大于 50 万吨的省份依次为内蒙古、江苏、山东、河南、河北和山西，占全国独立火电厂排放量的 42.1%。安装了 274 套脱硝设施，共去除氮氧化物 75 万吨，氮氧化物去除率为 6.5%。

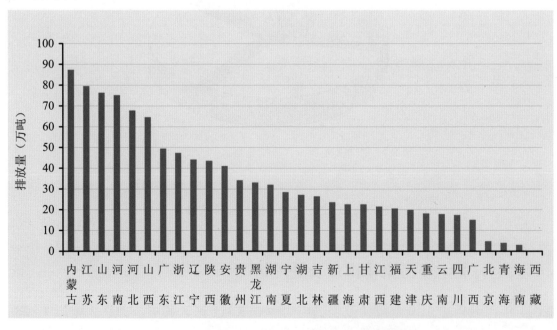

图 30　各地区独立火电厂氮氧化物排放情况

26

1.4 工业固体废物

1.4.1 一般工业固体废物产生及利用情况

2011 年，全国一般工业固体废物产生量 32.3 亿吨，一般工业固体废物综合利用量 19.5 亿吨，一般工业固体废物贮存量 6.0 亿吨，一般工业固体废物处置量 7.0 亿吨，一般工业固体废物倾倒丢弃量 0.04 亿吨。

2011 年，全国工业固体废物综合利用率为 59.9%。天津、上海、江苏、山东和浙江等地区工业固体废物综合利用率高于 90%。

表 14　全国工业固体废物产生及处理情况　　　　　　　　　单位：万吨

项目	产生量	倾倒丢弃量	综合利用量	贮存量	处置量
一般工业固体废物	322 772.3	433.3	195 214.6	60 424.3	70 465.3
危险废物	3 431.2	0.01	1 773.1	823.7	916.5

注：1. "综合利用量"包括综合利用往年贮存量，"处置量"包括处置往年贮存量；

2. 工业固体废物综合利用率=工业固体废物综合利用量/（工业固体废物产生量＋综合利用往年贮存量）；

3. 危险废物处理率=（危险废物综合利用量＋处置量）/（危险废物产生量＋综合利用往年贮存量＋处置往年贮存量）。

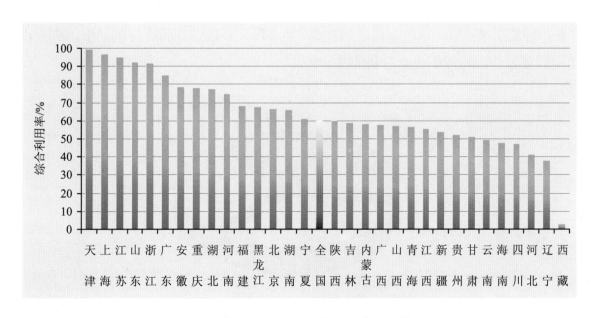

图 31　各地区工业固体废物综合利用情况

1.4.2 各地区一般工业固体废物倾倒丢弃情况

2011 年，工业固体废物倾倒丢弃量超过 20 万吨的省份有云南、新疆、山西、贵州和重庆，5 个省份的工业固体废物排放量占全国工业固体废物倾倒丢弃量的 78.4%。

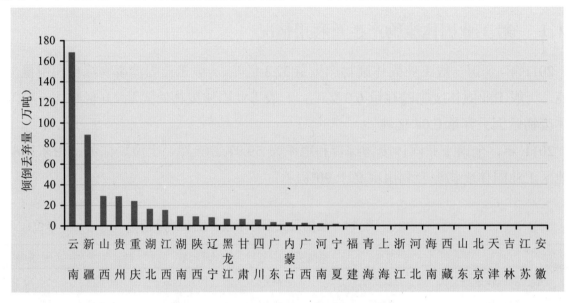

图 32　各地区工业固体废物倾倒丢弃情况

1.4.3 工业行业固体废物倾倒丢弃情况

2011 年，一般工业固体废物倾倒丢弃量超过 50 万吨的行业依次为煤炭开采和洗选业、有色金属矿采选业、黑色金属矿采选业。3 个行业一般工业固体废物倾倒丢弃量占统计工业行业固体废物倾倒丢弃总量的 71.0%。

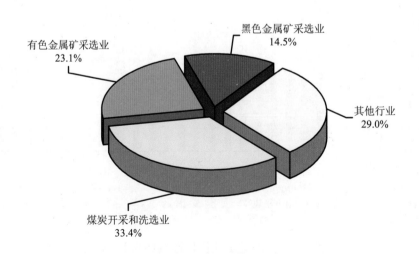

图 33　工业行业一般工业固体废物倾倒丢弃情况

1.4.4 工业危险废物产生及利用情况

2011 年，全国工业危险废物产生量 3 431.2 万吨，工业危险废物综合利用量 1 773.1 万吨，工业危险废物贮存量 823.7 万吨，工业危险废物处置量 916.5 万吨，工业危险废物倾倒丢弃量 0.01 万吨。全国工业危险废物综合利用处置率达到 76.5%。

1.5 集中式污染治理设施情况

1.5.1 城镇生活污水集中处理厂情况

2011 年，全国共调查统计 3 974 座污水处理厂，设计处理能力为 13 990.9 万吨/日，运行费用为 307.2 亿元。全年共处理废水 402.9 亿吨，其中，生活污水 349.7 亿吨，占总处理水量的 86.8%。再生水生产量 12.9 亿吨，再生水利用量 9.6 亿吨。共去除化学需氧量 1 014.9 万吨，氨氮 85.8 万吨，油类 5.5 万吨，总氮 67.4 万吨，总磷 15.0 万吨。污水处理厂的污泥处置量为 2 267.2 万吨。

1.5.2 生活垃圾处理厂（场）情况

2011 年，全国共调查统计了生活垃圾处理厂（场）2 039 座，填埋设计容量达 358 555 万米^3，堆肥设计处理能力达到 20 688 吨/日，焚烧设计处理能力达到 21 625 吨/日，运行费为 59.1 亿元。全年共处理生活垃圾 63.6 亿吨。

1.5.3 危险废物（医疗废物）集中处理（置）厂（场）情况

2011 年，全国共调查统计 644 座危险废物集中处理（置）厂（场），260 座医疗废物集中处理（置）厂（场），危险废物设计处置能力达到 352 541.4 吨/日，运行费用为 48.2 亿元。全年共处置危险废物 260.0 万吨，其中工业危险废物 193.4 万吨，医疗废物 49.8 万吨。采用填埋方式处置的危险废物共 121.5 万吨，采用焚烧方式处置的 133.6 万吨。

1.6 环境污染治理投资情况

2011 年，环境污染治理投资为 6 026.2 亿元，占当年国内生产总值（GDP）的 1.27%。其中，城市环境基础设施建设投资 3 469.4 亿元，比上年减少 17.9%；工业污染源治理投资 444.4 亿元, 比上年增加 11.9%; 建设项目"三同时"环保投资 2 112.4 亿元, 比上年增加 3.9%。

表 15　全国环境污染治理投资情况　　　　　　　　　　　　单位: 亿元

年度	城市环境基础设施建设投资	工业污染源治理投资	建设项目"三同时"环保投资	投资总额
2001	595.7	174.5	336.4	1 106.6
2005	1 289.7	458.2	640.1	2 388.0
2010	4 224.2	397.0	2 033.0	6 654.2
2011	3 469.4	444.4	2 112.4	6 026.2
增长率 (%)	−17.9%	11.9%	3.9%	−9.4%

1.6.1 城市环境基础设施建设

2011 年, 城市环境基础设施建设投资中, 燃气工程建设投资 331.4 亿元, 比上年增加 14.0%; 集中供热工程建设投资 437.6 亿元, 比上年增加 1.0%; 排水工程建设投资 770.1 亿元, 比上年减少 14.6%; 园林绿化工程建设投资 1 546.2 亿元, 比上年减少 32.7%; 市容环境卫生工程建设投资 384.1 亿元, 比上年增加 27.3%。

燃气、集中供热、排水、园林绿化和市容环境卫生投资分别占城市环境基础设施建设总投资的 9.6%、12.6%、22.2%、44.6% 和 11.1%, 排水设施和园林绿化投资为城市环境基础设施建设投资的重点。

表 16　全国近年城市环境基础设施建设投资构成　　　　　　单位: 亿元

年度	投资总额	燃气	集中供热	排水	园林绿化	市容环境卫生
2001	595.7	75.5	82.0	224.5	163.2	50.6
2005	1 289.7	142.4	220.2	368.0	411.3	147.8
2010	4 224.2	290.8	433.2	901.6	2 297.0	301.6
2011	3 469.4	331.4	437.6	770.1	1 546.2	384.1

1.6.2 工业污染源治理投资

2011 年, 工业污染源污染治理投资中, 废水治理资金 157.7 亿元, 比上年增加 21.2%; 废气治理资金 211.7 亿元, 比上年增加 12.1%。其中工业废气脱硫治理项目投资 112.7 亿元, 其中工业废气脱硝治理项目投资 12.7 亿元; 工业固体废物治理资金 31.4 亿元, 比上年增加 120.0%; 噪声治理资金 2.2 亿元, 比上年增加 42.3%。

废水、废气、固废、噪声以及其他污染要素治理投资，分别占工业源治理总投资的 35.5%、47.6%、7.1%、0.5%和9.3%，废水和废气仍是工业污染治理的重点。

表 17　全国近年工业源污染治理投资构成　　　　单位：万元

年度	废水	废气	固废	噪声	其他
2001	729 214.3	657 940.4	186 967.2	6 424.4	164 733.7
2005	1 337 146.9	2 129 571.3	274 181.3	30 613.3	810 395.9
2010	1 301 148.7	1 888 456.5	142 692.2	15 193.2	621 777.6
2011	1 577 471.1	2 116 810.6	313 875.3	21 622.5	413 830.7

1.6.3 建设项目"三同时"环保投资

2011 年，建设项目"三同时"环保投资 2 112.4 亿元，比上年增加 3.9%。建设项目"三同时"环保投资占环境污染治理投资总额的比例为 35.1%，占建设项目投资总额的 3.1%。

表 18　建设项目"三同时"投资情况

年　度	环保投资额（亿元）	占建设项目投资总额（%）	占全社会固定资产投资总额（%）	占环境治理投资总额（%）
2001	336.4	3.6	0.9	30.4
2005	640.1	4.0	0.7	26.8
2010	2 033.0	4.1	0.7	30.6
2011	2 112.4	3.1	0.7	35.1

1.7 全国辐射环境水平

2011 年，全国辐射环境质量总体良好。环境电离辐射水平保持稳定，核设施、核技术利用项目周围环境电离辐射水平总体未见明显变化；环境电磁辐射水平总体情况较好，电磁辐射发射设施周围环境电磁辐射水平总体未见明显变化。辐射监测数据表明，日本福岛核事故未对中国环境及公众健康产生影响。

1.7.1 环境电离辐射水平

全国地级及以上城市环境γ辐射空气吸收剂量率，省会城市及直辖市气溶胶、沉降物总α和总β活度浓度，省会城市及直辖市空气中氡活度浓度均为正常环境水平。长江、黄河、珠江、松花江、淮河、海河、辽河七大水系以及浙闽片河流、西南诸河与内陆诸河70个地表水国控监测断面，15个国控重点湖泊（水库）放射性核素活度浓度与历年相比未见明显变化，其中天然放射性核素活度浓度与1983—1990年全国环境天然放射性水平调查值处于同一水平。12个集中式饮用水水源地总α和总β活度浓度均低于《生活饮用水卫生标准》（GB 5749—2006）规定的限值。近岸海域10个海水国控点人工放射性核素锶-90和铯-137活度浓度均在《海水水质标准》（GB 3097—1997）规定限值内。省会城市、直辖市及部分地级城市土壤放射性核素活度浓度与历年相比未见明显变化，其中天然放射性核素活度浓度与1983—1990年全国环境天然放射性水平调查值处于同一水平。

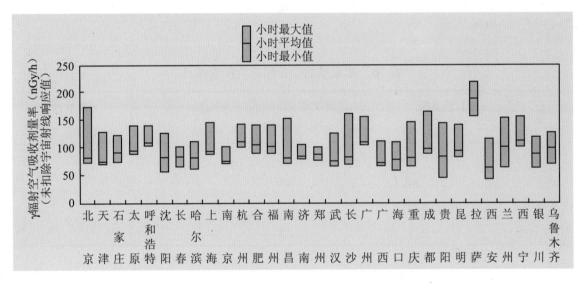

图34 省会城市和直辖市γ辐射空气吸收剂量率

1.7.2 运行核电厂周围环境电离辐射水平

秦山核电基地各核电厂、大亚湾/岭澳核电厂、田湾核电站外围各辐射环境自动监测站实时连续γ辐射空气吸收剂量率（未扣除宇宙射线响应值）年均值分别为101.0 nGy/h、123.8 nGy/h和100.9 nGy/h，在当地的天然本底水平涨落范围内。秦山核电基地周围关键居民点空气、降水、地表水及部分生物样品中氚活度浓度，大亚湾/岭澳核电厂和田湾核电站排放口附近海域海水氚活度浓度与核电站运行前本底值相比有所升高，但对公众附加的剂量远低于国家规定的限值。核电厂外围各种环境介质中除氚外其余放射性核素活度浓度与历年相比未见明显变化。

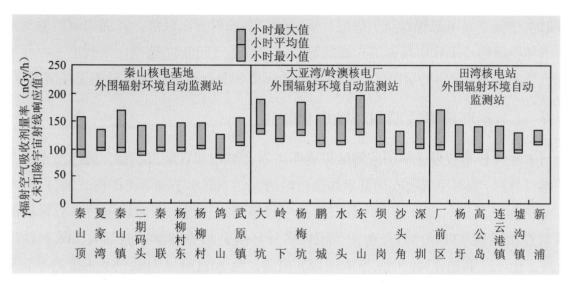

图 35　运行核电厂外围γ辐射空气吸收剂量率（未扣除宇宙射线响应值）

1.7.3　其他反应堆及核燃料循环设施周围环境电离辐射水平

中国原子能科学研究院、清华大学核能与新能源技术研究院、中国核动力研究设计院、陕西省西北核技术研究所等研究设施外围环境γ辐射剂量率，气溶胶、沉降物、地表水、土壤和生物样品中放射性核素活度浓度与历年相比未见明显变化；饮用地下水总α和总β活度浓度低于《生活饮用水卫生标准》（GB 5749—2006）规定的限值。

1.7.4　核燃料循环设施和废物处置设施周围环境电离辐射

兰州铀浓缩有限公司、陕西铀浓缩有限公司、包头核燃料元件厂、中核建中核燃料元件公司、中核四〇四有限公司等核燃料循环设施及西北低中放废物处置场、北龙低中放废物处置场外围环境γ辐射剂量率为正常环境水平，环境介质中也未监测到由企业生产、加工、贮存、处理、运输等活动引起的放射性核素活度浓度升高。

1.7.5　铀矿冶及伴生放射性矿周围环境电离辐射

铀矿冶设施周围环境空气中氡活度浓度，气溶胶、沉降物总α和总β活度浓度，地下水和生物样品中放射性核素铀和镭-226活度浓度未见异常。白云鄂博矿等部分伴生放射性矿的开采、冶炼、加工活动对企业周围局部环境产生了一定程度影响。

1.7.6 电磁辐射设施周围环境电磁辐射

环境电磁辐射水平总体情况较好。开展监测的移动通信基站天线周围环境敏感点电磁辐射水平低于《电磁辐射防护规定》规定的公众照射导出限值；开展监测的输变电设施周围环境敏感点工频电场强度和磁感应强度均低于《500kV 超高压送变电工程环境影响评价技术规范》规定的居民区工频电场评价标准和公众全天候辐射时的工频磁场限值。

1.7.7 日本福岛核事故期间环境电离辐射

日本福岛核事故发生后的监测结果表明，省会城市和直辖市、部分地级城市以及运行核电厂外围γ辐射空气吸收剂量率均在当地的天然本底水平涨落范围内。2011 年 3 月 25 日，中国东北地区部分城市气溶胶样品中监测到来自日本福岛核电站事故释放的人工放射性核素碘-131。此后在全国范围内多种环境介质中陆续监测到人工放射性核素碘-131、铯-137 和铯-134，但活度浓度极其微量，至 2011 年 4 月底已基本监测不到来自日本福岛核事故的人工放射性核素。根据全国范围内持续监测结果，中国管辖海域海水、海洋生物的放射性水平和海洋大气γ辐射空气吸收剂量率未见异常。

简 要 说 明

一、本年报资料根据全国 31 个省、自治区、直辖市及新疆生产建设兵团环境统计资料汇总整理而成，未包括香港特别行政区、澳门特别行政区以及台湾省数据。

二、本年报主要反映中国环境污染排放、治理及环境管理情况。主要内容包括废水及污染物的排放与治理情况，废气及污染物排放与治理情况，一般工业固体废物、危险废物（医疗废物）、生活垃圾的产生、综合利用及处理处置情况，环境污染治理投资以及环境管理等情况。

三、调查范围

本年报数据根据"十二五"环境统计报表制度调查收集，调查范围较"十一五"有所扩大。"十二五"环境统计报表制度调查范围包括工业污染源、农业污染源、城镇生活污染源、机动车、集中式污染治理设施和环境管理 6 方面内容。

1. 工业污染源调查范围：辖区内有污染物排放的所有工业企业。

2. 城镇生活污染源调查范围：城市和集镇内居民在日常生活及各种活动中产生、排放的污染物情况。

3. 机动车调查范围：辖区内所有机动车。

4. 农业污染源调查范围：畜禽养殖业、水产养殖业和种植业。

5. 集中式污染治理设施调查范围：辖区内所有集中式污染治理设施，包括污水处理厂、垃圾处理厂（场）、危险废物（医疗废物）处理（处置）厂。

6. 环境管理反映环保系统自身能力建设、业务工作进展及成果等情况，是在以往环境统计报表制度中专业年报基础上简化而来，调查范围主要包括环保机构数/人数、环境信访与环境法制、环境保护能力建设投资、环境污染源控制与管理、环境监测、污染源自动监控、排污费征收、自然生态保护与建设、环境影响评价、建设项目竣工环境保护验收情况、突发环境事件、环境宣教十二个方面的内容。

与"十一五"环境统计报表制度相比，本年报特点是：

1. 新增了农业污染源调查内容。农业污染源调查内容包括畜禽养殖业、水产养殖业和种植业；

2．细化了机动车污染调查统计。调查载客汽车、载货汽车、三轮汽车及低速载货汽车、摩托车的总颗粒物、氮氧化物、一氧化碳、碳氢化合物等污染物排放量；

3．新增了生活垃圾处理厂（场）调查内容。调查范围为垃圾填埋厂（场）、垃圾堆肥厂（场）、垃圾焚烧厂（场）和其他方式处理垃圾的处理厂（场）；

4．删除了医院污染排放情况调查表。因城镇生活污染源报表中的人均污染物排放量指标均已包含医院污染物排放。

四、调查及核算方法

1．主要污染物排放总量核算

废水污染物排放总量=工业污染源排放量＋农业污染源排放量＋城镇生活污染源排放量＋集中式污染治理设施排放量（集中式污水处理厂除外）；

废气污染物排放总量=工业污染源排放量＋城镇生活污染源排放量＋机动车排放量（仅限 NO_x 指标）＋集中式污染治理设施排放量。

2．工业污染源

工业污染源采取对重点调查工业企业逐个发表调查汇总，非重点调查工业企业采用比率估算法的方式核算。

工业污染排放总量为重点调查工业企业与非重点调查工业企业排放量之和。

工业污染治理投资的调查对象为调查年度正式施工的，且没有纳入"三同时"项目管理的老工业源污染治理投资项目，以及调查年度内完成"三同时"环境保护竣工验收的工业类建设项目。

3．农业污染源

农业污染源包括畜禽养殖业、水产养殖业和种植业，以县（区）为基本单位进行调查。

畜禽养殖业中的规模化养殖场（小区）采用逐场（小区）发表调查，根据饲养量、产污系数及去除率等测算污染物排放量；养殖专业户的污染物排放量根据饲养量和排污强度核算。

水产养殖业污染物排放量根据第一次全国污染源普查及总量减排核定减少水产围网养殖面积核算。

种植业污染物排放量与第一次全国污染源普查数据保持一致。

4．城镇生活污染源：以地市为基本调查单位，污染物产生量依据相关部门的统计数据和产污系数核算，排放量为产生量与集中式污水处理厂生活污染物的去除量之差。

5．机动车：以地市为基本调查单位，根据"遵循基数、核清增量、核实减量"原则核算污染物排放量。

6．集中式污染治理设施：逐家发表调查汇总。

五、其他需要说明的问题

1．自 2011 年起，环境统计年报中采用水利部流域代码对流域进行汇总，流域由松花江、辽河、海河、黄河、淮河、长江、珠江、东南诸河、西南诸河和西北诸河十大水系组成。

2．本年报中所指集中式污染治理设施的排放量，仅指生活垃圾处理厂（场）和危险（医疗）废物集中处置厂（场）的渗滤液和焚烧废气中的污染物。污水处理厂仅作为污染治理设施，不产生和排放污染物。

主要环境统计指标解释附后。

2

各地区环境统计

GEDIQU HUANJING TONGJI

ANNUAL STATISTIC REPORT ON ENVIRONMENT IN CHINA

2011

各地区主要污染物排放情况（一）
Discharge of Key Pollutants by Region（1）

单位：亿吨　　　　　　　　　　　　　　（2011）　　　　　　　　　　（100 million tons）

年　份　Year 地　区　Region	废水排放总量 Total Volume of Waste Water Discharged	工业 Industrial	生活 Household	集中式 Centralized Treatment
2011	659.19	230.87	427.92	0.402
北　京　Beijing	14.55	0.86	13.67	0.009
天　津　Tianjin	6.71	1.98	4.73	0.003
河　北　Hebei	27.86	11.85	16.00	0.008
山　西　Shanxi	11.61	3.97	7.64	0.003
内蒙古　Inner Mongolia	10.04	3.94	6.10	0.001
辽　宁　Liaoning	23.22	9.05	14.17	0.009
吉　林　Jilin	11.62	4.19	7.42	0.006
黑龙江　Heilongjiang	15.07	4.41	10.66	0.001
上　海　Shanghai	21.42	4.46	16.93	0.024
江　苏　Jiangsu	59.28	24.63	34.63	0.022
浙　江　Zhejiang	42.01	18.22	23.75	0.040
安　徽　Anhui	24.33	7.07	17.24	0.016
福　建　Fujian	31.62	17.72	13.87	0.026
江　西　Jiangxi	19.44	7.12	12.30	0.024
山　东　Shandong	44.33	18.72	25.59	0.022
河　南　Henan	37.88	13.87	24.00	0.013
湖　北　Hubei	29.31	10.44	18.84	0.025
湖　南　Hunan	27.88	9.72	18.12	0.037
广　东　Guangdong	78.56	17.86	60.66	0.037
广　西　Guangxi	22.24	10.12	12.11	0.010
海　南　Hainan	3.57	0.68	2.89	0.005
重　庆　Chongqing	13.14	3.40	9.74	0.014
四　川　Sichuan	27.99	8.04	19.92	0.019
贵　州　Guizhou	7.79	2.06	5.73	0.004
云　南　Yunnan	14.75	4.72	10.02	0.009
西　藏　Tibet	0.46	0.04	0.43	...
陕　西　Shaanxi	12.18	4.08	8.09	0.008
甘　肃　Gansu	5.92	1.97	3.95	0.002
青　海　Qinghai	2.13	0.87	1.26	0.001
宁　夏　Ningxia	3.94	1.93	2.01	0.001
新　疆　Xinjiang	8.33	2.88	5.45	0.004

各地区主要污染物排放情况（二）
Discharge of Key Pollutants by Region（2）

单位：万吨 　　　　　　　　　　　　　　　（2011）　　　　　　　　　（10 000 million tons）

年 份 Year 地 区 Region	化学需氧量 排放总量 Total Volume of COD Discharged	工业 Industrial	农业 Agricultural	生活 Household	集中式 Centralized Treatment
2011	2 499.86	354.80	1 186.11	938.84	20.114
北 京 Beijing	19.32	0.71	8.17	9.91	0.528
天 津 Tianjin	23.58	2.43	11.47	9.64	0.044
河 北 Hebei	138.88	19.40	94.08	24.61	0.787
山 西 Shanxi	48.96	9.11	18.45	21.14	0.258
内蒙古 Inner Mongolia	91.90	9.23	64.73	17.71	0.231
辽 宁 Liaoning	134.34	11.04	88.35	34.16	0.795
吉 林 Jilin	82.47	7.87	52.83	20.48	1.292
黑龙江 Heilongjiang	157.65	10.24	110.71	36.51	0.190
上 海 Shanghai	24.90	2.74	3.39	18.19	0.583
江 苏 Jiangsu	124.62	23.93	39.93	60.23	0.530
浙 江 Zhejiang	81.83	19.11	21.30	40.69	0.716
安 徽 Anhui	95.33	9.41	39.88	45.05	0.996
福 建 Fujian	67.94	9.59	22.52	35.42	0.412
江 西 Jiangxi	76.79	11.71	24.99	38.97	1.111
山 东 Shandong	198.25	14.08	137.97	45.73	0.471
河 南 Henan	143.67	19.36	82.30	41.02	0.979
湖 北 Hubei	110.47	14.47	48.12	46.13	1.745
湖 南 Hunan	130.52	17.09	58.57	53.19	1.669
广 东 Guangdong	188.45	24.40	62.03	100.63	1.398
广 西 Guangxi	79.33	20.49	22.28	36.17	0.390
海 南 Hainan	19.99	1.25	10.72	7.89	0.136
重 庆 Chongqing	41.68	5.80	12.74	23.08	0.062
四 川 Sichuan	130.23	12.91	54.92	61.04	1.353
贵 州 Guizhou	34.22	6.55	5.58	21.49	0.599
云 南 Yunnan	55.47	17.54	7.29	29.35	1.298
西 藏 Tibet	2.68	0.10	0.46	2.12	...
陕 西 Shaanxi	55.77	10.65	20.13	24.33	0.660
甘 肃 Gansu	39.66	9.61	14.66	15.23	0.163
青 海 Qinghai	10.32	4.06	2.15	3.92	0.191
宁 夏 Ningxia	23.37	11.15	10.28	1.86	0.076
新 疆 Xinjiang	67.29	18.76	35.10	12.97	0.450

各地区主要污染物排放情况（三）
Discharge of Key Pollutants by Region（3）

（2011）

单位：万吨　　　　　　　　　　　　　　　　　　　　　　　　　（10 000 tons）

年　份　Year 地　区　Region	氨氮排放总量 Total Volume of Ammonia Nitrogen Discharged	工业 Industrial	农业 Agricultural	生活 Household	集中式 Centralized Treatment
2011	260.44	28.12	82.65	147.66	2.005
北　京　Beijing	2.13	0.04	0.48	1.55	0.057
天　津　Tianjin	2.64	0.33	0.60	1.71	0.004
河　北　Hebei	11.43	1.77	4.59	5.01	0.058
山　西　Shanxi	5.91	0.89	1.26	3.73	0.025
内蒙古　Inner Mongolia	5.39	1.13	1.26	2.98	0.019
辽　宁　Liaoning	11.11	0.91	3.55	6.54	0.112
吉　林　Jilin	5.82	0.45	1.82	3.41	0.138
黑龙江　Heilongjiang	9.65	0.63	3.56	5.44	0.022
上　海　Shanghai	5.04	0.26	0.35	4.40	0.030
江　苏　Jiangsu	15.72	1.67	3.99	9.99	0.068
浙　江　Zhejiang	11.54	1.26	2.77	7.45	0.059
安　徽　Anhui	10.98	0.90	3.93	6.06	0.096
福　建　Fujian	9.54	0.77	3.40	5.33	0.040
江　西　Jiangxi	9.34	1.21	3.13	4.91	0.090
山　东　Shandong	17.29	1.18	7.58	8.49	0.050
河　南　Henan	15.38	1.39	6.62	7.25	0.118
湖　北　Hubei	13.12	1.71	4.75	6.46	0.204
湖　南　Hunan	16.50	2.79	6.44	7.12	0.152
广　东　Guangdong	23.09	1.56	5.93	15.46	0.143
广　西　Guangxi	8.39	0.85	2.78	4.73	0.031
海　南　Hainan	2.27	0.08	0.94	1.24	0.012
重　庆　Chongqing	5.50	0.32	1.36	3.81	0.016
四　川　Sichuan	14.37	0.57	5.87	7.86	0.077
贵　州　Guizhou	3.98	0.35	0.75	2.82	0.068
云　南　Yunnan	5.93	0.50	1.18	4.08	0.165
西　藏　Tibet	0.33	0.01	0.05	0.27	...
陕　西　Shaanxi	6.34	0.87	1.56	3.81	0.092
甘　肃　Gansu	4.26	1.40	0.59	2.25	0.011
青　海　Qinghai	0.96	0.19	0.09	0.67	0.013
宁　夏　Ningxia	1.80	0.93	0.23	0.64	0.005
新　疆　Xinjiang	4.68	1.21	1.23	2.21	0.032

各地区主要污染物排放情况（四）
Discharge of Key Pollutants by Region（4）

单位：万吨 （2011） （10 000 tons）

年 份 Year 地 区 Region	二氧化硫 排放总量 Total Volume of Sulphur Dioxide Discharged	工业 Industrial	生活 Household	集中式 Centralized Treatment
2011	2 217.91	2 017.23	200.39	0.287
北 京 Beijing	9.79	6.13	3.65	0.005
天 津 Tianjin	23.09	22.19	0.90	0.004
河 北 Hebei	141.21	131.71	9.49	0.013
山 西 Shanxi	139.91	129.43	10.47	0.004
内蒙古 Inner Mongolia	140.94	125.02	15.92	...
辽 宁 Liaoning	112.62	104.89	7.72	0.005
吉 林 Jilin	41.32	36.34	4.97	...
黑龙江 Heilongjiang	52.19	41.52	10.67	0.001
上 海 Shanghai	24.01	21.01	2.99	0.015
江 苏 Jiangsu	105.38	102.50	2.85	0.024
浙 江 Zhejiang	66.20	64.70	1.45	0.047
安 徽 Anhui	52.95	48.72	4.18	0.040
福 建 Fujian	38.92	37.03	1.88	0.012
江 西 Jiangxi	58.41	56.81	1.60	0.001
山 东 Shandong	182.74	162.86	19.86	0.016
河 南 Henan	137.05	122.92	14.13	0.004
湖 北 Hubei	66.56	59.50	7.07	0.001
湖 南 Hunan	68.55	63.62	4.93	0.002
广 东 Guangdong	84.77	82.60	2.13	0.045
广 西 Guangxi	52.10	48.87	3.23	0.003
海 南 Hainan	3.26	3.11	0.15	...
重 庆 Chongqing	58.69	53.13	5.56	...
四 川 Sichuan	90.20	82.89	7.28	0.024
贵 州 Guizhou	110.43	90.31	20.12	...
云 南 Yunnan	69.12	64.25	4.87	0.001
西 藏 Tibet	0.42	0.14	0.28	0
陕 西 Shaanxi	91.68	83.12	8.56	...
甘 肃 Gansu	62.39	52.77	9.60	0.017
青 海 Qinghai	15.66	13.43	2.23	...
宁 夏 Ningxia	41.04	38.79	2.25	...
新 疆 Xinjiang	76.31	66.91	9.40	...

各地区主要污染物排放情况（五）
Discharge of Key Pollutants by Region（5）

单位：万吨　　　　　　　　　　　　　　　　（2011）　　　　　　　　　　　　　　　（10 000 tons）

年　份　Year 地　区　Region	氮氧化物排放总量 Total Volume of Nitrogen Oxide Discharged	工业 Industrial	生活 Household	机动车 Vehicle	集中式 Centralized Treatment
2011	2 404. 27	1 729. 71	36. 62	637. 59	0. 350
北　京　Beijing	18. 83	9. 03	1. 21	8. 55	0. 032
天　津　Tianjin	35. 89	30. 04	0. 44	5. 40	0. 005
河　北　Hebei	180. 11	122. 09	1. 84	56. 18	0. 007
山　西　Shanxi	128. 60	98. 65	2. 90	27. 04	0. 011
内蒙古　Inner Mongolia	142. 19	115. 05	2. 45	24. 69	0. 001
辽　宁　Liaoning	106. 28	76. 78	1. 37	28. 13	0. 001
吉　林　Jilin	60. 47	41. 00	1. 13	18. 34	0. 001
黑龙江　Heilongjiang	78. 38	48. 49	4. 25	25. 63	0. 002
上　海　Shanghai	43. 54	32. 69	0. 91	9. 90	0. 033
江　苏　Jiangsu	153. 57	119. 56	0. 62	33. 35	0. 044
浙　江　Zhejiang	85. 91	69. 08	0. 37	16. 43	0. 024
安　徽　Anhui	95. 91	72. 27	0. 85	22. 76	0. 031
福　建　Fujian	49. 45	39. 13	0. 24	10. 06	0. 026
江　西　Jiangxi	61. 23	39. 35	0. 27	21. 61	0. 002
山　东　Shandong	179. 03	127. 36	3. 10	48. 56	0. 013
河　南　Henan	166. 54	114. 34	2. 32	49. 85	0. 019
湖　北　Hubei	66. 96	47. 76	1. 21	18. 00	0. 001
湖　南　Hunan	66. 64	48. 77	0. 76	17. 10	0. 006
广　东　Guangdong	138. 82	89. 59	1. 22	47. 97	0. 041
广　西　Guangxi	49. 40	34. 78	0. 35	14. 26	0. 005
海　南　Hainan	9. 54	6. 54	0. 03	2. 96	0. 001
重　庆　Chongqing	40. 26	29. 38	0. 48	10. 40	0. 001
四　川　Sichuan	67. 49	46. 52	1. 00	19. 93	0. 035
贵　州　Guizhou	55. 32	44. 69	1. 18	9. 45	...
云　南　Yunnan	54. 85	34. 83	0. 60	19. 42	0. 001
西　藏　Tibet	4. 06	0. 38	0. 02	3. 66	0
陕　西　Shaanxi	83. 17	63. 37	1. 92	17. 88	0. 001
甘　肃　Gansu	48. 09	35. 60	1. 17	11. 30	0. 005
青　海　Qinghai	12. 41	8. 76	0. 49	3. 17	...
宁　夏　Ningxia	45. 82	38. 59	0. 27	6. 96	...
新　疆　Xinjiang	75. 51	45. 23	1. 64	28. 64	0. 001

各地区主要污染物排放情况（六）
Discharge of Key Pollutants by Region（6）

单位：万吨 　　　　　　　　　　　　　　（2011）　　　　　　　　　（10 000 tons）

年份 地区	Year Region	烟（粉）尘排放 总量 Total Volume of Soot and Dust Discharged	工业 Industrial	生活 Household	机动车 Vehicle	集中式 Centralized Treatment
	2011	1 278.83	1 100.88	114.76	62.93	0.247
北 京	Beijing	6.58	2.94	3.20	0.44	0.003
天 津	Tianjin	7.59	6.53	0.41	0.65	0.002
河 北	Hebei	132.25	122.35	4.28	5.61	0.004
山 西	Shanxi	112.99	100.84	9.79	2.32	0.037
内蒙古	Inner Mongolia	73.99	60.45	10.63	2.90	0.001
辽 宁	Liaoning	69.32	59.12	7.05	3.15	0.004
吉 林	Jilin	43.22	36.13	5.20	1.89	0.002
黑龙江	Heilongjiang	65.59	41.98	20.99	2.61	...
上 海	Shanghai	8.98	6.64	1.49	0.85	0.004
江 苏	Jiangsu	52.74	48.64	1.20	2.88	0.017
浙 江	Zhejiang	32.33	30.28	0.39	1.64	0.023
安 徽	Anhui	45.22	41.06	1.55	2.52	0.090
福 建	Fujian	22.53	20.99	0.58	0.96	0.010
江 西	Jiangxi	39.60	35.91	1.00	2.69	...
山 东	Shandong	78.38	61.27	11.82	5.29	0.004
河 南	Henan	66.82	57.57	4.12	5.14	0.003
湖 北	Hubei	34.61	30.74	2.33	1.55	0.002
湖 南	Hunan	38.44	35.52	1.44	1.48	0.003
广 东	Guangdong	32.43	26.39	1.25	4.78	0.012
广 西	Guangxi	28.83	26.00	1.17	1.65	0.003
海 南	Hainan	1.58	1.11	0.06	0.42	...
重 庆	Chongqing	18.10	17.12	0.25	0.73	0.001
四 川	Sichuan	38.59	35.76	1.28	1.54	0.014
贵 州	Guizhou	30.35	25.72	3.69	0.94	...
云 南	Yunnan	38.22	35.28	1.40	1.54	0.001
西 藏	Tibet	1.00	0.45	0.12	0.43	0
陕 西	Shaanxi	46.34	39.70	4.33	2.31	...
甘 肃	Gansu	23.62	18.78	4.06	0.79	0.001
青 海	Qinghai	13.83	11.76	1.76	0.31	...
宁 夏	Ningxia	21.55	19.82	0.96	0.77	...
新 疆	Xinjiang	53.19	44.05	6.98	2.16	0.006

各地区工业废水排放及处理情况（一）
Discharge and Treatment of Industrial Waste Water by Region（1）

单位：万吨　　　　　　　　　　　　　　（2011）　　　　　　　　　（10 000 tons）

年 份　Year 地 区　Region	工业废水排放量 Total Volume of Industrial Waste Water Discharged	直接排入环境的 Direct discharged to environment	排入污水处理厂的 Discharged to centralized Waste water treatment	工业废水处理量 Industrial Waste Water treated
2011	2 308 743	1 839 635	469 108	5 805 511
北　京　Beijing	8 633	2 827	5 806	11 294
天　津　Tianjin	19 795	8 413	11 382	35 198
河　北　Hebei	118 505	85 164	33 342	933 519
山　西　Shanxi	39 665	36 063	3 603	242 477
内蒙古　InnerMongolia	39 409	24 432	14 976	95 283
辽　宁　Liaoning	90 457	75 930	14 527	351 993
吉　林　Jilin	41 884	34 704	7 180	65 570
黑龙江　Heilongjiang	44 072	33 055	11 017	109 965
上　海　Shanghai	44 626	20 366	24 260	71 099
江　苏　Jiangsu	246 298	167 348	78 951	404 070
浙　江　Zhejiang	182 240	82 202	100 038	304 258
安　徽　Anhui	70 720	62 203	8 516	213 335
福　建　Fujian	177 186	163 337	13 848	251 318
江　西　Jiangxi	71 196	69 735	1 461	138 830
山　东　Shandong	187 245	126 900	60 344	423 600
河　南　Henan	138 654	127 626	11 028	207 377
湖　北　Hubei	104 434	93 061	11 373	325 234
湖　南　Hunan	97 197	93 558	3 639	239 481
广　东　Guangdong	178 626	151 662	26 963	258 541
广　西　Guangxi	101 234	97 950	3 283	350 689
海　南　Hainan	6 820	6 231	589	8 817
重　庆　Chongqing	33 954	32 777	1 176	37 328
四　川　Sichuan	80 420	73 875	6 545	286 354
贵　州　Guizhou	20 626	19 917	709	100 259
云　南　Yunnan	47 228	45 885	1 343	145 135
西　藏　Tibet	363	348	15	498
陕　西　Shaanxi	40 806	34 348	6 458	73 091
甘　肃　Gansu	19 720	18 371	1 349	22 546
青　海　Qinghai	8 677	8 677	0	16 777
宁　夏　Ningxia	19 285	17 996	1 289	26 135
新　疆　Xinjiang	28 769	24 673	4 096	55 440

各地区工业废水排放及处理情况（二）
Discharge and Treatment of Industrial Waste Water by Region（2）

（2011）

年　份　Year 地　区　Region	废水治理设施 数（套） Number of Facilities for Treatment of Waste Water （set）	废水治理设施 治理能力/ （万吨/日） Capacity of Facilities forTreatment of Waste Water （10 000 tons/day）	废水治理设施 运行费用/ 万元 Annual Expenditure for Operation （10 000 yuan）
2011	91 506	31 406	7 321 459.5
北　京　Beijing	508	60	37 124.0
天　津　Tianjin	957	136	134 315.1
河　北　Hebei	4 769	3 522	468 524.1
山　西　Shanxi	3 401	1 000	358 729.6
内蒙古　InnerMongolia	1 020	474	90 305.0
辽　宁　Liaoning	2 287	6 391	302 283.4
吉　林　Jilin	685	350	63 066.9
黑龙江　Heilongjiang	1 171	465	241 107.4
上　海　Shanghai	5 872	305	398 206.8
江　苏　Jiangsu	7 255	3 042	715 131.2
浙　江　Zhejiang	8 462	1 422	466 564.0
安　徽　Anhui	2 321	903	183 885.1
福　建　Fujian	3 429	962	159 360.1
江　西　Jiangxi	2 948	747	146 285.3
山　东　Shandong	5 444	1 859	811 853.5
河　南　Henan	3 299	1 224	222 815.0
湖　北　Hubei	5 296	1 322	344 256.0
湖　南　Hunan	3 111	670	267 415.2
广　东　Guangdong	10 143	1 408	468 752.7
广　西　Guangxi	2 329	1 064	163 188.1
海　南　Hainan	299	40	35 788.1
重　庆　Chongqing	1 507	338	67 100.1
四　川　Sichuan	4 073	1 116	622 493.0
贵　州　Guizhou	1 648	530	168 344.5
云　南　Yunnan	4 857	933	121 650.7
西　藏　Tibet	30	2	231.7
陕　西　Shaanxi	2 274	388	120 108.4
甘　肃　Gansu	553	226	30 269.1
青　海　Qinghai	176	64	9 309.4
宁　夏　Ningxia	397	165	35 041.1
新　疆　Xinjiang	985	279	67 954.8

各地区工业废水排放及处理情况（三）
Discharge and Treatment of Industrial Waste Water by Region（3）

单位：吨 　　　　　　　　　　　　　　　　　　　（2011） 　　　　　　　　　　　　（tons）

年　份　Year 地　区　Region	工业废水中污染物产生量 Amount of Pollutants Produced in the Industrial Waste Water				
	化学需氧量 COD	氨氮 Ammonia Nitrogen	石油类 Petroleum	挥发酚 Volatile Hydroxy - benzene	氰化物 Cyanide
2011	34 475 339.7	1 734 151.3	375 986.196	70 560.5	6 220.9
北　京　Beijing	81 321.5	3 190.8	2 730.4	877.0	6.9
天　津　Tianjin	157 653.5	6 814.7	1 251.3	3.2	1.4
河　北　Hebei	1 750 630.9	77 435.6	22 088.7	11 626.9	696.6
山　西　Shanxi	497 636.2	147 252.0	9 831.5	11 802.6	596.6
内蒙古　Inner Mongolia	420 139.2	42 426.0	3 464.0	1 090.0	31.7
辽　宁　Liaoning	554 312.7	61 129.1	18 210.6	1 958.2	563.1
吉　林　Jilin	624 519.9	23 117.9	1 970.5	860.3	19.7
黑龙江　Heilongjiang	1 590 226.6	69 270.5	48 926.8	5 602.1	102.1
上　海　Shanghai	320 510.9	12 564.1	9 509.3	1 119.2	243.9
江　苏　Jiangsu	2 246 235.6	89 494.3	16 528.1	2 019.7	440.3
浙　江　Zhejiang	2 668 596.6	77 458.2	22 840.5	381.9	704.5
安　徽　Anhui	777 487.8	98 328.5	8 187.8	1 989.2	300.6
福　建　Fujian	1 049 746.9	44 873.3	12 261.0	262.0	141.6
江　西　Jiangxi	559 503.8	51 141.8	6 153.2	1 821.3	60.7
山　东　Shandong	2 848 762.7	132 060.1	31 036.4	5 377.1	274.1
河　南　Henan	2 018 709.3	82 589.1	58 217.8	2 365.1	137.3
湖　北　Hubei	8 798 415.5	50 135.7	4 459.6	928.8	69.0
湖　南　Hunan	587 853.5	67 418.8	4 280.7	1 255.0	364.8
广　东　Guangdong	1 665 255.3	58 416.8	5 008.8	157.1	643.6
广　西　Guangxi	1 119 356.4	55 134.2	3 417.1	1 568.0	26.8
海　南　Hainan	112 930.9	1 918.7	232.1	0.1	0.2
重　庆　Chongqing	260 752.3	28 728.3	3 536.0	1 219.5	88.4
四　川　Sichuan	871 294.1	67 464.1	6 217.0	1 861.6	106.6
贵　州　Guizhou	196 708.5	79 450.2	1 935.9	644.0	139.4
云　南　Yunnan	757 279.8	72 925.6	4 364.0	3 292.2	149.7
西　藏　Tibet	6 137.8	157.5	4.9	0	0
陕　西　Shaanxi	558 781.6	40 534.2	11 169.5	7 534.6	235.9
甘　肃　Gansu	232 357.2	26 279.6	2 491.2	289.2	1.8
青　海　Qinghai	73 404.3	4 072.2	1 091.8	5.3	4.9
宁　夏　Ningxia	510 864.4	90 838.7	4 250.0	564.6	24.4
新　疆　Xinjiang	557 953.9	71 530.9	50 319.8	2 084.6	44.3

各地区工业废水排放及处理情况（四）
Discharge and Treatment of Industrial Waste Water by Region（4）
（2011）

单位：吨 （tons）

年　份　Year 地　区　Region	工业废水中污染物产生量 Amount of Pollutants Produced in the Industrial Waste Water				
	汞 Mercury	镉 Cadmium	总铬 Total Chrome	铅 Lead	砷 Arsenic
2011	23.916	2 580.017	8 456.407	3 793.586	9 977.714
北　京　Beijing	…	0.001	14.134	0.337	3.821
天　津　Tianjin	0.004	0.011	21.094	20.290	0.054
河　北　Hebei	0.051	16.732	220.003	34.432	79.796
山　西　Shanxi	0.166	15.598	16.661	35.589	88.496
内蒙古　Inner Mongolia	0.912	92.858	7.253	94.630	257.114
辽　宁　Liaoning	0.168	56.314	20.190	56.704	48.902
吉　林　Jilin	0.344	0.023	12.013	0.421	2.955
黑龙江　Heilongjiang	0.063	0.014	12.683	1.602	17.345
上　海　Shanghai	0.003	0.069	216.112	1.042	0.828
江　苏　Jiangsu	0.571	2.176	401.630	29.240	5.919
浙　江　Zhejiang	0.010	2.817	2 223.643	9.215	32.686
安　徽　Anhui	2.325	56.218	30.578	58.023	570.832
福　建　Fujian	0.387	26.587	888.826	94.996	32.654
江　西　Jiangxi	0.370	376.236	57.683	86.217	4 022.833
山　东　Shandong	1.004	241.309	149.120	265.690	865.854
河　南　Henan	1.352	260.460	224.062	304.212	138.032
湖　北　Hubei	0.459	3.425	98.825	23.669	108.610
湖　南　Hunan	4.484	204.093	93.165	497.696	520.024
广　东　Guangdong	2.146	15.383	3 213.163	42.083	86.908
广　西　Guangxi	0.527	104.384	127.607	155.141	207.438
海　南　Hainan	0.001	0.005	16.896	0.040	0.027
重　庆　Chongqing	…	0.004	220.530	6.095	3.529
四　川　Sichuan	1.601	52.321	73.041	78.529	326.686
贵　州　Guizhou	0.139	4.349	8.035	7.844	22.471
云　南　Yunnan	0.696	311.129	5.336	529.757	719.504
西　藏　Tibet	…	0.376	3.140	2.316	15.589
陕　西　Shaanxi	0.399	112.826	10.147	403.077	93.424
甘　肃　Gansu	1.261	576.151	49.237	913.870	1 494.115
青　海　Qinghai	0.435	30.328	7.704	23.097	18.924
宁　夏　Ningxia	0.723	5.254	1.512	6.331	152.012
新　疆　Xinjiang	3.315	12.565	12.383	11.400	40.329

各地区工业废水排放及处理情况（五）

Discharge and Treatment of Industrial Waste Water by Region（5）

单位：吨 　　　　　　　　　　　　　　（2011） 　　　　　　　　　　　　　　（tons）

| 年　份　Year
地　区　Region | 工业废水中污染物排放量
Amount of Pollutants Discharged in the Industrial Waste Water | | | | |
	化学需氧量 COD	氨氮 Ammonia Nitrogen	石油类 Petroleum	挥发酚 Volatile Hydroxy - benzene	氰化物 Cyanide
2011	3 547 997.6	281 201.7	20 589.1	2 410.5	215.4
北　京　Beijing	7 117.7	433.8	77.5	0.2	0.2
天　津　Tianjin	24 294.2	3 253.1	199.4	1.3	0.0
河　北　Hebei	194 002.4	17 737.1	1 294.6	307.3	16.6
山　西　Shanxi	91 108.3	8 868.7	1 484.8	1 047.7	41.2
内蒙古　Inner Mongolia	92 274.3	11 298.4	803.1	5.3	1.3
辽　宁　Liaoning	110 369.4	9 142.4	880.4	69.1	4.3
吉　林　Jilin	78 701.2	4 499.8	274.2	2.4	2.2
黑龙江　Heilongjiang	102 372.0	6 251.2	1 251.4	25.6	2.1
上　海　Shanghai	27 357.0	2 611.4	773.8	5.7	3.5
江　苏　Jiangsu	239 318.7	16 710.2	1 573.0	63.0	16.4
浙　江　Zhejiang	191 148.6	12 597.1	976.0	23.0	11.1
安　徽　Anhui	94 088.9	8 971.0	776.9	6.8	6.7
福　建　Fujian	95 915.3	7 700.2	441.2	16.5	7.4
江　西　Jiangxi	117 139.9	12 134.6	696.6	75.7	10.3
山　东　Shandong	140 786.2	11 779.7	718.0	48.0	5.5
河　南　Henan	193 601.8	13 927.8	1 259.6	158.6	17.8
湖　北　Hubei	144 718.5	17 091.3	1 136.4	20.8	9.1
湖　南　Hunan	170 946.6	27 868.5	867.6	102.5	12.8
广　东　Guangdong	243 959.2	15 559.7	830.5	13.2	14.8
广　西　Guangxi	204 914.0	8 472.9	311.3	64.4	5.1
海　南　Hainan	12 499.6	811.9	4.3	0.1	0.0
重　庆　Chongqing	58 028.2	3 205.3	521.1	35.0	2.7
四　川　Sichuan	129 125.7	5 680.3	547.5	55.4	1.8
贵　州　Guizhou	65 506.8	3 455.1	481.5	1.0	8.1
云　南　Yunnan	175 355.6	5 045.2	486.3	1.9	1.5
西　藏　Tibet	1 021.0	65.7	0.4	0.0	0.0
陕　西　Shaanxi	106 489.6	8 694.5	848.0	199.1	7.9
甘　肃　Gansu	96 072.0	14 040.2	332.1	10.2	0.3
青　海　Qinghai	40 585.6	1 906.8	320.3	1.5	0.0
宁　夏　Ningxia	111 549.6	9 254.4	145.8	9.5	1.5
新　疆　Xinjiang	187 629.6	12 133.2	275.4	39.9	3.3

各地区工业废水排放及处理情况（六）
Discharge and Treatment of Industrial Waste Water by Region（6）
（2011）

单位：吨　　　　　　　　　　　　　　　　　　　　　　　　　　　　　　　　　（tons）

年　份　Year 地　区　Region	工业废水中污染物排放量 Amount of Pollutants Discharged in the Industrial Waste Water				
	汞 Mercury	镉 Cadmium	总铬 Total Chrome	铅 Lead	砷 Arsenic
2011	1. 214	35. 096	290. 264	150. 831	145. 169
北　京　Beijing	0	0. 001	0. 446	0. 068	0. 013
天　津　Tianjin	0. 001	0. 010	0. 285	1. 454	0. 023
河　北　Hebei	0	0. 007	8. 429	0. 440	0. 045
山　西　Shanxi	0. 004	0. 825	0. 508	0. 638	0. 750
内蒙古　Inner Mongolia	0. 041	0. 536	0. 089	3. 033	4. 912
辽　宁　Liaoning	0. 002	0. 072	0. 624	0. 889	0. 459
吉　林　Jilin	0. 003	0. 004	0. 136	0. 123	1. 001
黑龙江　Heilongjiang	0	0. 001	0. 867	0. 023	0. 072
上　海　Shanghai	0	0. 003	2. 505	0. 104	0. 010
江　苏　Jiangsu	0. 093	0. 126	12. 221	3. 493	0. 769
浙　江　Zhejiang	0. 002	0. 261	21. 586	0. 454	0. 187
安　徽　Anhui	0. 004	0. 746	6. 852	2. 873	7. 064
福　建　Fujian	0. 045	0. 332	15. 276	5. 034	1. 484
江　西　Jiangxi	0. 088	2. 759	22. 541	9. 223	10. 795
山　东　Shandong	0. 029	0. 988	13. 770	0. 686	2. 080
河　南　Henan	0. 021	2. 755	37. 327	7. 019	1. 775
湖　北　Hubei	0. 212	0. 810	17. 365	4. 024	11. 921
湖　南　Hunan	0. 257	14. 409	33. 987	41. 007	55. 393
广　东　Guangdong	0. 045	1. 104	74. 665	11. 164	2. 126
广　西　Guangxi	0. 078	2. 487	4. 632	15. 596	9. 058
海　南　Hainan	0	0. 005	0. 131	0. 025	0. 027
重　庆　Chongqing	0	0	0. 592	0. 075	1. 377
四　川　Sichuan	0. 065	0. 160	1. 680	1. 575	3. 500
贵　州　Guizhou	0. 038	0. 114	0. 237	0. 494	0. 608
云　南　Yunnan	0. 052	3. 322	0. 057	27. 844	15. 431
西　藏　Tibet	0	0	0. 001	0	2. 721
陕　西　Shaanxi	0. 031	0. 954	1. 624	4. 201	0. 954
甘　肃　Gansu	0. 072	1. 419	4. 660	6. 858	5. 729
青　海　Qinghai	0. 010	0. 144	1. 025	1. 100	2. 063
宁　夏　Ningxia	0. 004	0. 028	0. 448	0. 102	0. 211
新　疆　Xinjiang	0. 019	0. 717	5. 697	1. 213	2. 615

各地区工业废气排放及处理情况（一）
Discharge and Treatment of Industrial Waste Gas by Region（1）
（2011）

年　份　Year 地　区　Region	工业废气排放总量 （标态）/亿米³ Total Volume of Industrial Waste Gas Emission （100 millon m^3）	工业废气中污染物排放量/吨 Volume of Pollutants Emission in the Industrial Waste Gas （tons）		
		二氧化硫 Sulphur Dioxide	氮氧化物 Nitrogen Oxide	烟（粉）尘 Soot and Dust
2011	674 509	20 172 265	17 297 077	11 008 839
北　京　Beijing	4 897	61 299	90 317	29 405
天　津　Tianjin	8 919	221 897	300 404	65 333
河　北　Hebei	77 185	1 317 099	1 220 888	1 223 502
山　西　Shanxi	42 195	1 294 321	986 503	1 008 410
内蒙古　InnerMongolia	30 063	1 250 180	1 150 486	604 534
辽　宁　Liaoning	31 701	1 048 914	767 838	591 152
吉　林　Jilin	10 637	363 443	410 046	361 279
黑龙江　Heilongjiang	10 377	415 209	484 901	419 810
上　海　Shanghai	13 704	210 092	326 949	66 446
江　苏　Jiangsu	48 182	1 025 013	1 195 592	486 389
浙　江　Zhejiang	24 790	647 048	690 811	302 752
安　徽　Anhui	30 411	487 228	722 700	410 590
福　建　Fujian	14 973	370 269	391 277	209 852
江　西　Jiangxi	16 102	568 058	393 521	359 085
山　东　Shandong	50 452	1 628 647	1 273 603	612 740
河　南　Henan	40 791	1 229 200	1 143 421	575 655
湖　北　Hubei	22 841	594 980	477 568	307 356
湖　南　Hunan	16 778	636 181	487 671	355 203
广　东　Guangdong	31 465	825 952	895 892	263 918
广　西　Guangxi	29 853	488 744	347 767	260 006
海　南　Hainan	1 676	31 058	65 449	11 054
重　庆　Chongqing	9 121	531 340	293 820	171 224
四　川　Sichuan	23 172	828 929	465 165	357 577
贵　州　Guizhou	10 820	903 132	446 933	257 164
云　南　Yunnan	17 545	642 523	348 291	352 817
西　藏　Tibet	114	1 369	3 802	4 452
陕　西　Shaanxi	15 704	831 222	633 708	396 997
甘　肃　Gansu	12 892	527 689	356 045	187 755
青　海　Qinghai	5 226	134 315	87 559	117 630
宁　夏　Ningxia	10 056	387 865	385 860	198 220
新　疆　Xinjiang	11 868	669 050	452 288	440 532

各地区工业废气排放及处理情况（二）
Discharge and Treatment of Industrial Waste Gas by Region（2）

单位：吨　　　　　　　　　　　　　　　　（2011）　　　　　　　　　　　　　　　　（tons）

年 份　Year 地 区　Region	工业废气中污染物产生量 Volume of Pollutants Emission in the Industrial Waste Gas		
	二氧化硫 Sulphur Dioxide	氮氧化物 Nitrogen Oxide	烟（粉）尘 Soot and Dust
2011	59 847 513	18 159 854	818 909 810
北 京　Beijing	160 817	100 549	3 615 113
天 津　Tianjin	578 843	309 578	7 699 268
河 北　Hebei	3 600 900	1 276 443	57 783 503
山 西　Shanxi	3 314 705	1 129 763	51 233 268
内蒙古　Inner Mongolia	4 231 951	1 183 519	69 813 492
辽 宁　Liaoning	2 224 153	787 899	36 530 104
吉 林　Jilin	569 467	418 226	18 933 319
黑龙江　Heilongjiang	568 285	490 436	19 303 245
上 海　Shanghai	545 040	357 279	7 329 122
江 苏　Jiangsu	3 040 900	1 286 012	38 302 383
浙 江　Zhejiang	2 297 462	748 456	26 035 283
安 徽　Anhui	2 324 137	735 207	38 489 287
福 建　Fujian	930 014	443 400	15 815 787
江 西　Jiangxi	2 241 362	415 200	19 255 497
山 东　Shandong	5 928 754	1 285 704	62 311 686
河 南　Henan	3 478 222	1 148 170	60 007 693
湖 北　Hubei	1 861 508	494 119	26 405 142
湖 南　Hunan	1 872 041	511 941	23 837 772
广 东　Guangdong	2 363 833	1 015 096	26 528 026
广 西　Guangxi	1 978 857	349 744	19 320 905
海 南　Hainan	104 620	90 093	1 516 007
重 庆　Chongqing	1 589 429	296 364	18 008 680
四 川　Sichuan	1 989 181	490 300	37 073 423
贵 州　Guizhou	2 693 459	447 648	24 212 198
云 南　Yunnan	2 233 594	350 355	26 551 600
西 藏　Tibet	1 369	3 802	312 027
陕 西　Shaanxi	2 172 337	676 374	22 940 053
甘 肃　Gansu	2 609 523	369 235	19 272 544
青 海　Qinghai	216 905	87 559	3 682 392
宁 夏　Ningxia	1 211 261	408 999	23 256 568
新 疆　Xinjiang	914 584	452 381	13 534 422

各地区工业废气排放及处理情况（三）
Discharge and Treatment of Industrial Waste Gas by Region（3）
（2011）

单位：套　　　　　　　　　　　　　　　　　　　　　　　　　　　　　　（sets）

年　份　Year 地　区　Region	废气治理设施数 Facilities for Treatment of Waste Gas	脱硫设施数 Desulfurization Facilities	脱硝设施数 Denitrification Facilities	除尘设施数 Dedusting Facilities
2011	216 457	20 539	497	169 788
北　京　Beijing	2 962	481	25	2 170
天　津　Tianjin	3 837	968	4	2 575
河　北　Hebei	16 602	1 199	23	12 279
山　西　Shanxi	11 557	2 916	23	8 555
内蒙古　Inner Mongolia	6 804	480	11	6 263
辽　宁　Liaoning	11 627	1 009	23	10 293
吉　林　Jilin	3 096	105	10	2 966
黑龙江　Heilongjiang	4 725	163	4	4 411
上　海　Shanghai	4 730	351	37	2 838
江　苏　Jiangsu	16 065	2 022	61	10 380
浙　江　Zhejiang	16 729	1 448	30	11 665
安　徽　Anhui	4 755	264	6	4 242
福　建　Fujian	7 592	145	23	6 649
江　西　Jiangxi	5 092	332	16	4 510
山　东　Shandong	15 308	2 738	17	11 920
河　南　Henan	9 944	964	16	8 886
湖　北　Hubei	7 355	432	13	4 866
湖　南　Hunan	5 282	469	13	4 668
广　东　Guangdong	17 211	1 605	71	10 901
广　西　Guangxi	6 028	426	9	5 425
海　南　Hainan	502	32	2	464
重　庆　Chongqing	4 134	249	3	3 275
四　川　Sichuan	9 411	475	26	7 576
贵　州　Guizhou	3 079	400	0	2 494
云　南　Yunnan	6 305	301	8	5 215
西　藏　Tibet	217	1	0	216
陕　西　Shaanxi	4 848	296	12	3 963
甘　肃　Gansu	3 288	108	6	3 112
青　海　Qinghai	1 099	13	0	1 085
宁　夏　Ningxia	1 719	70	5	1 630
新　疆　Xinjiang	4 554	77	0	4 296

各地区工业废气排放及处理情况（四）
Discharge and Treatment of Industrial Waste Gas by Region（4）

单位：万米³/小时 　　　　　　　　（2011）　　　　　　　　（10 000 m³/h）

年　份　Year 地　区　Region	废气治理设施处理能力（标态）Capacity of Facilities for Treatment of Waste Gas	脱硫设施处理能力 Capacity of Desulfurization Facilities	脱硝设施处理能力 Capacity of Denitrification Facilities	除尘设施处理能力 Capacity of Dedusting Facilities
2011	1 568 592	301 219	37 505	1 106 962
北　京　Beijing	10 947	2 696	1 375	6 075
天　津　Tianjin	23 588	6 164	260	15 687
河　北　Hebei	174 476	22 738	1 487	140 259
山　西　Shanxi	85 627	16 293	1 740	58 988
内蒙古　InnerMongolia	73 690	20 755	2 095	50 131
辽　宁　Liaoning	86 842	15 109	997	68 560
吉　林　Jilin	22 888	4 165	694	17 998
黑龙江　Heilongjiang	42 441	5 303	157	36 167
上　海　Shanghai	30 581	6 556	1 537	16 998
江　苏　Jiangsu	105 172	24 276	4 264	64 302
浙　江　Zhejiang	68 986	17 630	4 353	38 167
安　徽　Anhui	44 419	9 610	781	26 828
福　建　Fujian	35 301	7 281	3 172	24 313
江　西　Jiangxi	27 281	4 095	1 157	20 476
山　东　Shandong	153 377	46 325	1 456	94 659
河　南　Henan	77 041	14 332	1 136	60 558
湖　北　Hubei	78 940	5 717	359	70 752
湖　南　Hunan	34 635	7 491	921	25 040
广　东　Guangdong	94 094	20 069	6 374	43 960
广　西　Guangxi	49 948	4 749	23	39 789
海　南　Hainan	2 517	374	120	2 023
重　庆　Chongqing	22 710	3 214	4	17 590
四　川　Sichuan	54 666	5 436	680	45 029
贵　州　Guizhou	26 147	6 728	0	17 921
云　南　Yunnan	31 440	3 776	222	23 444
西　藏　Tibet	283	3	0	273
陕　西　Shaanxi	28 079	5 520	667	20 685
甘　肃　Gansu	18 935	3 300	493	13 492
青　海　Qinghai	6 670	556	0	6 110
宁　夏　Ningxia	21 111	6 077	980	12 970
新　疆　Xinjiang	35 760	4 881	0	27 720

各地区工业废气排放及处理情况（五）
Discharge and Treatment of Industrial Waste Gas by Region（5）

单位：万元　　　　　　　　　　　　　（2011）　　　　　　　　　　　　　　（10 000 yuan）

年　份　Year 地　区　Region	废气治理设施 运行费用 Annul Expenditure for Operation	脱硫设施运行费用 Annul Expenditure for Desulfurization Facilities	脱硝设施运行费用 Annul Expenditure for Denitrification Facilities	除尘设施运行费用 Annul Expenditure for Dedusting Facilities
2011	15 794 757.8	5 872 661.3	444 155.0	8 186 504.2
北　京　Beijing	82 698.9	25 091.6	16 714.0	32 008.2
天　津　Tianjin	436 180.8	232 651.6	11 050.5	176 325.7
河　北　Hebei	1 439 026.2	330 613.3	14 810.5	903 154.0
山　西　Shanxi	884 117.2	370 902.2	20 402.5	332 368.2
内蒙古　Inner Mongolia	520 124.2	234 274.5	4 401.8	265 833.8
辽　宁　Liaoning	580 509.8	159 523.6	5 626.8	369 434.7
吉　林　Jilin	139 339.6	41 734.3	3 756.1	85 910.0
黑龙江　Heilongjiang	1 672 870.2	43 846.5	1 570.4	1 612 885.6
上　海　Shanghai	416 225.5	147 895.9	33 807.9	163 831.7
江　苏　Jiangsu	1 139 761.9	456 183.0	75 240.2	506 475.4
浙　江　Zhejiang	796 157.5	335 886.2	79 506.2	273 611.0
安　徽　Anhui	430 479.1	150 938.5	9 014.0	239 937.0
福　建　Fujian	283 903.9	96 220.3	30 333.4	127 637.8
江　西　Jiangxi	313 726.1	127 386.0	10 569.4	169 211.3
山　东　Shandong	1 168 727.6	505 303.4	12 870.3	501 684.0
河　南　Henan	540 172.8	247 856.4	2 389.8	285 601.4
湖　北　Hubei	1 225 056.3	779 693.2	5 762.1	425 796.5
湖　南　Hunan	338 839.5	144 492.5	11 060.1	167 228.7
广　东　Guangdong	889 076.1	440 662.0	63 607.2	275 978.2
广　西　Guangxi	254 498.1	77 298.9	570.0	151 612.6
海　南　Hainan	37 135.6	15 481.0	4 000.0	17 605.0
重　庆　Chongqing	166 041.2	76 208.3	39.0	80 694.2
四　川　Sichuan	566 793.8	181 257.5	2 494.4	353 068.1
贵　州　Guizhou	327 679.9	187 694.2	0	65 018.1
云　南　Yunnan	301 815.5	98 784.0	379.0	184 642.4
西　藏　Tibet	906.6	10.0	0	835.6
陕　西　Shaanxi	223 923.0	116 913.9	4 972.4	92 379.3
甘　肃　Gansu	201 258.2	110 234.2	5 195.0	83 392.4
青　海　Qinghai	64 001.9	9 017.7	0	54 945.2
宁　夏　Ningxia	209 692.9	97 927.8	14 012.0	94 423.8
新　疆　Xinjiang	144 017.9	30 678.6	0	92 974.3

各地区一般工业固体废物产生及处置利用情况

Generation and Utilization of Industrial Solid Wastes by Region

单位：万吨　　　　　　　　　　　　　　　（2011）　　　　　　　　　　　　　（10 000 tons）

年　份　Year 地　区　Region	一般工业固体废物产生量 Industrial Solid Wastes Generated	一般工业固体废物综合利用量 Industrial Solid Wastes Utilized	一般工业固体废物处置量 Industrial Solid Wastes Disposed	一般工业固体废物贮存量 Stock of Industrial Solid Wastes	一般工业固体废物倾倒丢弃量 Industrial Solid Wastes Discharged	一般工业固体废物综合利用率（%） Industrial Solid Wastes Utilization Rate (%)
2011	322 772	195 215	70 465	60 424	433	59.9
北　京　Beijing	1 126	749	349	28	0	66.5
天　津　Tianjin	1 752	1 749	9	0	0	99.5
河　北　Hebei	45 129	18 821	6 806	20 184	0	41.1
山　西　Shanxi	27 556	15 818	9 187	2 578	29	57.2
内蒙古　InnerMongolia	23 584	13 701	7 429	2 647	3	58.0
辽　宁　Liaoning	28 270	10 748	13 394	4 335	8	37.9
吉　林　Jilin	5 379	3 171	920	1 290	0	58.9
黑龙江　Heilongjiang	6 017	4 139	643	1 308	7	67.6
上　海　Shanghai	2 442	2 358	75	11	0	96.5
江　苏　Jiangsu	10 475	9 997	335	220	0	94.9
浙　江　Zhejiang	4 446	4 092	310	51	0	91.9
安　徽　Anhui	11 473	9 366	1 761	1 096	0	78.7
福　建　Fujian	4 415	3 024	1 304	98	1	68.4
江　西　Jiangxi	11 372	6 305	652	4 420	15	55.3
山　东　Shandong	19 533	18 298	1 106	350	0	92.4
河　南　Henan	14 574	10 964	2 602	1 200	2	74.8
湖　北　Hubei	7 596	6 007	1 424	268	17	77.6
湖　南　Hunan	8 487	5 679	2 215	696	9	66.2
广　东　Guangdong	5 849	5 119	810	75	3	84.9
广　西　Guangxi	7 438	4 292	2 050	1 516	3	57.5
海　南　Hainan	421	201	182	42	0	47.7
重　庆　Chongqing	3 299	2 585	518	199	24	77.8
四　川　Sichuan	12 684	6 002	3 988	2 773	6	47.1
贵　州　Guizhou	7 598	4 015	2 033	1 552	29	52.3
云　南　Yunnan	17 335	8 728	4 969	3 687	169	49.4
西　藏　Tibet	301	8	16	284	0	2.7
陕　西　Shaanxi	7 118	4 266	1 836	1 008	9	59.9
甘　肃　Gansu	6 524	3 342	1 143	1 143	7	51.2
青　海　Qinghai	12 017	6 785	14	5 226	0	56.4
宁　夏　Ningxia	3 344	2 048	865	435	2	61.1
新　疆　Xinjiang	5 219	2 838	621	1 702	89	54.1

各地区工业危险废物产生及处置利用情况
Generation and Utilization of Hazardous Wastes by Region
（2011）

单位：吨   （tons）

年 份 Year 地 区 Region	危险废物产生量 Hazardous Wastes Generated	危险废物综合利用量 Hazardous Wastes Utilized	危险废物处置量 Hazardous Wastes Disposed	危险废物贮存量 Stock of Hazardous Wastes	危险废物倾倒丢弃量 Hazardous Wastes Discharged
2011	34 312 203	17 730 523	9 164 762	8 237 304	96
北 京 Beijing	119 190	50 545	68 645	0	0
天 津 Tianjin	102 690	30 935	71 762	12	0
河 北 Hebei	520 375	332 671	187 931	0	0
山 西 Shanxi	222 055	180 632	41 103	364	0
内蒙古 InnerMongolia	1 118 759	544 308	499 320	116 507	0
辽 宁 Liaoning	784 945	598 590	252 989	919	2
吉 林 Jilin	900 435	601 254	299 015	271	0
黑龙江 Heilongjiang	195 633	36 996	151 560	7 280	0
上 海 Shanghai	563 600	301 257	260 051	3 424	0
江 苏 Jiangsu	1 889 447	952 939	933 328	10 136	0
浙 江 Zhejiang	784 707	318 319	460 814	13 055	0
安 徽 Anhui	242 811	212 707	29 489	1 918	0
福 建 Fujian	103 082	43 764	57 128	2 717	0
江 西 Jiangxi	232 396	178 909	51 992	1 861	32
山 东 Shandong	9 378 395	6 575 217	2 909 617	2 610	0
河 南 Henan	476 202	394 231	70 452	13 660	0
湖 北 Hubei	405 281	268 499	208 472	3 188	36
湖 南 Hunan	2 466 776	1 943 783	442 171	231 461	0
广 东 Guangdong	1 267 933	701 988	569 051	4 558	0
广 西 Guangxi	229 360	134 949	57 129	74 443	0
海 南 Hainan	7 149	1 247	5 177	725	0
重 庆 Chongqing	464 961	56 373	436 102	5 066	0
四 川 Sichuan	1 158 495	779 693	467 482	11 238	0
贵 州 Guizhou	388 906	129 486	10 482	256 865	25
云 南 Yunnan	1 340 964	877 395	195 074	270 492	0
西 藏 Tibet	0	0	0	0	0
陕 西 Shaanxi	331 539	173 220	98 614	60 631	0
甘 肃 Gansu	246 773	108 801	179 185	80 569	0
青 海 Qinghai	3 559 836	563 585	60 967	2 980 558	0
宁 夏 Ningxia	108 139	98 117	6 388	3 647	0
新 疆 Xinjiang	4 701 369	540 113	83 272	4 079 129	0

各地区汇总工业企业概况（一）

Summarization of Industrial Enterprises Investigeted by Region（1）

（2011）

年　份 Year 地　区 Region	汇总工业企业数/个 Number of Industrial Enterprises Investigated（unit）	工业总产值（现价）/亿元 Gross Industrial Output Value（current rate）（100 million yuan）	工业炉窑数/台 Number of Industrial Furnaces（unit）	工业锅炉数/台 Number of Industrial Boilers（unit）	35 蒸吨及以上的 Beyond 35 tons of Steam	20（含）～35 蒸吨的 Between 20 and 35 tons of Steam	10（含）～20 蒸吨的 Between 20 and 35 tons of Steam	10 蒸吨以下的 Below 10 tons of Steam
2011	153 027	450 660.2	101 705	104 009	13 013	7 619	13 984	69 393
北　京 Beijing	914	6 471.6	330	1 751	301	288	304	858
天　津 Tianjin	1 988	10 596.8	718	2 150	412	288	336	1 114
河　北 Hebei	10 524	73 120.6	6 145	6 209	732	325	804	4 348
山　西 Shanxi	6 464	11 971.1	5 521	5 778	701	419	863	3 795
内蒙古 Inner Mongolia	3 187	7 721.5	3 847	4 051	747	418	835	2 051
辽　宁 Liaoning	6 586	18 001.9	6 602	7 211	1 120	971	1 343	3 777
吉　林 Jilin	1 489	7 367.5	690	2 682	596	238	305	1 543
黑龙江 Heilongjiang	2 083	14 591.0	3 443	4 852	727	454	706	2 965
上　海 Shanghai	2 283	16 561.2	1 503	2 254	158	77	185	1 834
江　苏 Jiangsu	11 291	40 929.5	4 439	7 137	987	298	661	5 191
浙　江 Zhejiang	13 931	25 243.2	5 168	8 227	601	241	864	6 521
安　徽 Anhui	8 552	10 542.7	4 482	2 728	249	105	373	2 001
福　建 Fujian	5 802	9 454.1	2 479	3 451	154	105	369	2 823
江　西 Jiangxi	5 197	7 038.9	4 130	2 068	127	75	219	1 647
山　东 Shandong	8 008	67 111.9	4 685	6 603	1 497	719	949	3 438
河　南 Henan	6 678	12 717.5	5 267	4 234	480	260	515	2 979
湖　北 Hubei	3 911	10 383.3	2 494	2 339	222	98	321	1 698
湖　南 Hunan	5 008	9 075.4	4 888	2 179	157	87	230	1 705
广　东 Guangdong	15 907	33 238.7	5 103	8 145	502	291	1 030	6 322
广　西 Guangxi	3 565	6 983.3	2 446	1 846	340	195	199	1 112
海　南 Hainan	483	1 310.4	194	289	50	45	11	183
重　庆 Chongqing	3 212	5 349.4	5 305	1 438	109	49	119	1 161
四　川 Sichuan	7 720	11 305.2	6 236	3 480	219	180	362	2 719
贵　州 Guizhou	4 131	3 250.7	3 415	1 542	130	51	100	1 261
云　南 Yunnan	4 112	6 776.3	2 194	1 698	253	165	191	1 089
西　藏 Tibet	93	44.6	20	28	0	0	2	26
陕　西 Shaanxi	4 070	8 067.6	3 693	3 123	287	191	446	2 199
甘　肃 Gansu	2 473	6 496.9	2 609	1 821	169	209	441	1 002
青　海 Qinghai	597	1 445.0	1 005	433	47	45	61	280
宁　夏 Ningxia	938	1 973.1	1 360	1 208	230	141	189	648
新　疆 Xinjiang	1 830	5 519.4	1 294	3 054	709	591	651	1 103

各地区汇总工业企业概况（二）

Summarization of Industrial Enterprises Investigeted by Region（2）

（2011）

年 份 地 区	Year Region	工业用水 总量/万吨 Quantity of Water Used by Industry （10 000 tons）	取水量 Fresh Water	重 复 用水量 Recycled Water	工业煤炭 消耗量/ 万吨 Total Amount of Industrial Coal Consumed （10 000 tons）	燃料煤 Coal Used as Fuel
	2011	39 622 524	6 678 557	32 943 967	385 802	288 960
北 京	Beijing	450 344	20 061	430 283	1 651	1 530
天 津	Tianjin	1 028 306	40 283	988 023	4 537	4 262
河 北	Hebei	4 392 566	267 161	4 125 404	27 287	18 689
山 西	Shanxi	1 881 496	125 813	1 755 683	31 641	18 823
内蒙古	InnerMongolia	1 523 065	101 760	1 421 305	32 471	23 803
辽 宁	Liaoning	1 912 752	406 398	1 506 354	16 718	13 364
吉 林	Jilin	913 444	105 430	808 014	9 526	7 761
黑龙江	Heilongjiang	816 593	107 176	709 417	13 440	8 074
上 海	Shanghai	833 695	85 828	747 867	5 935	4 390
江 苏	Jiangsu	3 543 317	651 022	2 892 295	25 581	23 514
浙 江	Zhejiang	2 445 866	901 053	1 544 814	14 826	13 772
安 徽	Anhui	1 348 663	179 070	1 169 593	12 965	10 188
福 建	Fujian	964 284	426 412	537 871	7 920	6 811
江 西	Jiangxi	697 649	160 814	536 835	6 446	4 798
山 东	Shandong	3 341 557	338 941	3 002 616	30 833	23 723
河 南	Henan	2 640 918	246 602	2 394 315	29 269	18 259
湖 北	Hubei	1 494 298	325 448	1 168 850	10 484	6 870
湖 南	Hunan	1 011 478	295 794	715 684	9 724	6 489
广 东	Guangdong	1 875 510	854 455	1 021 055	18 096	16 381
广 西	Guangxi	1 411 357	343 214	1 068 143	6 806	5 256
海 南	Hainan	214 839	11 041	203 797	797	796
重 庆	Chongqing	645 054	172 196	472 858	4 657	3 626
四 川	Sichuan	894 302	132 114	762 188	8 949	6 369
贵 州	Guizhou	784 545	41 158	743 386	8 577	7 354
云 南	Yunnan	949 043	93 320	855 724	8 480	6 331
西 藏	Tibet	2 760	1 478	1 282	39	34
陕 西	Shaanxi	422 468	78 861	343 608	11 699	9 274
甘 肃	Gansu	230 932	47 469	183 463	5 945	4 993
青 海	Qinghai	137 085	14 792	122 293	1 360	1 099
宁 夏	Ningxia	216 387	40 315	176 072	10 843	6 662
新 疆	Xinjiang	597 953	63 078	534 876	8 298	5 667

各地区汇总工业企业概况（三）

Summarization of Industrial Enterprises Investigeted by Region（3）

（2011）

年 份 地 区	Year Region	燃料油消耗量/万吨 Total Amount of Fuel Oil Consumed （10 000 tons）	焦炭消耗量/万吨 Total Amount of Coke Consumed （10 000 tons）	天然气消耗量/ 亿米³ Total Amount of Natural Gas Consumed （100 million m³）	其他燃料消耗量/ 万吨标煤 Total Amount of other Fuel Consumed （10 000 tons of standard coal）
	2011	1 613	28 698	1 542	16 466
北 京	Beijing	21	1	29	94
天 津	Tianjin	15	428	18	262
河 北	Hebei	19	6 898	265	2 001
山 西	Shanxi	4	2 120	10	456
内蒙古	InnerMongolia	11	1 264	28	276
辽 宁	Liaoning	221	2 172	19	1 958
吉 林	Jilin	28	556	18	154
黑龙江	Heilongjiang	24	286	25	372
上 海	Shanghai	69	152	37	645
江 苏	Jiangsu	150	1 763	133	1 822
浙 江	Zhejiang	82	334	51	187
安 徽	Anhui	14	926	41	700
福 建	Fujian	74	395	30	475
江 西	Jiangxi	30	872	3	556
山 东	Shandong	93	2 365	361	961
河 南	Henan	34	1 157	67	479
湖 北	Hubei	44	345	11	150
湖 南	Hunan	11	497	8	836
广 东	Guangdong	368	488	94	663
广 西	Guangxi	56	705	6	737
海 南	Hainan	3	0	28	30
重 庆	Chongqing	3	301	30	258
四 川	Sichuan	10	840	72	1 234
贵 州	Guizhou	12	339	4	44
云 南	Yunnan	13	1 208	0	245
西 藏	Tibet	12	0	0	0
陕 西	Shaanxi	166	427	47	142
甘 肃	Gansu	12	289	13	115
青 海	Qinghai	2	620	24	70
宁 夏	Ningxia	2	212	5	43
新 疆	Xinjiang	9	739	68	502

各地区工业污染防治投资情况（一）

Treatment Investment for Industrial Pollution by Region（1）

单位：个 　　　　　　　　　　　　　　　　（2011）　　　　　　　　　　　　　　　　（unit）

年　份 Year 地　区 Region	汇总工业企业数 Number of Industrial Enterprises Collected	本年施工项目总数 Numer of Projects under Construction	工业废水治理项目 Treatment of Waste Water	工业废气治理项目 Treatment of Waste Gas	脱硫治理项目 Treatment of desulfurization	脱硝治理项目 Treatment of Denitration	工业固体废物治理项目 Treatment of Solid Wastes	噪声治理项目 Treatment of Noise Pollution	其他治理项目 Treatment of Other Pollution
2011	6 500	9 257	3 738	3 478	956	98	577	175	1 289
北　京 Beijing	48	66	25	30	14	0	5	2	4
天　津 Tianjin	90	117	40	51	24	1	4	2	20
河　北 Hebei	212	340	90	203	74	9	10	1	36
山　西 Shanxi	309	528	110	207	82	7	66	13	132
内蒙古 Inner Mongolia	151	245	60	108	25	3	26	5	46
辽　宁 Liaoning	139	177	75	64	20	1	11	7	20
吉　林 Jilin	96	117	36	60	26	2	7	0	14
黑龙江 Heilongjiang	65	73	25	35	13	1	5	0	8
上　海 Shanghai	77	120	33	58	5	2	3	6	20
江　苏 Jiangsu	572	783	382	270	59	13	42	13	76
浙　江 Zhejiang	703	913	463	294	58	7	22	13	121
安　徽 Anhui	133	183	91	63	12	4	4	6	19
福　建 Fujian	298	399	193	121	23	5	26	10	49
江　西 Jiangxi	139	234	103	63	12	1	16	1	51
山　东 Shandong	737	959	353	452	138	10	41	9	104
河　南 Henan	261	373	117	167	61	6	25	9	55
湖　北 Hubei	112	182	74	74	16	4	5	5	24
湖　南 Hunan	154	243	113	56	21	5	34	8	32
广　东 Guangdong	631	936	365	369	83	10	25	10	167
广　西 Guangxi	170	290	139	73	29	1	36	4	38
海　南 Hainan	33	56	28	16	1	0	5	1	6
重　庆 Chongqing	174	227	115	51	18	0	25	10	26
四　川 Sichuan	255	344	170	104	20	1	19	15	36
贵　州 Guizhou	114	174	95	37	14	0	12	3	27
云　南 Yunnan	404	529	190	213	29	3	66	10	50
西　藏 Tibet	7	8	4	2	0	0	1	0	1
陕　西 Shaanxi	171	248	100	75	19	2	17	8	48
甘　肃 Gansu	76	114	43	43	6	0	11	1	16
青　海 Qinghai	25	54	11	16	2	0	4	0	23
宁　夏 Ningxia	61	89	17	54	10	0	4	2	12
新　疆 Xinjiang	83	136	78	49	42	0	0	1	8

各地区工业污染防治投资情况（二）
Treatment Investment for Industrial Pollution by Region（2）

单位：个 　　　　　　　　　　　　　　　　（2011）　　　　　　　　　　　　　（unit）

年　份 Year 地　区 Region	本年竣工项目总数 Number of Projects Completed	工业废水治理项目 Treatment of Waste Water	工业废气治理项目 Treatment of Waste Gas	脱硫治理项目 Treatment of Desulfuri-zation	脱硝治理项目 Treatment of Denitration	工业固体废物治理项目 Treatment of Solid Wastes	噪声治理项目 Treatment of Noise Pollution	其他治理项目 Treatment of Other Pollution
2011	7 005	2 633	2 787	753	62	420	146	1 019
北　京 Beijing	59	21	28	12	0	5	2	3
天　津 Tianjin	99	33	42	18	0	3	1	20
河　北 Hebei	280	71	168	64	6	9	0	32
山　西 Shanxi	424	79	183	69	4	49	12	101
内蒙古 Inner Mongolia	136	29	71	19	2	14	3	19
辽　宁 Liaoning	123	41	52	10	1	5	5	20
吉　林 Jilin	88	30	43	21	2	5	0	10
黑龙江 Heilongjiang	50	13	26	8	0	4	0	7
上　海 Shanghai	93	22	47	3	2	1	5	18
江　苏 Jiangsu	660	316	234	51	7	37	11	62
浙　江 Zhejiang	712	336	249	52	3	17	11	99
安　徽 Anhui	158	79	54	11	3	2	6	17
福　建 Fujian	233	111	71	15	1	12	8	31
江　西 Jiangxi	176	67	46	7	1	14	1	48
山　东 Shandong	770	250	396	114	9	27	8	89
河　南 Henan	286	93	122	43	2	25	8	38
湖　北 Hubei	149	59	64	13	4	4	5	17
湖　南 Hunan	196	99	42	16	3	23	6	26
广　东 Guangdong	690	243	274	65	7	19	7	147
广　西 Guangxi	198	98	51	19	0	24	4	21
海　南 Hainan	44	19	15	1	0	4	1	5
重　庆 Chongqing	169	73	42	14	0	21	9	24
四　川 Sichuan	274	129	84	19	1	15	12	34
贵　州 Guizhou	129	73	26	9	0	4	3	23
云　南 Yunnan	388	129	172	16	3	50	8	29
西　藏 Tibet	4	1	2	0	0	1	0	0
陕　西 Shaanxi	165	61	52	11	1	9	6	37
甘　肃 Gansu	88	30	35	5	0	9	1	13
青　海 Qinghai	35	7	11	1	0	4	0	13
宁　夏 Ningxia	65	11	39	6	0	4	2	9
新　疆 Xinjiang	64	10	46	41	0	0	1	7

各地区工业污染防治投资情况（三）

Treatment Investment for Industrial Pollution by Region（3）

单位：万元　　　　　　　　　　（2011）　　　　　　　　　　（10 000 yuan）

年　份 Year 地　区 Region	施工项目本年完成投资 Investment Completed in the Treatment of Industrial Pollution This Year	工业废水治理项目 Treatment of Waste Water	工业废气治理项目 Treatment of Waste Gas	脱硫治理项目 Treatment of Desulfurization	脱硝治理项目 Treatment of Denitration	工业固体废物治理项目 Treatment of Solid Wastes	噪声治理项目 Treatment of Noise Pollution	其他治理项目 Treatment of Other Pollution
2011	4 443 610.1	1 577 471.1	2 116 810.6	1 127 347.8	126 755.7	313 875.3	21 622.5	413 830.7
北　京 Beijing	10 945.7	5 892.6	4 514.9	2 831.1	0	104.2	16.0	418.0
天　津 Tianjin	152 848.5	31 317.9	46 799.6	26 056.5	0	33 245.0	14.0	41 472.0
河　北 Hebei	243 398.6	70 720.7	156 827.6	112 720.7	4 211.0	7 765.8	595.0	7 489.5
山　西 Shanxi	279 450.2	48 857.5	145 000.7	81 762.7	14 498.6	28 717.9	2 304.5	54 569.6
内蒙古 Inner Mongolia	310 164.0	40 807.5	214 616.6	163 448.6	5 600.0	17 299.3	658.5	36 782.0
辽　宁 Liaoning	116 031.8	63 953.5	44 931.8	31 359.7	190.0	2 869.4	573.3	3 703.8
吉　林 Jilin	65 624.1	25 059.2	35 434.4	23 533.6	5.0	815.3	0	4 315.2
黑龙江 Heilongjiang	100 891.1	18 565.3	78 077.1	68 151.3	270.0	693.5	0	3 555.2
上　海 Shanghai	63 601.6	4 929.7	51 244.0	9 845.3	2 360.0	59.9	878.5	6 489.5
江　苏 Jiangsu	310 062.0	125 480.5	134 460.7	63 489.3	23 572.6	13 905.5	1 420.6	34 794.7
浙　江 Zhejiang	178 372.8	88 097.8	65 062.7	16 231.1	22 271.1	2 232.5	1 486.0	21 493.8
安　徽 Anhui	92 793.4	26 865.0	29 532.0	7 838.8	2 195.4	2 073.5	273.2	34 049.7
福　建 Fujian	142 599.2	54 267.8	62 600.9	5 921.3	16 028.3	10 482.4	579.5	14 668.6
江　西 Jiangxi	66 234.7	24 440.6	18 401.2	1 849.9	8.2	4 745.0	12.0	18 635.9
山　东 Shandong	624 465.6	295 540.2	244 688.4	135 121.2	6 153.0	53 923.0	204.4	30 109.6
河　南 Henan	213 727.8	50 859.4	146 766.2	46 903.5	9 610.0	5 374.5	658.0	10 069.8
湖　北 Hubei	92 872.6	35 041.4	38 972.4	25 378.5	514.0	10 085.8	474.9	8 298.2
湖　南 Hunan	97 038.5	50 215.4	31 816.2	21 427.4	3 870.0	8 852.5	1 744.7	4 409.7
广　东 Guangdong	166 419.8	55 886.1	76 909.4	19 066.2	6 401.1	7 972.9	146.0	25 505.8
广　西 Guangxi	86 229.8	45 998.3	34 460.3	26 076.3	576.0	3 241.6	46.4	2 483.1
海　南 Hainan	27 533.6	18 403.8	7 038.5	15.0	0	1 218.0	120.0	753.3
重　庆 Chongqing	49 384.5	24 915.1	8 241.4	5 333.5	0	14 281.3	353.8	1 592.9
四　川 Sichuan	166 536.7	66 136.2	84 726.0	20 677.0	5 000.0	3 655.6	6 150.2	5 868.7
贵　州 Guizhou	131 969.6	7 872.9	64 249.4	43 628.0	0	56 022.9	89.3	3 735.2
云　南 Yunnan	137 331.1	40 869.6	67 798.9	24 831.0	1 668.5	12 015.9	1 214.9	15 431.8
西　藏 Tibet	1 628.1	660.0	580.0	0	0	380.0	0	8.1
陕　西 Shaanxi	237 247.8	157 712.0	60 408.9	50 081.4	1 753.0	5 582.5	1 417.9	12 126.5
甘　肃 Gansu	105 337.5	29 085.0	69 925.5	37 468.2	0	388.7	98.3	5 840.0
青　海 Qinghai	27 858.0	8 510.0	10 091.0	7 727.0	0	5 705.0	0	3 552.0
宁　夏 Ningxia	38 735.4	17 416.9	20 529.2	4 522.4	0	166.0	17.6	605.7
新　疆 Xinjiang	106 276.2	43 093.3	62 104.9	44 050.9	0	0	75.0	1 003.0

各地区工业污染防治投资情况（四）

Treatment Investment for Industrial Pollution by Region（4）

（2011）

年 份　Year 地 区　Region	本年竣工项目新增设计处理能力 Capacity for Treatment of Industrial Pollution New-added		
	治理废水/ （万吨/日） Treatment of Waste Water （10 000 tons/day）	治理废气（标态）/ （万米³/时） Treatment of Waste Gas （100 00 m³/hour）	治理固体废物/ （万吨/日） Treatment of Solid Wastes （10 000 tons/day）
2011	1 493.5	52 513.0	495.7
北 京　Beijing	2.4	63.5	4.9
天 津　Tianjin	9.7	513.9	0.1
河 北　Hebei	14.9	3 727.5	1.6
山 西　Shanxi	24.5	2 233.4	21.1
内蒙古　Inner Mongolia	68.4	747.2	90.7
辽 宁　Liaoning	19.1	1 235.8	12.1
吉 林　Jilin	5.9	1 133.7	0.3
黑龙江　Heilongjiang	5.1	207.3	11.1
上 海　Shanghai	1.2	312.0	0.7
江 苏　Jiangsu	113.0	6 455.8	10.9
浙 江　Zhejiang	59.8	491.6	0.1
安 徽　Anhui	8.9	560.5	⋯
福 建　Fujian	41.6	536.9	0.9
江 西　Jiangxi	47.8	394.4	1.0
山 东　Shandong	306.1	3 755.8	10.6
河 南　Henan	37.2	1 291.1	31.5
湖 北　Hubei	28.1	387.1	0.3
湖 南　Hunan	33.1	1 186.8	10.0
广 东　Guangdong	92.1	3 041.1	113.0
广 西　Guangxi	31.4	1 868.8	0.7
海 南　Hainan	7.3	9.5	0
重 庆　Chongqing	5.4	431.8	0.2
四 川　Sichuan	45.3	1 520.8	21.8
贵 州　Guizhou	5.8	725.5	29.4
云 南　Yunnan	219.1	17 072.5	15.3
西 藏　Tibet	30.0	0.2	6.5
陕 西　Shaanxi	21.9	1 524.2	27.0
甘 肃　Gansu	37.2	639.8	73.8
青 海　Qinghai	2.1	23.3	0.1
宁 夏　Ningxia	17.6	227.1	0
新 疆　Xinjiang	151.5	194.4	0

各地区农业污染排放情况（一）
Discharge of Agricultural Pollution by Region
（2011）

单位：万吨 （10 000 tons）

年 份 Year 地 区 Region	农业污染物排放（流失）总量 Total Amount of Discharge of Agricultural Pollution			
	化学需氧量 COD	氨氮 Ammonial Nitrogen	总氮 Total Nitrogen	总磷 Total Phosphorus
2011	1 186.1	82.7	454.7	56.3
北 京 Beijing	8.2	0.5	3.3	0.4
天 津 Tianjin	11.5	0.6	3.7	0.5
河 北 Hebei	94.1	4.6	45	6.8
山 西 Shanxi	18.4	1.3	8.8	1
内蒙古 Inner Mongolia	64.7	1.3	15.9	1.7
辽 宁 Liaoning	88.3	3.6	20	2.8
吉 林 Jilin	52.8	1.8	11	1.4
黑龙江 Heilongjiang	110.7	3.6	26.9	2.9
上 海 Shanghai	3.4	0.3	1.5	0.2
江 苏 Jiangsu	39.9	4	17.7	1.9
浙 江 Zhejiang	21.3	2.8	9.2	1.1
安 徽 Anhui	39.9	3.9	18.9	3
福 建 Fujian	22.5	3.4	9.5	1.3
江 西 Jiangxi	25.0	3.1	9.5	1.4
山 东 Shandong	138.0	7.6	56	6.5
河 南 Henan	82.3	6.6	41.5	4.9
湖 北 Hubei	48.1	4.7	19.8	2.5
湖 南 Hunan	58.6	6.4	23.1	2.8
广 东 Guangdong	62.0	5.9	20.1	3.1
广 西 Guangxi	22.3	2.8	11.6	1.4
海 南 Hainan	10.7	0.9	3.9	0.6
重 庆 Chongqing	12.7	1.4	5.5	0.7
四 川 Sichuan	54.9	5.9	23	2.7
贵 州 Guizhou	5.6	0.8	4.4	0.5
云 南 Yunnan	7.3	1.2	8.1	0.9
西 藏 Tibet	0.5	0	0.5	0.1
陕 西 Shaanxi	20.1	1.6	10.2	1
甘 肃 Gansu	14.7	0.6	5.2	0.5
青 海 Qinghai	2.2	0.1	0.8	0.1
宁 夏 Ningxia	10.3	0.2	3.1	0.3
新 疆 Xinjiang	35.1	1.2	17.1	1.4

各地区农业污染排放情况（二）
Discharge of Agricultural Pollution by Region（2）
（2011）

单位：吨 　　　　　　　　　　　　　　　　　　　　　　　　　　　　　　　　（tons）

年 份 地 区	Year Region	化学需氧量排放 总量 Total Amount of COD Discharged	水产养殖业 Aquiculture	畜禽养殖业 Livestock and Poultry	规模化畜禽养 殖场/小区 Large-scale Farms	养殖专业户 Small-scale Farms
	2011	1 1861 057	556 485	11 304 573	3 884 218	7 420 355
北 京	Beijing	81 692	3 833	77 859	41 379	36 479
天 津	Tianjin	114 674	15 405	99 268	34 484	64 784
河 北	Hebei	940 830	8 811	932 019	382 002	550 017
山 西	Shanxi	184 473	246	184 227	73 473	110 755
内蒙古	Inner Mongolia	647 266	677	646 589	122 841	523 748
辽 宁	Liaoning	883 488	13 115	870 373	257 941	612 432
吉 林	Jilin	528 297	1 022	527 275	96 481	430 795
黑龙江	Heilongjiang	1107 126	4 743	1 102 383	238 714	863 668
上 海	Shanghai	33 932	5 561	28 371	14 712	13 659
江 苏	Jiangsu	399 292	55 986	343 306	111 229	232 077
浙 江	Zhejiang	213 002	24 087	188 915	79 601	109 315
安 徽	Anhui	398 791	13 738	385 053	193 188	191 865
福 建	Fujian	225 214	32 828	192 386	86 822	105 563
江 西	Jiangxi	249 935	12 978	236 956	119 680	117 276
山 东	Shandong	1 379 733	9 368	1 370 365	377 874	992 491
河 南	Henan	823 045	8 908	814 138	460 947	353 191
湖 北	Hubei	481 159	142 614	338 545	170 910	167 634
湖 南	Hunan	585 668	39 979	545 690	191 559	354 131
广 东	Guangdong	620 268	114 508	505 760	283 338	222 422
广 西	Guangxi	222 786	12 678	210 109	64 348	145 761
海 南	Hainan	107 203	13 517	93 686	25 731	67 955
重 庆	Chongqing	127 357	2 821	124 536	43 582	80 954
四 川	Sichuan	549 240	8 571	540 669	168 190	372 479
贵 州	Guizhou	55 802	1 089	54 713	15 078	39 635
云 南	Yunnan	72 896	3 167	69 729	18 986	50 743
西 藏	Tibet	4 635	0	4 635	402	4 233
陕 西	Shaanxi	201 321	1 457	199 864	73 579	126 285
甘 肃	Gansu	146 599	377	146 222	23 151	123 071
青 海	Qinghai	21 501	5	21 496	7 479	14 018
宁 夏	Ningxia	102 803	2 501	100 303	31 484	68 819
新 疆	Xinjiang	351 028	1 895	349 133	75 035	274 098

各地区农业污染排放情况（三）
Discharge of Agricultural Pollution by Region（3）

单位：吨　　　　　　　　　　　　　　　　（2011）　　　　　　　　　　　　　　　　（tons）

年　份　Year 地　区　Region	氨氮排放总量 Total Amount of Ammonia Nitrogen Discharged	种植业 Crops	水产养殖业 Aquiculture	畜禽养殖业 Livestock and Poultry	规模化畜禽养殖场/小区 Large-scale Farms	养殖专业户 Small-scale Farms
2011	826 529	151 424	23 143	651 962	362 832	289 130
北　京　Beijing	4 813	372	188	4 252	2 874	1 379
天　津　Tianjin	5 961	488	713	4 760	2 070	2 690
河　北　Hebei	45 871	5 947	771	39 154	20 225	18 929
山　西　Shanxi	12 644	2 818	24	9 802	5 309	4 492
内蒙古　Inner Mongolia	12 634	3 152	44	9 437	3 092	6 345
辽　宁　Liaoning	35 510	2 487	907	32 116	13 797	18 319
吉　林　Jilin	18 237	2 717	65	15 456	5 794	9 662
黑龙江　Heilongjiang	35 635	7 644	202	27 790	13 352	14 438
上　海　Shanghai	3 500	908	53	2 538	1 593	945
江　苏　Jiangsu	39 925	13 723	1 819	24 383	10 737	13 646
浙　江　Zhejiang	27 702	5 719	1 498	20 485	12 990	7 495
安　徽　Anhui	39 299	8 693	464	30 143	19 995	10 148
福　建　Fujian	34 027	3 876	6 983	23 168	14 642	8 526
江　西　Jiangxi	31 290	5 590	164	25 536	19 051	6 485
山　东　Shandong	75 800	8 940	774	66 086	30 030	36 056
河　南　Henan	66 202	4 900	437	60 865	47 149	13 716
湖　北　Hubei	47 451	13 734	1 254	32 463	20 931	11 532
湖　南　Hunan	64 370	8 855	410	55 105	30 357	24 748
广　东　Guangdong	59 265	9 165	3 631	46 469	33 778	12 691
广　西　Guangxi	27 825	9 333	779	17 714	8 985	8 728
海　南　Hainan	9 445	2 121	826	6 498	2 446	4 052
重　庆　Chongqing	13 612	2 425	60	11 126	6 588	4 538
四　川　Sichuan	58 700	9 758	298	48 644	20 337	28 307
贵　州　Guizhou	7 501	3 332	373	3 796	1 393	2 403
云　南　Yunnan	11 822	7 533	109	4 180	2 361	1 819
西　藏　Tibet	485	428	0	57	22	35
陕　西　Shaanxi	15 625	3 263	68	12 295	5 894	6 401
甘　肃　Gansu	5 914	1 641	20	4 253	1 202	3 052
青　海　Qinghai	856	154	1	702	452	250
宁　夏　Ningxia	2 302	412	120	1 770	814	956
新　疆　Xinjiang	12 305	1 297	87	10 921	4 575	6 346

各地区农业污染排放情况（四）
Discharge of Agricultural Pollution by Region（4）
（2011）

单位：吨   (tons)

年　份 地　区 Year Region	总氮排放总量 Total Amount of Total Nitrogen Discharged	种植业 Crops	水产养殖业 Aquiculture	畜禽养殖业 Livestock and Poultry	规模化畜禽养殖场/小区 Large-scale Farms	养殖专业户 Small-scale Farms
2011	4 546 677	1 495 332	85 589	2 965 756	1 503 427	1 462 329
北　京 Beijing	32 840	8 262	629	23 948	17 881	6 067
天　津 Tianjin	36 713	9 098	2 298	25 317	14 133	11 183
河　北 Hebei	450 082	114 624	1 344	334 115	143 152	190 963
山　西 Shanxi	87 879	35 274	85	52 520	32 544	19 976
内蒙古 Inner Mongolia	159 214	20 411	77	138 725	55 274	83 451
辽　宁 Liaoning	200 073	18 140	2 870	179 063	74 723	104 340
吉　林 Jilin	110 260	20 126	177	89 957	36 779	53 178
黑龙江 Heilongjiang	269 293	27 441	378	241 474	95 674	145 800
上　海 Shanghai	15 292	5 885	383	9 024	6 377	2 646
江　苏 Jiangsu	176 502	88 580	7 215	80 707	38 704	42 002
浙　江 Zhejiang	92 274	38 075	3 706	50 493	28 479	22 015
安　徽 Anhui	188 509	82 822	2 943	102 743	56 512	46 232
福　建 Fujian	95 450	28 143	11 240	56 066	32 760	23 307
江　西 Jiangxi	94 647	21 844	1 023	71 780	50 250	21 529
山　东 Shandong	559 805	228 213	6 881	324 711	163 169	161 543
河　南 Henan	414 561	149 892	1 629	263 039	200 081	62 958
湖　北 Hubei	197 703	76 668	14 592	106 444	71 695	34 749
湖　南 Hunan	230 748	68 584	4 509	157 655	83 928	73 727
广　东 Guangdong	201 430	66 403	13 512	121 516	82 428	39 087
广　西 Guangxi	115 629	66 648	2 325	46 656	19 544	27 112
海　南 Hainan	38 607	16 783	2 010	19 814	6 595	13 219
重　庆 Chongqing	55 088	22 089	355	32 644	17 510	15 134
四　川 Sichuan	230 228	78 594	1 898	149 736	74 969	74 767
贵　州 Guizhou	44 468	30 465	1 044	12 958	4 525	8 433
云　南 Yunnan	80 972	59 414	1 521	20 038	9 777	10 260
西　藏 Tibet	5 330	4 299	0	1 031	170	861
陕　西 Shaanxi	102 092	49 540	214	52 337	28 188	24 150
甘　肃 Gansu	51 853	20 493	67	31 293	10 538	20 755
青　海 Qinghai	7 913	1 537	1	6 376	3 448	2 928
宁　夏 Ningxia	30 608	5 431	384	24 793	12 209	12 583
新　疆 Xinjiang	170 617	31 554	280	138 783	31 410	107 373

各地区农业污染排放情况（五）
Discharge of Agricultural Pollution by Region（5）

单位：吨 　　　　　　　　　　　　　　　　　　　　　（2011） 　　　　　　　　　　　　　　　　　　　　　（tons）

年 份 Year 地 区 Region	总磷排放总量 Total Amount of Total phosphorus Discharged	种植业 Crops	水产养殖业 Aquiculture	畜禽养殖业 Livestock and Poultry	规模化畜禽养 殖场/小区 Large-scale Farms	养殖 专业户 Small-scale Farms
2011	563 340	115 987	16 388	430 965	249 224	181 740
北 京 Beijing	4 471	388	124	3 959	3 133	827
天 津 Tianjin	4 689	498	452	3 739	2 485	1 254
河 北 Hebei	68 133	8 298	267	59 568	24 318	35 251
山 西 Shanxi	9 631	1 817	19	7 795	5 533	2 262
内蒙古 Inner Mongolia	17 285	2 229	56	15 000	8 564	6 436
辽 宁 Liaoning	27 817	1 176	326	26 314	12 529	13 785
吉 林 Jilin	13 680	1 054	27	12 599	5 598	7 001
黑龙江 Heilongjiang	29 058	1 980	214	26 864	13 819	13 045
上 海 Shanghai	1 971	427	76	1 468	1 073	394
江 苏 Jiangsu	18 975	4 353	1 593	13 028	7 206	5 822
浙 江 Zhejiang	11 459	3 231	597	7 631	4 827	2 804
安 徽 Anhui	29 732	13 265	899	15 568	9 771	5 797
福 建 Fujian	12 968	2 975	1 925	8 069	5 410	2 658
江 西 Jiangxi	13 622	2 405	211	11 007	8 199	2 807
山 东 Shandong	64 768	11 797	1 017	51 954	28 151	23 803
河 南 Henan	48 802	7 530	295	40 977	32 765	8 212
湖 北 Hubei	24 984	5 473	2 948	16 563	12 343	4 221
湖 南 Hunan	28 202	6 287	844	21 071	13 459	7 612
广 东 Guangdong	31 484	7 383	2 851	21 250	14 453	6 797
广 西 Guangxi	14 233	6 491	434	7 308	3 340	3 968
海 南 Hainan	5 808	2 792	312	2 704	1 146	1 558
重 庆 Chongqing	6 917	2 178	68	4 671	2 867	1 804
四 川 Sichuan	26 923	6 554	353	20 016	12 130	7 886
贵 州 Guizhou	4 529	2 710	223	1 596	786	810
云 南 Yunnan	8 817	5 856	67	2 893	1 645	1 248
西 藏 Tibet	795	701	0	94	28	65
陕 西 Shaanxi	10 235	3 202	43	6 989	4 592	2 398
甘 肃 Gansu	4 748	1 388	14	3 346	1 601	1 746
青 海 Qinghai	812	85	1	727	506	221
宁 夏 Ningxia	3 322	300	76	2 946	1 955	991
新 疆 Xinjiang	14 470	1 162	56	13 252	4 992	8 259

各地区城镇生活污染排放及处理情况（一）

Discharge and Treatment of Household Pollution by Region（1）

（2011）

年　份　Year 地　区　Region	城镇人口/ 万人 Urban Population （10 000 people）	生活煤炭消费量/ 万吨 Total Amount of Household Coal Consumed （10 000 tons）	生活用水总量/ 万吨 Quantity of Water Consumed by Household （10 000 tons）	居民家庭 用水总量 Water Consumed in Household	公共服务 用水总量 Water Consumed by Public Services
2011	69 258.2	15 909.6	4 983 749.9	4 115 515.5	868 234.4
北　京　Beijing	1 740.7	462.9	151 935.0	85 447.0	66 488.0
天　津　Tianjin	1 091.0	203.5	52 580.3	35 980.3	16 600.0
河　北　Hebei	3 260.0	814.0	187 873.3	159 982.2	27 891.1
山　西　Shanxi	1 759.1	1 329.3	108 641.7	80 063.0	28 578.7
内蒙古　Inner Mongolia	1 405.0	1 210.2	75 942.4	61 232.0	14 710.4
辽　宁　Liaoning	2 807.2	762.8	176 838.4	140 826.6	36 011.8
吉　林　Jilin	1 468.3	593.9	89 059.8	64 149.2	24 910.6
黑龙江　Heilongjiang	2 176.3	1 825.1	126 892.0	112 387.8	14 504.2
上　海　Shanghai	2 096.0	165.0	188 100.0	173 052.0	15 048.0
江　苏　Jiangsu	4 891.9	198.7	349 009.8	276 734.3	72 275.5
浙　江　Zhejiang	3 403.4	126.7	262 581.4	217 443.3	45 138.1
安　徽　Anhui	2 673.7	402.3	190 771.1	166 586.9	24 184.2
福　建　Fujian	2 160.2	115.8	159 954.7	141 449.1	18 505.6
江　西　Jiangxi	2 050.8	80.0	142 366.8	123 415.9	18 950.9
山　东　Shandong	4 909.7	1 428.3	292 657.0	253 714.6	38 942.4
河　南　Henan	3 808.7	1 308.3	287 736.6	243 412.7	44 323.9
湖　北　Hubei	2 913.6	734.8	214 147.0	186 566.0	27 581.0
湖　南　Hunan	2 975.0	356.8	219 352.3	176 346.7	43 005.6
广　东　Guangdong	6 985.9	193.8	716 000.9	625 050.4	90 950.5
广　西　Guangxi	1 942.0	165.9	137 807.1	114 924.3	22 882.8
海　南　Hainan	443.1	9.1	37 086.8	28 992.9	8 093.9
重　庆　Chongqing	1 606.0	149.8	114 536.0	101 873.7	12 662.3
四　川　Sichuan	3 404.0	318.4	235 503.4	170 432.6	65 070.8
贵　州　Guizhou	1 213.1	552.5	84 232.9	66 796.7	17 436.2
云　南　Yunnan	1 706.1	233.7	117 447.8	95 617.9	21 829.9
西　藏　Tibet	99.1	12.4	5 167.9	4 638.6	529.3
陕　西　Shaanxi	1 770.4	581.6	104 556.8	86 165.8	18 391.0
甘　肃　Gansu	953.5	518.6	50 174.7	42 136.2	8 038.5
青　海　Qinghai	263.0	232.2	15 579.4	10 979.3	4 600.1
宁　夏　Ningxia	319.0	107.2	22 687.4	17 070.7	5 616.7
新　疆　Xinjiang	962.5	716.3	66 529.1	52 046.7	14 482.4

各地区城镇生活污染排放及处理情况（二）

Discharge and Treatment of Household Pollution by Region（2）

（2011）

年　份　Year 地　区　Region	城镇生活污水 排放量/ 万吨 Amount of Household Waste Water Discharged （10 000 tons）	城镇生活化学需 氧量产生量/吨 Amount of Household COD Generated（tons）	城镇生活氨氮 产生量/吨 Amount of Household Ammonia Nitrogen Generated（tons）	城镇生活化学需 氧量排放量/吨 Amount of Household COD Discharged （tons）	城镇生活氨氮 排放量/吨 Amount of Household Ammonia Nitrogen Discharged（tons）
2011	4 279 159.1	16 984 202.1	2 166 719.0	9 388 423.1	1 476 626.1
北　京　Beijing	136 741.5	501 931.0	61 629.0	99 097.0	15 513.0
天　津　Tianjin	47 322.3	274 010.3	37 356.4	96 422.4	17 128.0
河　北　Hebei	159 968.9	724 496.3	96 342.1	246 119.9	50 119.2
山　西　Shanxi	76 439.1	418 153.3	55 467.3	211 404.7	37 302.9
内蒙古　Inner Mongolia	60 970.8	310 290.3	38 604.1	177 124.2	29 760.5
辽　宁　Liaoning	141 699.0	641 690.3	88 513.4	341 578.0	65 354.1
吉　林　Jilin	74 213.7	335 539.1	43 103.2	204 821.1	34 085.8
黑龙江　Heilongjiang	106 574.2	486 353.7	64 692.7	365 134.3	54 365.1
上　海　Shanghai	169 290.0	581 430.4	74 208.9	181 866.3	43 962.8
江　苏　Jiangsu	346 252.3	1 246 228.7	161 063.2	602 252.0	99 853.4
浙　江　Zhejiang	237 493.9	860 362.5	113 542.8	406 940.0	74 535.5
安　徽　Anhui	172 383.8	634 346.0	75 135.2	450 508.9	60 575.5
福　建　Fujian	138 728.9	528 432.3	68 630.5	354 193.4	53 259.7
江　西　Jiangxi	122 996.4	645 236.1	74 312.9	389 701.3	49 082.4
山　东　Shandong	255 869.5	1 155 123.1	153 887.1	457 270.3	84 869.1
河　南　Henan	239 998.6	862 890.9	112 707.4	410 241.0	72 492.4
湖　北　Hubei	188 384.1	729 004.7	85 902.2	461 322.7	64 645.5
湖　南　Hunan	181 243.2	716 226.0	86 967.1	531 850.7	71 212.4
广　东　Guangdong	606 589.4	1 784 935.6	229 524.1	1 006 295.5	154 633.0
广　西　Guangxi	121 107.8	468 101.9	59 426.0	361 672.3	47 304.8
海　南　Hainan	28 858.3	122 194.2	14 542.1	78 857.8	12 364.6
重　庆　Chongqing	97 355.6	398 609.2	50 998.5	230 769.5	38 069.0
四　川　Sichuan	199 244.9	871 057.1	106 724.7	610 360.4	78 572.9
贵　州　Guizhou	57 261.5	275 752.1	34 955.2	214 885.3	28 170.9
云　南　Yunnan	100 208.3	435 209.3	51 411.0	293 472.8	40 798.9
西　藏　Tibet	4 271.7	21 704.5	2 785.4	21 172.0	2 745.4
陕　西　Shaanxi	80 932.7	405 599.4	53 815.7	243 270.9	38 140.4
甘　肃　Gansu	39 491.5	204 003.9	26 400.9	152 303.4	22 512.2
青　海　Qinghai	12 606.7	55 965.2	7 481.2	39 202.6	6 714.7
宁　夏　Ningxia	20 138.0	71 650.0	9 591.0	18 568.1	6 397.8
新　疆　Xinjiang	54 522.4	217 674.7	26 997.7	129 744.5	22 084.3

各地区城镇生活污染排放及处理情况（三）
Discharge and Treatment of Household Pollution by Region（3）

单位：吨 （2011） （tons）

年 份 Year 地 区 Region	城镇生活二氧化硫 排放量 Amount of Household Sulphur Dioxide Emission	城镇生活氮氧化物 排放量 Amount of Household Nitrogen Oxide Emission	城镇生活烟尘 排放量 Amount of Household Soot Emission
2011	2 003 949.5	366 236.5	1 147 647.1
北 京 Beijing	36 538.9	12 145.1	31 966.7
天 津 Tianjin	8 959.0	4 447.0	4 071.0
河 北 Hebei	94 900.3	18 414.2	42 807.2
山 西 Shanxi	104 694.4	29 005.4	97 910.9
内蒙古 Inner Mongolia	159 220.4	24 502.6	106 339.3
辽 宁 Liaoning	77 208.4	13 700.9	70 499.2
吉 林 Jilin	49 747.2	11 262.1	52 003.5
黑龙江 Heilongjiang	106 678.7	42 508.0	209 921.8
上 海 Shanghai	29 860.0	9 079.2	14 850.0
江 苏 Jiangsu	28 541.5	6 236.4	11 963.4
浙 江 Zhejiang	14 526.9	3 729.6	3 929.5
安 徽 Anhui	41 841.0	8 502.4	15 498.7
福 建 Fujian	18 790.9	2 364.0	5 777.5
江 西 Jiangxi	15 988.1	2 715.7	10 001.6
山 东 Shandong	198 588.6	30 952.3	118 185.0
河 南 Henan	141 261.3	23 229.8	41 166.6
湖 北 Hubei	70 653.6	12 089.3	23 316.9
湖 南 Hunan	49 324.3	7 632.9	14 435.2
广 东 Guangdong	21 326.2	12 225.3	12 458.9
广 西 Guangxi	32 250.3	3 545.2	11 705.1
海 南 Hainan	1 512.0	289.1	565.0
重 庆 Chongqing	55 585.4	4 816.9	2 463.3
四 川 Sichuan	72 835.6	9 984.3	12 780.5
贵 州 Guizhou	201 151.4	11 800.0	36 929.3
云 南 Yunnan	48 696.0	5 984.7	14 002.3
西 藏 Tibet	2 807.1	247.2	1 235.5
陕 西 Shaanxi	85 613.5	19 204.8	43 342.2
甘 肃 Gansu	96 040.1	11 721.1	40 570.0
青 海 Qinghai	22 286.0	4 863.1	17 616.0
宁 夏 Ningxia	22 520.4	2 672.3	9 584.8
新 疆 Xinjiang	94 002.1	16 365.7	69 750.4

各地区机动车污染排放情况
Discharge of Motor Vehicle Pollution by Region
（2011）

单位：万吨 （10 000 tons）

年　份　Year 地　区　Region	机动车污染物排放总量 Total Amount of Discharge of Motor Vehicle Pollution			
	总颗粒物 Total Particulate	氮氧化物 Nitrogen Oxide	一氧化碳 Carbon Monoxide	碳氢化合物 Hydrocarbon
2011	62.924	637.585	3 466.570	440.475
北　京　Beijing	0.445	8.555	89.436	9.858
天　津　Tianjin	0.649	5.400	43.481	4.959
河　北　Hebei	5.613	56.177	255.901	33.178
山　西　Shanxi	2.316	27.040	132.812	16.673
内蒙古　Inner Mongolia	2.899	24.690	120.684	16.619
辽　宁　Liaoning	3.147	28.126	121.458	16.762
吉　林　Jilin	1.888	18.340	107.303	13.883
黑龙江　Heilongjiang	2.613	25.633	151.522	19.178
上　海　Shanghai	0.849	9.904	46.074	6.737
江　苏　Jiangsu	2.884	33.347	194.842	23.997
浙　江　Zhejiang	1.641	16.427	120.470	14.289
安　徽　Anhui	2.523	22.760	85.352	11.852
福　建　Fujian	0.961	10.061	68.828	8.148
江　西　Jiangxi	2.695	21.608	80.298	11.456
山　东　Shandong	5.288	48.557	236.183	31.372
河　南　Henan	5.137	49.853	204.754	27.007
湖　北　Hubei	1.546	17.998	104.308	12.585
湖　南　Hunan	1.477	17.101	82.049	10.293
广　东　Guangdong	4.771	47.961	330.720	37.818
广　西　Guangxi	1.655	14.264	89.637	11.056
海　南　Hainan	0.419	2.964	21.930	2.765
重　庆　Chongqing	0.727	10.397	62.949	7.869
四　川　Sichuan	1.541	19.935	116.156	13.953
贵　州　Guizhou	0.937	9.445	52.412	7.050
云　南　Yunnan	1.541	19.423	127.005	18.000
西　藏　Tibet	0.435	3.657	27.097	3.242
陕　西　Shaanxi	2.309	17.883	105.055	13.575
甘　肃　Gansu	0.785	11.304	101.368	12.082
青　海　Qinghai	0.309	3.169	24.122	3.067
宁　夏　Ningxia	0.766	6.964	29.189	4.242
新　疆　Xinjiang	2.157	28.643	133.179	16.909

各地区城镇污水处理情况（一）
Urban Waste Water Treatment by Region（1）

（2011）

年 份 Year 地 区 Region	污水处理厂数/座 Number of Urban Waste Water Treatment Plants （unit）	污水处理厂设计处理能力/（万吨/日） Treatment Capacity （10 000 tons/day）	本年运行费用/万元 Annul Expenditure for Operation （10 000 yuan）	污水处理厂累计完成投资/万元 Total Investment of Urban Waste Water Treatment （10 000 yuan）	新增固定资产/万元 Newly-added Fixed Assets （10 000 yuan）
2011	39 74	139 91	3 071 639.1	29 021 719.6	2 309 446.1
北 京 Beijing	116	410	106 592.9	826 604.6	10 395.8
天 津 Tianjin	52	235	47 655.9	367 651.3	3 692.8
河 北 Hebei	235	809	156 514.0	1 843 447.0	155 028.5
山 西 Shanxi	160	275	62 438.8	692 381.7	61 558.8
内蒙古 Inner Mongolia	98	267	51 601.0	861 698.1	57 546.7
辽 宁 Liaoning	121	568	117 229.1	950 010.9	49 185.6
吉 林 Jilin	47	251	46 148.1	369 577.5	10 248.7
黑龙江 Heilongjiang	52	260	47 490.7	574 233.9	7 494.1
上 海 Shanghai	55	670	136 142.7	1 233 708.5	26 472.9
江 苏 Jiangsu	560	13 20	323 007.1	3 329 085.5	256 684.2
浙 江 Zhejiang	224	977	314 250.1	2 235 560.3	115 263.5
安 徽 Anhui	105	445	87 567.2	869 125.6	58 713.0
福 建 Fujian	121	382	64 764.6	485 682.2	21 697.3
江 西 Jiangxi	112	285	57 996.0	559 342.7	32 283.6
山 东 Shandong	270	10 66	275 838.1	1 907 585.5	162 946.5
河 南 Henan	161	707	116 950.4	1 109 224.1	60 606.3
湖 北 Hubei	135	542	116 038.3	1 116 290.5	205 969.1
湖 南 Hunan	128	489	95 013.8	1 326 495.0	26 153.8
广 东 Guangdong	342	17 42	357 649.6	3 315 262.7	164 033.8
广 西 Guangxi	110	367	61 881.1	954 720.9	578 923.4
海 南 Hainan	44	129	20 394.4	286 171.7	1 203.1
重 庆 Chongqing	117	261	93 180.7	675 037.1	65 046.7
四 川 Sichuan	227	455	109 882.1	928 047.4	38 841.3
贵 州 Guizhou	103	166	26 387.6	417 658.5	19 804.0
云 南 Yunnan	56	208	55 537.0	505 824.4	43 077.9
西 藏 Tibet	2	1	162.0	0.0	0.0
陕 西 Shaanxi	107	271	50 199.9	507 901.1	21 822.4
甘 肃 Gansu	28	125	22 900.5	236 892.2	4 941.2
青 海 Qinghai	18	37	7 855.7	107 728.9	192.4
宁 夏 Ningxia	18	74	11 393.1	117 342.9	2 582.0
新 疆 Xinjiang	50	200	30 976.6	311 427.3	47 037.0

各地区城镇污水处理情况（二）
Urban Waste Water Treatment by Region（2）

单位：万吨 （2011） （10 000 tons）

年 份 Year 地 区 Region	污水实际 处理量 Quantity of Waste Water Treated	污水再生 利用量 Waste Water Recycled	工业用水量 Waste Water Recycled for Industry	市政用水量 Waste Water Recycled for Municipal Services	景观用水量 Waste Water Recycled for Landscape
2011	4 028 972	96 262	60 402	12 461	23 399
北 京 Beijing	119 624	8 926	4 180	1 688	3 059
天 津 Tianjin	60 344	210	152	58	0
河 北 Hebei	209 567	16 672	11 588	1 338	3 747
山 西 Shanxi	65 202	7 262	6 511	278	472
内蒙古 InnerMongolia	58 072	11 399	8 883	825	1 690
辽 宁 Liaoning	146 713	9 177	6 541	21	2 614
吉 林 Jilin	59 270	0	0	0	0
黑龙江 Heilongjiang	53 392	48	0	47	1
上 海 Shanghai	204 982	619	274	179	167
江 苏 Jiangsu	348 497	2 343	1 884	303	156
浙 江 Zhejiang	279 804	9 885	5 459	3 586	839
安 徽 Anhui	125 611	0	0	0	0
福 建 Fujian	109 732	219	175	36	8
江 西 Jiangxi	85 844	328	108	15	205
山 东 Shandong	516 281	7 221	5 082	554	1 585
河 南 Henan	203 003	3 950	3 837	0	113
湖 北 Hubei	153 308	686	74	553	59
湖 南 Hunan	132 206	120	67	32	20
广 东 Guangdong	512 013	4 974	1 563	364	3 047
广 西 Guangxi	86 697	204	178	13	12
海 南 Hainan	35 775	624	0	49	575
重 庆 Chongqing	76 060	0	0	0	0
四 川 Sichuan	132 280	2 054	7	530	1 517
贵 州 Guizhou	43 250	2 899	0	187	2 712
云 南 Yunnan	61 537	0	0	0	0
西 藏 Tibet	236	0	0	0	0
陕 西 Shaanxi	62 183	2 006	1 126	819	61
甘 肃 Gansu	25 148	3 065	2 452	605	8
青 海 Qinghai	7 722	0	0	0	0
宁 夏 Ningxia	18 010	402	0	0	402
新 疆 Xinjiang	36 619	969	258	381	330

各地区城镇污水处理情况（三）
Urban Waste Water Treatment by Region（3）

单位：万吨　　　　　　　　　　　　　　（2011）　　　　　　　　　　　　（10 000 tons）

年 份 Year 地 区 Region	污泥产生量 Quantity of Sludge Generated	污泥处置量 Quantity of Sludge Disposed	土地利用量 Landuse	填埋处置量 Landfill	建筑材料利用量 As Building Material	焚烧处置量 Incineration	污泥倾倒丢弃量 Quantity of Sludge Discharged
2011	2 267.54	2 267.24	354.60	1 270.38	228.59	413.67	0.12
北 京 Beijing	113.27	113.27	79.20	8.42	5.66	19.99	0
天 津 Tianjin	21.72	21.72	1.27	19.28	0.91	0.26	0
河 北 Hebei	140.58	140.58	17.53	120.40	1.01	1.64	0
山 西 Shanxi	36.56	36.56	17.27	17.25	0.02	2.02	0
内蒙古 Inner Mongolia	23.36	23.36	0.49	22.58	0.10	0.20	0
辽 宁 Liaoning	82.58	82.58	0.48	69.89	9.28	2.93	0
吉 林 Jilin	27.62	27.62	0.16	27.47	0	0	0
黑龙江 Heilongjiang	25.62	25.62	0.57	24.98	0	0.07	0
上 海 Shanghai	115.69	115.69	4.88	101.00	4.36	5.44	0
江 苏 Jiangsu	246.11	246.11	16.38	49.38	38.63	141.72	0
浙 江 Zhejiang	343.01	342.98	35.42	95.42	87.06	125.09	0.027 5
安 徽 Anhui	65.94	65.94	15.55	32.15	4.67	13.56	0
福 建 Fujian	67.69	67.69	10.69	45.50	4.46	7.05	0
江 西 Jiangxi	29.64	29.64	3.06	19.99	2.84	3.74	0
山 东 Shandong	193.60	193.60	57.09	80.54	16.06	39.91	0
河 南 Henan	121.11	121.11	14.09	102.32	3.81	0.89	0
湖 北 Hubei	39.90	39.90	1.17	35.90	0.02	2.81	0
湖 南 Hunan	41.02	41.02	0.55	37.26	3.20	0	0
广 东 Guangdong	253.07	253.07	34.38	159.23	23.22	36.23	…8
广 西 Guangxi	21.24	21.24	6.52	9.80	4.92	0	0
海 南 Hainan	6.68	6.68	6.50	0.18	0	0	0
重 庆 Chongqing	36.44	36.43	5.04	17.05	13.17	1.18	0.004 4
四 川 Sichuan	49.23	49.05	3.29	40.64	4.88	0.23	0
贵 州 Guizhou	16.15	16.15	0.03	16.06	0.05	0	0
云 南 Yunnan	55.80	55.80	5.17	50.63	0	0	0
西 藏 Tibet	0.13	0.13	0.05	0.09	0	0	0
陕 西 Shaanxi	48.21	48.21	7.55	36.69	0.02	3.96	0
甘 肃 Gansu	18.74	18.74	2.95	15.78	0	0	0
青 海 Qinghai	5.52	5.52	0	5.52	0	0	0
宁 夏 Ningxia	13.92	13.92	5.60	3.34	0.24	4.75	0
新 疆 Xinjiang	7.39	7.30	1.68	5.63	0	0	0.084 7

各地区城镇污水处理情况（四）
Urban Waste Water Treatment by Region（4）

单位：吨 　　　　　　　　　　　　　　　　（2011）　　　　　　　　　　　　　　　　（tons）

年　份　Year	污染物去除量 Quantity of Pollutants Removed by Urban Waste Water Treatment						
地　区　Region	化学需氧量 COD	氨氮 Ammonia Nitrogen	油类 Oil	总氮 Total Nitrogen	总磷 Total Phosphorus	挥发酚 Volatile Phenols	氰化物 Cyanide
2011	10 148 983.5	857 544.0	54 812.8	674 199.6	150 369.9	966.47	1 222.60
北　京　Beijing	387 310.7	51 995.8	6 482.5	42 280.0	4 473.5	21.17	1.41
天　津　Tianjin	231 509.0	22 592.8	291.4	39 507.0	2 533.0	0.14	0.07
河　北　Hebei	669 398.1	59 201.1	498.1	23 235.1	24 926.4	24.55	0.02
山　西　Shanxi	215 372.3	17 911.0	3 983.2	14 543.6	16 554.1	30.34	0.13
内蒙古　Inner Mongolia	164 995.5	11 798.0	1 035.9	8 325.8	1 343.7	3.00	8.92
辽　宁　Liaoning	341 610.6	24 276.1	1 612.8	18 844.4	7 677.4	12.65	0.00
吉　林　Jilin	168 497.4	10 053.4	598.0	2 137.5	364.9	77.62	9.83
黑龙江　Heilongjiang	135 640.6	11 620.8	1 552.0	11 328.5	1 288.1	6.12	1.28
上　海　Shanghai	519 251.6	33 778.6	2 115.3	41 740.9	6 632.2	149.36	2.59
江　苏　Jiangsu	876 566.3	78 283.4	3 392.2	48 062.1	7 666.1	45.37	0.12
浙　江　Zhejiang	1 156 024.6	64 101.2	6 652.3	35 628.1	8 316.0	0	307.54
安　徽　Anhui	201 535.0	16 429.8	561.7	15 810.2	1 927.3	35.14	0.06
福　建　Fujian	278 617.4	19 829.9	918.6	18 278.3	2 040.9	0	728.30
江　西　Jiangxi	121 794.4	10 203.1	405.6	6 036.6	1 025.9	0	0
山　东　Shandong	1 470 883.1	146 396.0	541.5	102 321.9	17 769.7	65.62	13.99
河　南　Henan	516 587.7	44 023.9	2 523.4	34 210.5	5 432.7	8.92	2.00
湖　北　Hubei	228 467.4	18 778.8	848.6	9 939.5	2 068.1	154.50	2.91
湖　南　Hunan	203 605.4	16 574.1	266.8	11 471.0	1 428.1	8.71	0.75
广　东　Guangdong	987 037.8	84 365.6	5 745.5	76 564.0	13 126.4	72.63	132.47
广　西　Guangxi	124 005.4	13 911.1	841.6	11 539.0	2 324.2	0.09	0
海　南　Hainan	59 042.2	2 289.0	54.4	3 788.6	1 105.0	0	0
重　庆　Chongqing	184 673.6	13 233.0	987.4	18 810.1	2 395.5	0	0
四　川　Sichuan	275 253.6	27 976.9	1 812.1	22 124.9	3 127.9	104.79	1.77
贵　州　Guizhou	65 171.2	7 925.6	177.0	3 283.8	511.2	0.24	0
云　南　Yunnan	135 441.4	11 142.4	730.6	11 977.9	3 582.3	0.10	0.01
西　藏　Tibet	532.6	40.1	0	0	3.8	0	0
陕　西　Shaanxi	176 301.7	17 447.9	676.8	16 082.3	3 627.9	120.86	0.17
甘　肃　Gansu	65 269.1	6 980.3	5 880.8	19 928.8	5 939.5	18.44	1.17
青　海　Qinghai	20 627.2	1 151.5	3.4	141.2	15.1	0	0
宁　夏　Ningxia	42 512.8	2 657.3	183.5	732.1	586.1	2.47	0
新　疆　Xinjiang	125 448.5	10 575.5	3 439.5	5 526.1	556.7	3.62	7.10

各地区生活垃圾处理情况（一）
Centralized Treatment of Garbage by Region（1）
（2011）

年　份　Year 地　区　Region	生活垃圾处理厂数/座 Number of Garbage Treatment Plants（unit）	本年运行费用/万元 Annul Expenditure for Operation （10 000 yuan）	生活垃圾处理厂累计 完成投资/万元 Total Investment of Garbage Treatment Plants （10 000 yuan）	新增固定资产/万元 Newly-added Fixed Assets （10 000 yuan）
2011	2 039	591 506.9	6 676 883.6	658 537.0
北　京　Beijing	19	68 433.2	293 802.6	3 401.2
天　津　Tianjin	8	6 905.6	50 024.9	0.0
河　北　Hebei	129	22 904.4	411 664.3	30 067.7
山　西　Shanxi	72	9 807.2	157 604.1	2 261.9
内蒙古　Inner Mongolia	56	12 693.0	208 115.1	11 529.4
辽　宁　Liaoning	36	13 665.1	147 510.1	14 350.7
吉　林　Jilin	47	14 997.1	157 744.1	3 361.2
黑龙江　Heilongjiang	23	4 930.9	95 940.0	3 575.0
上　海　Shanghai	25	44 712.3	223 284.1	26 758.6
江　苏　Jiangsu	55	29 453.3	396 096.9	31 017.9
浙　江　Zhejiang	85	44 762.4	489 731.8	40 084.3
安　徽　Anhui	125	22 624.2	193 634.1	27 300.3
福　建　Fujian	90	18 625.1	355 360.4	10 413.2
江　西　Jiangxi	85	7 986.1	117 554.3	7 286.1
山　东　Shandong	82	24 965.7	338 587.6	26 244.1
河　南　Henan	121	26 011.9	409 543.0	31 190.7
湖　北　Hubei	95	22 119.3	311 519.3	3 882.6
湖　南　Hunan	94	25 613.2	354 696.9	47 681.3
广　东　Guangdong	131	62 642.4	447 072.1	58 279.3
广　西　Guangxi	52	10 803.1	241 861.9	53 804.7
海　南　Hainan	18	5 957.2	79 012.9	1 035.3
重　庆　Chongqing	43	15 649.1	225 896.3	160 851.5
四　川　Sichuan	89	25 379.2	229 876.1	15 327.4
贵　州　Guizhou	29	5 078.7	116 588.0	10 439.0
云　南　Yunnan	93	13 780.6	160 718.2	22 513.2
西　藏　Tibet	9	1 090.8	20 834.3	380.0
陕　西　Shaanxi	67	9 058.9	133 730.9	1 387.7
甘　肃　Gansu	63	6 571.2	79 544.6	4 169.7
青　海　Qinghai	52	2 012.4	55 885.0	2 648.0
宁　夏　Ningxia	20	3 175.2	47 002.9	1 292.2
新　疆　Xinjiang	126	9 098.2	126 446.9	6 002.9

各地区生活垃圾处理情况（二）
Centralized Treatment of Garbage by Region（2）
（2011）

单位：吨 (tons)

年 份 Year 地 区 Region	渗滤液中污染物排放量 Amount of Pollutants Discharged in Landfill Leachate					
	化学需氧量 COD	氨氮 Ammonia Nitrogen	油类 Oil	总磷 Total Phosphorus	挥发酚 Volatile Phenols	氰化物 Cyanide
2011	199 699.907	20 008.380	410.006	288.898	20.024	2.509
北 京 Beijing	5 274.570	563.052	4.448	4.886	0.693	0.025
天 津 Tianjin	434.683	35.694	0.200	0.001	0.010	0.001
河 北 Hebei	7 872.381	580.089	7.594	12.290	1.563	0.033
山 西 Shanxi	2 577.370	247.030	4.304	4.644	0.321	0.008
内蒙古 Inner Mongolia	2 308.400	190.926	4.962	4.171	0.574	0.007
辽 宁 Liaoning	7 005.690	1 105.780	7.187	8.476	0.957	0.029
吉 林 Jilin	12 915.260	1 376.980	13.289	14.741	2.231	0.052
黑龙江 Heilongjiang	1 898.130	219.850	0.744	1.143	0.126	0.005
上 海 Shanghai	5 826.660	295.870	3.165	5.030	0.112	0.025
江 苏 Jiangsu	5 290.100	672.347	5.829	6.591	0.363	0.025
浙 江 Zhejiang	7 029.769	590.245	5.243	10.959	0.653	0.047
安 徽 Anhui	9 959.468	958.267	8.337	13.401	0.258	0.069
福 建 Fujian	4 117.251	398.876	13.796	7.466	0.230	0.038
江 西 Jiangxi	11 110.817	897.284	10.275	18.160	1.129	0.101
山 东 Shandong	4 703.160	495.670	5.475	6.783	0.624	0.917
河 南 Henan	9 794.799	1 177.901	6.930	14.821	1.532	0.029
湖 北 Hubei	17 432.550	2 042.397	7.830	19.310	1.943	0.064
湖 南 Hunan	16 686.120	1 520.502	12.733	25.677	1.987	0.426
广 东 Guangdong	13 687.658	1 428.055	6.528	12.652	0.039	0.400
广 西 Guangxi	3 897.120	306.725	252.505	34.053	0.104	0.027
海 南 Hainan	1 357.830	123.045	0.289	0.703	0.030	0.003
重 庆 Chongqing	601.500	155.450	1.189	1.697	0.001	0.010
四 川 Sichuan	13 530.360	767.990	5.270	10.475	0.129	0.036
贵 州 Guizhou	5 994.777	679.950	2.304	6.080	0.080	0.017
云 南 Yunnan	12 983.350	1 651.234	8.062	22.022	2.036	0.034
西 藏 Tibet	3.150	0.210	0	0.002	0.001	0
陕 西 Shaanxi	6 596.926	917.564	3.505	6.129	0.118	0.021
甘 肃 Gansu	1 631.391	113.509	1.550	1.839	0.963	0.004
青 海 Qinghai	1 912.361	126.636	0.831	1.497	0.137	0.025
宁 夏 Ningxia	763.304	51.331	0.488	0.775	0.026	0.002
新 疆 Xinjiang	4 503.002	317.921	5.144	12.424	1.053	0.028

各地区生活垃圾处理情况（三）
Centralized Treatment of Garbage by Region（3）

单位：千克 　　　　　　　　　　　　　　　　（2011）　　　　　　　　　　　　　　　　（kg）

年　份　Year 地　区　Region	渗滤液中污染物排放量 Amount of Pollutants Discharged in Landfill Leachate				
	汞 Hydrargyrum	镉 Cadmium	总铬 Total Chromium	铅 Plumbum	砷 Arsenic
2011	164.803	792.959	2 687.263	4 299.244	1 404.626
北　京　Beijing	1.218	9.469	27.174	103.836	12.494
天　津　Tianjin	0.010	0.030	0.090	0.350	0.170
河　北　Hebei	4.203	26.796	50.992	126.733	33.555
山　西　Shanxi	1.341	5.970	11.930	25.010	5.797
内蒙古　Inner Mongolia	2.225	13.319	25.478	53.747	17.709
辽　宁　Liaoning	8.152	23.720	66.126	92.465	20.068
吉　林　Jilin	5.089	31.565	63.206	144.498	27.222
黑龙江　Heilongjiang	1.663	4.295	8.124	14.614	6.147
上　海　Shanghai	3.285	15.609	43.739	71.575	27.589
江　苏　Jiangsu	4.599	21.512	69.207	115.429	35.488
浙　江　Zhejiang	6.589	20.437	73.385	107.684	27.957
安　徽　Anhui	5.793	32.200	113.039	152.613	101.671
福　建　Fujian	4.807	89.859	49.702	68.030	18.905
江　西　Jiangxi	7.379	33.202	123.035	175.818	54.686
山　东　Shandong	19.519	110.019	188.946	410.863	97.897
河　南　Henan	6.729	25.959	52.555	121.568	35.740
湖　北　Hubei	11.611	36.179	91.638	142.858	40.537
湖　南　Hunan	22.395	109.327	619.642	1 459.301	312.034
广　东　Guangdong	21.416	42.699	584.777	343.840	160.908
广　西　Guangxi	3.342	11.482	34.951	43.508	12.721
海　南　Hainan	0.379	1.464	6.270	7.159	2.338
重　庆　Chongqing	1.175	5.908	13.253	37.760	11.312
四　川　Sichuan	6.090	22.962	129.045	185.700	96.656
贵　州　Guizhou	2.211	9.922	31.400	42.980	11.858
云　南　Yunnan	5.187	21.789	77.224	102.673	149.316
西　藏　Tibet	0.003	0.063	0.105	0.210	0.063
陕　西　Shaanxi	2.093	11.519	31.203	41.141	11.150
甘　肃　Gansu	0.965	5.447	12.354	27.184	6.906
青　海　Qinghai	0.750	4.091	8.102	20.416	5.084
宁　夏　Ningxia	0.256	1.772	3.930	8.513	2.319
新　疆　Xinjiang	4.329	44.374	76.641	51.168	58.329

各地区生活垃圾处理情况（四）
Centralized Treatment of Garbage by Region（4）
（2011）

单位：吨 (tons)

年 份 Year		焚烧废气中污染物排放量 Volume of Pollutants Emission in the Incineration Gas		
地 区 Region		二氧化硫 Sulfur Dioxide	氮氧化物 Nitrogen Oxide	烟尘 Soot
	2011	1 375.74	2 089.58	1 460.98
北 京	Beijing	44.64	300.76	25.97
天 津	Tianjin	0	0	0
河 北	Hebei	81.42	0	23.43
山 西	Shanxi	32.42	103.91	24.15
内蒙古	Inner Mongolia	0	0	0
辽 宁	Liaoning	37.96	7.72	32.84
吉 林	Jilin	0	0	0
黑龙江	Heilongjiang	0	0	0
上 海	Shanghai	0	0.01	0.01
江 苏	Jiangsu	162.34	192.95	35.42
浙 江	Zhejiang	83.63	109.49	139.07
安 徽	Anhui	394.95	296	874.14
福 建	Fujian	107.92	237.72	92.63
江 西	Jiangxi	0	0	0
山 东	Shandong	27.25	60.73	14.9
河 南	Henan	23.5	136	22.2
湖 北	Hubei	0.01	0.03	0.01
湖 南	Hunan	0	0	0
广 东	Guangdong	113.44	260.33	56.47
广 西	Guangxi	24.02	42.01	4.93
海 南	Hainan	1.14	5.09	0.84
重 庆	Chongqing	0	0	0
四 川	Sichuan	234.92	336.62	106.71
贵 州	Guizhou	0	0	0
云 南	Yunnan	5.96	0.03	2.63
西 藏	Tibet	0	0	0
陕 西	Shaanxi	0	0	0
甘 肃	Gansu	0	0	0
青 海	Qinghai	0	0	0
宁 夏	Ningxia	0	0	0
新 疆	Xinjiang	0.23	0.18	4.61

各地区危险（医疗）废物集中处置情况（一）

Centralized Treatment of Hazardous (Medical) Wastes by Region（1）

（2011）

年　份　Year 地　区　Region	危险废物集中处置厂数/个 Number of Centralized Hazardous Wastes Treatment Plants（unit）	医疗废物集中处置厂数/个 Number of Centralized Medical Wastes Treatment Plants（unit）	本年运行费用/万元 Annul Expenditure for Operation（10 000 yuan）	危险（医疗）废物集中处置厂累计完成投资/万元 Total Investment of Hazardous/ Medical Wastes Treatment Plants（10 000 yuan）	新增固定资产/万元 Newly-added Fixed Assets（10 000 yuan）
2011	644	260	481 785. 9	1 908 958. 4	189 987. 1
北　京　Beijing	2	1	25 574. 2	72 726. 0	1 391. 0
天　津　Tianjin	13	3	19 298. 5	27 480. 0	443. 1
河　北　Hebei	30	13	7 884. 0	30 537. 4	1 877. 2
山　西　Shanxi	6	10	2 055. 8	13 551. 9	2 146. 6
内蒙古　Inner Mongolia	1	14	2 240. 0	29 367. 6	49. 0
辽　宁　Liaoning	55	8	33 709. 7	147 754. 9	20 170. 0
吉　林　Jilin	2	4	3 922. 0	11 552. 0	148. 0
黑龙江　Heilongjiang	5	8	3 035. 2	44 244. 5	6 302. 6
上　海　Shanghai	47	2	57 042. 1	160 987. 7	23 325. 2
江　苏　Jiangsu	126	11	66 329. 1	243 569. 5	19 395. 0
浙　江　Zhejiang	63	11	42 357. 5	165 773. 7	20 169. 6
安　徽　Anhui	8	11	6 122. 6	25 963. 9	4 005. 9
福　建　Fujian	7	8	5 247. 2	25 129. 6	1 539. 3
江　西　Jiangxi	20	8	6 399. 4	57 501. 5	380. 0
山　东　Shandong	40	16	21 599. 8	74 102. 7	4 917. 1
河　南　Henan	3	17	8 580. 0	23 756. 4	1 229. 6
湖　北　Hubei	33	13	11 028. 9	115 918. 4	16 810. 1
湖　南　Hunan	2	11	3 330. 4	18 627. 0	397. 5
广　东　Guangdong	92	19	110 903. 6	251 368. 4	23 861. 6
广　西　Guangxi	13	8	3 821. 1	14 200. 8	1 220. 0
海　南　Hainan	1	3	1 078. 3	20 159. 0	17 507. 2
重　庆　Chongqing	6	4	7 710. 1	48 987. 0	2 103. 5
四　川　Sichuan	17	18	11 984. 4	32 293. 5	3 042. 8
贵　州　Guizhou	4	3	1 194. 0	21 336. 4	280. 0
云　南　Yunnan	1	6	4 300. 9	19 305. 9	0. 9
陕　西　Shaanxi	13	6	6 122. 0	26 833. 6	2 576. 5
甘　肃　Gansu	7	10	3 988. 8	38 997. 1	11 987. 8
青　海　Qinghai	3	1	928. 1	5 041. 6	544. 0
宁　夏　Ningxia	7	2	1 265. 2	26 653. 0	466. 2
新　疆　Xinjiang	17	11	2 733. 0	115 237. 6	1 700. 0

各地区危险（医疗）废物集中处置情况（二）

Centralized Treatment of Hazardous (Medical) Wastes by Region（2）

（2011）

年 份 Year 地 区 Region	危险废物设计处置能力/（吨/日）Treatment Capacity（tons/day）	危险废物实际处置量/吨 Volume of Hazardous Wastes Disposed（tons）	工业危险废物处置量 Industrial Hazardous Wastes	医疗废物处置量 Medical Hazardous Wastes	其他危险废物处置量 Other Hazardous Wastes	危险废物综合利用量/吨 Volume of Hazardous Wastes Utilized（tons）
2011	352 541	2 600 177	1 933 930	497 782	168 465	3 330 420
北 京 Beijing	622	71 096	60 547	10 549	0	33 852
天 津 Tianjin	1 671	78 322	52 789	16 983	8 550	112 571
河 北 Hebei	6 238	144 600	67 833	69 435	7 332	28 153
山 西 Shanxi	164	11 708	83	11 623	2	2 556
内蒙古 InnerMongolia	187	7 589	2	7 587	0	0
辽 宁 Liaoning	3 188	193 643	158 322	14 761	20 560	316 898
吉 林 Jilin	227	32 678	22 469	8 221	1 988	373
黑龙江 Heilongjiang	224	7 731	0	7 495	235	356
上 海 Shanghai	2245	169 472	140 577	22 975	5 920	145 595
江 苏 Jiangsu	91 586	437 009	336 805	29 461	70 744	943 552
浙 江 Zhejiang	11 178	276 593	227 184	37 562	11 847	165 217
安 徽 Anhui	3 013	47 117	38 848	8 268	0	11 206
福 建 Fujian	2 610	16 765	5 204	11 561	0	19 020
江 西 Jiangxi	641	11 579	2 879	8 700	0	73 332
山 东 Shandong	8 523	165 759	129 270	34 088	2 400	248 982
河 南 Henan	166	28 412	0	28 412	0	3 634
湖 北 Hubei	9 189	59 608	40 919	16 204	2 485	50 361
湖 南 Hunan	301	15 532	3 647	11 886	0	290
广 东 Guangdong	42 526	292 798	227 649	56 825	8 324	565 276
广 西 Guangxi	287	13 606	6 515	7 091	0	6 848
海 南 Hainan	82	5 944	0	5 944	0	0
重 庆 Chongqing	1 374	27 761	18 661	7 143	1 957	255 893
四 川 Sichuan	1 114	34 803	12 214	20 688	1 901	181 390
贵 州 Guizhou	41 321	5 691	431	5 235	25	3 365
云 南 Yunnan	50	10 552	0	10 483	69	0
陕 西 Shaanxi	4 074	153 031	136 401	13 331	3 300	9 690
甘 肃 Gansu	100 730	184 936	181 039	3 893	5	43 025
青 海 Qinghai	169	2 404	0	2 404	0	20 811
宁 夏 Ningxia	9 555	3 502	214	1 178	2 110	2 011
新 疆 Xinjiang	9 286	89 938	63 431	7 797	18 710	86 163

各地区危险（医疗）废物集中处置情况（三）
Centralized Treatment of Hazardous (Medical) Wastes by Region（3）
（2011）

年 份 Year 地 区 Region	渗滤液中污染物排放量 Amount of Pollutants Discharged in Landfill Leachate					
	化学需氧量/吨 COD （tons）	氨氮/吨 Ammonia Nitrogen （tons）	油类/吨 Oil （tons）	总磷/吨 Total Phosphorus （tons）	挥发酚/千克 Volatile Phenols（kg）	氰化物/千克 Cyanide （kg）
2011	1 436.773	39.385	12.946	0.104	6.768	39.502
北 京 Beijing	2.700	2.475	0.090	0.005	2.000	4.000
天 津 Tianjin	6.787	0.120	0	0	0	0
河 北 Hebei	0.052	0.005	0	0.001	0	0
山 西 Shanxi	0	0	0	0	0	0
内蒙古 Inner Mongolia	0.530	0.350	0.010	0	0	0
辽 宁 Liaoning	946.400	10.781	1…	0.001	0.001	0.001
吉 林 Jilin	0.130	0.040	0.001	0	0.026	0.042
黑龙江 Heilongjiang	0.340	0.010	0	0	0	0
上 海 Shanghai	0	0	0	0	0	0
江 苏 Jiangsu	13.681	6.414	0.061	0.010	0.017	0.065
浙 江 Zhejiang	129.525	3.003	0.006	0.051	0.002	0.002
安 徽 Anhui	0.550	0.090	0	0	0	0
福 建 Fujian	7.722	0.031	0	0	0.001	4.711
江 西 Jiangxi	0.538	0.340	0.023	0	0.335	0.345
山 东 Shandong	4.170	0.370	2.210	0	0	0.050
河 南 Henan	0	0	0	0	0	0
湖 北 Hubei	19.171	1.953	0.020	0	0	0
湖 南 Hunan	1.003	0.191	0.006	0	0.030	0.127
广 东 Guangdong	288.183	5.657	0.067	0.031	4.314	0.099
广 西 Guangxi	0	0	0	0	0	0
海 南 Hainan	0	0	0	0	0	0
重 庆 Chongqing	15.223	7.544	0.450	0.006	0.003	30.003
四 川 Sichuan	0.002	0	0	0	0	0
贵 州 Guizhou	0	0	0.001	0	0.026	0.042
云 南 Yunnan	0	0	0	0	0	0
陕 西 Shaanxi	0	0	0	0	0.004	0.010
甘 肃 Gansu	0.051	0.002	0	0	0	0
青 海 Qinghai	0	0	0	0	0	0
宁 夏 Ningxia	0.012	0.006	0	0	0.008	0
新 疆 Xinjiang	0.004	0.003	0	0	0.001	0.005

各地区危险（医疗）废物集中处置情况（四）
Centralized Treatment of Hazardous (Medical) Wastes by Region（4）

单位：千克 　　　　　　　　　　　　　　（2011） 　　　　　　　　　　　　　　（kg）

年　份　Year 地　区　Region		渗滤液中污染物排放量 Amount of Pollutants Discharged in Landfill Leachate				
		汞 Hydrargyrum	镉 Cadmium	总铬 Total Chromium	铅 Plumbum	砷 Arsenic
	2011	7.953	10.303	215.437	111.454	42.445
北　京	Beijing	0.500	2.000	36.000	14.000	3.000
天　津	Tianjin	0	0	0	4.900	0.010
河　北	Hebei	0	0	0.003	0	0
山　西	Shanxi	0	0	0	0	0
内蒙古	Inner Mongolia	0	0	0	0	0
辽　宁	Liaoning	0.023	0.850	2.253	0.616	0.150
吉　林	Jilin	0.006	0.026	0.425	0.159	0.037
黑龙江	Heilongjiang	0	0	0	0	0
上　海	Shanghai	0	0	0	0	0
江　苏	Jiangsu	0.011	0.042	27.635	0.280	0.274
浙　江	Zhejiang	0.055	0.097	0.928	6.676	3.405
安　徽	Anhui	0	0	0	0	0
福　建	Fujian	4.880	0.341	2.266	2.449	1.859
江　西	Jiangxi	0.032	0.115	4.975	2.730	2.325
山　东	Shandong	0	0.010	0.160	0.100	0.030
河　南	Henan	0	0	0	0	0
湖　北	Hubei	0	0	0	0	0
湖　南	Hunan	0.010	0.027	0.087	0.319	0.139
广　东	Guangdong	0.030	0.761	5.009	4.034	1.160
广　西	Guangxi	0	0	0	0	0
海　南	Hainan	0	0	0	0	0
重　庆	Chongqing	2.401	6.003	135.240	75.016	30.016
四　川	Sichuan	0	0	0	0	0
贵　州	Guizhou	0.005	0.030	0.420	0.160	0.036
云　南	Yunnan	0	0	0	0	0
陕　西	Shaanxi	0	0	0	0.005	0
甘　肃	Gansu	0	0	0	0	0
青　海	Qinghai	0	0	0	0	0
宁　夏	Ningxia	0	0	0	0	0
新　疆	Xinjiang	0	0.001	0.036	0.01	0.004

各地区危险（医疗）废物集中处置情况（五）
Centralized Treatment of Hazardous (Medical) Wastes by Region（5）

单位：吨
（2011）
（tons）

年 份 Year 地 区 Region	焚烧废气中污染物排放量 Volume of Pollutants Emission in the Incineration Gas		
	二氧化硫 Sulfur Dioxide	氮氧化物 Nitrogen Oxide	烟尘 Soot
2011	1 491.408	1 406.524	1 011.210
北 京 Beijing	0.470	16.800	1.770
天 津 Tianjin	44.480	45.151	24.712
河 北 Hebei	48.391	70.017	15.669
山 西 Shanxi	3.270	6.980	350.132
内蒙古 Inner Mongolia	4.373	7.616	14.438
辽 宁 Liaoning	9.700	6.580	8.944
吉 林 Jilin	0.510	8.345	20.729
黑龙江 Heilongjiang	8.169	17.130	2.965
上 海 Shanghai	149.859	331.592	39.107
江 苏 Jiangsu	82.582	245.587	129.724
浙 江 Zhejiang	389.328	126.923	94.089
安 徽 Anhui	9.779	15.508	27.998
福 建 Fujian	7.276	19.910	3.529
江 西 Jiangxi	14.794	21.883	4.267
山 东 Shandong	134.512	66.128	23.537
河 南 Henan	19.065	51.163	6.295
湖 北 Hubei	6.570	11.400	16.680
湖 南 Hunan	24.400	56.940	33.405
广 东 Guangdong	336.201	148.635	58.911
广 西 Guangxi	3.980	10.830	20.180
海 南 Hainan	1.180	4.670	3.030
重 庆 Chongqing	3.545	14.405	9.716
四 川 Sichuan	7.071	17.917	28.775
贵 州 Guizhou	0.440	1.380	0.450
云 南 Yunnan	1.158	13.260	4.320
陕 西 Shaanxi	3.288	10.935	1.288
甘 肃 Gansu	173.380	49.740	14.170
青 海 Qinghai	0.790	2.310	0.250
宁 夏 Ningxia	0.218	0.840	1.680
新 疆 Xinjiang	2.629	5.949	50.450

各地区环境污染治理投资情况
Investment in the Treatment of Environmental Pollution by Region

单位：亿元　　　　　　　　　　　　　　　　　（2011）　　　　　　　　　　　　　　　（100 million yuan）

年　份　Year 地　区　Region	环境污染治理投资总额 Total Investment in Treatment of Environmental Pollution	城市环境基础设施建设投资 Investment in Urban Environment Infrastructure Facilities	工业污染源治理投资 Investment in Treatment of Industrial Pollution Sources	建设项目"三同时"环保投资 Investments in Environment Components for New Construction Projects
2011	6 026. 2	3 469. 4	444. 4	2 112. 4
国家级　State Level	433. 5	—	—	433. 5
北　京　Beijing	213. 2	192. 3	1. 1	19. 8
天　津　Tianjin	167. 3	97. 6	15. 3	54. 4
河　北　Hebei	443. 6	254. 0	24. 3	165. 3
山　西　Shanxi	185. 7	100. 0	27. 9	57. 8
内蒙古　Inner Mongolia	315	206. 7	31. 0	77. 3
辽　宁　Liaoning	352. 1	289. 1	11. 6	51. 4
吉　林　Jilin	92. 2	65. 4	6. 6	20. 2
黑龙江　Heilongjiang	125. 1	59. 5	10. 1	55. 5
上　海　Shanghai	144. 8	68. 9	6. 4	69. 5
江　苏　Jiangsu	527. 1	301. 9	31. 0	194. 2
浙　江　Zhejiang	210. 4	105. 0	17. 8	87. 6
安　徽　Anhui	212. 2	155. 7	9. 3	47. 2
福　建　Fujian	180. 4	113. 0	14. 3	53. 1
江　西　Jiangxi	173	121. 0	6. 6	45. 4
山　东　Shandong	527. 6	336. 5	62. 4	128. 7
河　南　Henan	137. 9	73. 7	21. 4	42. 8
湖　北　Hubei	184. 3	121. 9	9. 3	53. 1
湖　南　Hunan	101. 1	63. 9	9. 7	27. 5
广　东　Guangdong	329. 6	172. 8	16. 6	140. 2
广　西　Guangxi	143. 9	110. 3	8. 6	25. 0
海　南　Hainan	27	14. 4	2. 8	9. 8
重　庆　Chongqing	210. 3	165. 7	4. 9	39. 7
四　川　Sichuan	108. 6	51. 2	16. 7	40. 7
贵　州　Guizhou	47. 1	20. 2	13. 2	13. 7
云　南　Yunnan	88. 4	27. 3	13. 7	47. 4
西　藏　Tibet	28. 3	0. 0	0. 2	2. 2
陕　西　Shaanxi	98. 1	51. 8	23. 7	28. 1
甘　肃　Gansu	42. 6	21. 6	10. 5	22. 6
青　海　Qinghai	22. 1	7. 1	2. 8	10. 5
宁　夏　Ningxia	51. 9	11. 9	3. 9	12. 2
新　疆　Xinjiang	101. 8	89. 0	10. 6	36. 1

各地区城市环境基础设施建设投资情况
Investment in Urban Environment Infrastructure by Region

单位：亿元　　　　　　　　　　　　　　　　（2011）　　　　　　　　　　　　（100 million yuan）

年 份 Year 地 区 Region	投资总额 Total Investment	燃气 Gas Supply	集中供热 Gerneral Heating	排水 Sewerage Projects	园林绿化 Gardening & Greening	市容环境 卫生 Sanitation
2011	3 469.4	331.4	437.6	770.1	1 546.2	384.1
北　京　Beijing	192.3	17.0	27.6	37.8	17.8	92.0
天　津　Tianjin	97.6	13.5	2.8	15.2	53.1	13.0
河　北　Hebei	254.0	23.0	66.6	35.7	118.3	10.4
山　西　Shanxi	100.0	13.6	29.7	16.4	32.9	7.4
内蒙古　Inner Mongolia	206.7	7.8	48.1	29.8	112.1	9.0
辽　宁　Liaoning	289.1	15.1	74.5	66.7	98.8	34.0
吉　林　Jilin	65.4	7.0	34.5	7.0	11.1	5.9
黑龙江　Heilongjiang	59.5	4.6	21.7	10.0	17.3	5.9
上　海　Shanghai	68.9	16.6	0.0	11.4	25.4	15.5
江　苏　Jiangsu	301.9	23.4	0.0	85.8	174.0	18.6
浙　江　Zhejiang	105.0	13.4	1.0	22.1	58.5	10.2
安　徽　Anhui	155.7	11.1	2.9	36.5	96.2	9.0
福　建　Fujian	113.0	3.1	0.0	42.8	49.1	17.9
江　西　Jiangxi	121.0	42.8	0.0	13.3	62.6	2.3
山　东　Shandong	336.5	30.0	56.9	81.6	149.3	18.8
河　南　Henan	73.7	13.3	17.8	15.7	24.3	2.7
湖　北　Hubei	121.9	10.9	1.4	53.7	41.1	14.7
湖　南　Hunan	63.9	6.1	0.0	20.8	25.4	11.6
广　东　Guangdong	172.8	16.7	0.0	59.4	39.8	56.8
广　西　Guangxi	110.3	6.7	0.0	17.9	79.3	6.4
海　南　Hainan	14.4	0.3	0.0	6.8	5.3	2.1
重　庆　Chongqing	165.7	3.5	0.8	17.5	141.8	2.0
四　川　Sichuan	51.2	4.1	0.0	16.9	26.2	4.0
贵　州　Guizhou	20.2	2.7	0.0	8.0	7.0	2.6
云　南　Yunnan	27.3	1.8	0.0	12.7	9.5	3.3
西　藏　Tibet	0.0	0.0	0.0	0.0	0.0	0.0
陕　西　Shaanxi	51.8	5.0	4.3	11.1	27.6	3.7
甘　肃　Gansu	21.6	2.9	5.4	4.9	7.1	1.3
青　海　Qinghai	7.1	0.8	0.4	3.2	1.7	1.0
宁　夏　Ningxia	11.9	1.0	3.9	2.2	4.3	0.6
新　疆　Xinjiang	89.0	13.6	37.2	7.1	29.4	1.7

重点城市环境统计

ZHONGDIAN CHENGSHI HUANJING TONGJI

ANNUAL STATISTIC REPORT ON ENVIRONMENT IN CHINA

2011

重点城市工业废水排放及处理情况（一）

（2011）

城 市 名 称	工业废水 排放量/万吨	直接排入 环境的	排入污水 处理厂的	工业废水 处理量/万吨	废水治理 设施数/套	废水治理设施 治理能力/ （吨/日）	废水治理设施 运行费用/万元
总 计	1 218 085	861 847	356 238	3344 270	49 589	202 354 151	3 966 372.5
北 京	8 633	2 827	5 806	11 294	508	603 149	37 124.0
天 津	19 795	8 413	11 382	35 198	957	1 355 029	134 315.1
石 家 庄	25 591	12 507	13 084	60 234	621	1 365 658	51 512.5
唐 山	17 308	11 847	5 461	485 071	878	18 289 766	206 304.2
秦 皇 岛	6 380	3 695	2 685	12 558	271	919 528	12 066.9
邯 郸	7 205	7 092	112	137 044	352	1 577 279	44 073.7
保 定	16 395	11 220	5 175	24 205	733	3 626 452	28 295.9
太 原	2 453	1 837	616	36 238	312	1 365 722	52 304.1
大 同	6 013	5 106	907	5 908	275	283 732	8 188.3
阳 泉	779	469	309	3 931	61	146 957	4 078.7
长 治	7 287	7 028	259	19 609	499	1 051 503	27 179.6
临 汾	4 382	4 327	54	16 366	360	982 376	62 588.8
呼和浩特	2 650	2 144	507	2 328	87	148 768	8 212.6
包 头	9 327	1 710	7 617	55 585	106	2 493 272	34 226.2
赤 峰	2 924	2 620	303	5 815	127	215 087	7 239.5
沈 阳	7 239	2 084	5 156	11 920	289	435 540	13 857.1
大 连	31 487	28 945	2 542	28 766	420	51 332 290	23 821.1
鞍 山	7 529	6 046	1 483	165 357	202	3 964 994	93 541.3
抚 顺	3 003	2 722	281	15 088	135	738 834	19 895.9
本 溪	5 103	4 930	173	67 110	195	3 718 091	38 126.0
锦 州	5 497	4 381	1 116	5 715	101	219 575	23 839.5
长 春	6 335	4 848	1 487	6 161	143	315 013	11 494.1
吉 林	11 291	6 691	4 600	13 608	126	1 474 087	17 314.2
哈 尔 滨	5 838	2 240	3 598	5 922	186	302 139	28 394.4
齐齐哈尔	4 991	4 940	50	10 372	121	506 931	6 624.4
牡 丹 江	3 157	2 110	1 047	3 777	70	206 048	3 802.0
上 海	44 626	20 366	24 260	71 099	5 872	3 050 430	398 206.8
南 京	25 379	18 985	6 394	118 597	620	2 840 173	110 103.6
无 锡	26 795	8 275	18 520	42 818	981	12 112 505	92 311.3
徐 州	10 987	9 299	1 688	19 598	306	3 688 829	32 143.6
常 州	17 371	9 654	7 717	21 221	572	721 226	50 436.0
苏 州	71 441	44 729	26 711	69 746	1 725	4 303 191	189 352.0
南 通	19 649	13 062	6 587	15 110	792	937 615	59 437.4
连 云 港	6 548	5 184	1 364	11 300	218	923 272	26 746.5
扬 州	9 132	7 323	1 809	5 641	334	471 817	20 528.5
镇 江	11 172	8 472	2 700	10 473	425	524 732	27 015.8
杭 州	47 890	16 370	31 520	139 178	1 300	6 166 522	107 169.3

重点城市工业废水排放及处理情况（一）（续表）

（2011）

城 市名 称	工业废水排放量/万吨	直接排入环境的	排入污水处理厂的	工业废水处理量/万吨	废水治理设施数/套	废水治理设施治理能力/（吨/日）	废水治理设施运行费用/万元
宁 波	19 797	12 471	7 326	28 077	860	1 095 228	80 048.0
温 州	8 038	5 598	2 441	7 191	1 003	724 426	30 548.2
湖 州	11 980	6 302	5 678	7 797	525	536 441	16 524.3
绍 兴	33 124	8 079	25 045	34 281	1 006	1 857 127	77 331.5
合 肥	6 039	4 545	1 494	26 612	258	853 138	14 604.0
芜 湖	2 691	1 414	1 277	6 706	216	331 553	10 718.6
马 鞍 山	7 447	6 872	575	92 395	197	3 102 181	61 952.8
福 州	5 492	4 002	1 490	20 645	424	1 283 543	22 697.1
厦 门	31 542	28 367	3 175	32 238	323	285 170	15 216.8
泉 州	20 716	12 651	8 065	26 886	723	1 162 035	49 704.9
南 昌	9 367	8 830	537	14 692	207	599 746	14 460.1
九 江	9 185	9 122	63	14 780	197	608 479	20 966.7
济 南	6 396	5 381	1 015	38 757	255	1 342 239	15 946.8
青 岛	11 289	3 104	8 185	32 335	494	1 137 757	29 527.2
淄 博	17 922	9 771	8 152	55 810	755	1 719 408	204 575.5
枣 庄	10 801	8 878	1 923	14 193	148	557 260	15 615.1
烟 台	8 875	5 392	3 483	13 352	484	1 078 792	24 282.5
潍 坊	28 191	12 479	15 712	35 772	641	3 670 264	66 023.9
济 宁	16 690	13 943	2 747	25 496	293	1 300 190	35 244.1
泰 安	8 019	6 648	1 371	18 450	215	915 255	18 602.4
日 照	8 361	6 752	1 610	41 922	203	426 537	18 610.0
郑 州	14 454	10 547	3 906	14 837	379	1 209 087	13 560.4
开 封	8 676	8 512	165	2 574	71	174 047	4 887.7
洛 阳	9 334	8 202	1 132	11 138	407	765 578	17 140.3
平 顶 山	6 252	5 886	366	15 196	130	930 383	14 799.7
安 阳	6 726	6 725	1	37 366	289	2 061 817	30 292.1
焦 作	11 489	10 142	1 347	12 580	183	615 852	14 837.6
三 门 峡	6 080	6 080	0	6 573	279	573 821	13 181.6
武 汉	23 389	20 299	3 090	121 231	272	3 740 031	35 569.3
宜 昌	19 988	18 595	1 393	18 030	375	806 113	21 087.1
荆 州	13 895	12 141	1 755	7 056	3093	285 994	23 812.2
长 沙	4 051	3 785	265	3 278	208	240 136	7 033.5
株 洲	7 152	7 152	0	11 558	450	629 909	13 786.4
湘 潭	9 688	9 311	377	6 647	170	387 233	6 774.8
岳 阳	13 214	12 077	1 137	13 058	183	496 114	28 935.5
常 德	9 973	9 853	120	9 034	154	580 399	10 914.2
张 家 界	564	564	0	50	13	9 220	49.2
广 州	24 579	15 265	9 314	23 696	986	1 975 107	61 354.9

重点城市工业废水排放及处理情况（一）（续表）

（2011）

城　市 名　称	工业废水 排放量/万吨	直接排入 环境的	排入污水处理 厂的	工业废水 处理量/万吨	废水治理 设施数/套	废水治理设施 治理能力/ （吨/日）	废水治理设施 运行费用/万元
韶　关	6 225	6 071	154	71 423	318	2 141 183	33 361.0
深　圳	11 715	8 501	3 214	12 586	1 508	454 975	65 268.8
珠　海	4 876	3 754	1 122	4 243	373	255 163	19 157.9
汕　头	5 789	5 111	677	5 210	318	445 760	11 896.7
湛　江	6 131	6 066	65	5 580	216	633 774	6 532.3
南　宁	14 690	14 204	485	18 036	279	1 324 432	21 699.9
柳　州	12 764	11 516	1 249	73 393	273	3 338 439	60 845.9
桂　林	5 153	4 649	504	7 925	310	485 711	4 673.7
北　海	1 749	1 201	547	11 079	44	426 847	6 010.7
海　口	620	108	512	526	57	47 737	2 004.7
重　庆	33 954	32 777	1 176	37 328	1 507	3 379 031	67 100.1
成　都	12 845	9 052	3 794	32 844	1 004	1 528 458	34 284.3
自　贡	2 540	2 345	195	1 647	128	167 632	2 639.4
攀枝花	2 087	1 341	745	139 315	251	2 004 949	57 178.6
泸　州	3 782	3 750	32	3 535	192	191 120	8 240.1
德　阳	6 797	6 336	461	8 793	297	496 660	30 587.6
绵　阳	9 265	8 882	383	9 700	255	424 807	8 855.3
南　充	2 862	2 479	383	930	37	206 272	2 492.4
宜　宾	7 612	7 592	20	7 520	251	557 616	13 234.6
贵　阳	1 986	1 573	412	17 403	122	1 064 608	10 516.6
遵　义	2 647	2 622	25	2 192	161	173 382	6 576.3
昆　明	6 264	5 789	475	30 265	487	1 430 824	34 457.3
曲　靖	4 558	4 545	13	17 602	445	1 006 019	14 959.0
玉　溪	5 482	5 462	20	43 658	270	1 542 795	19 259.5
拉　萨	329	320	9	227	19	13 760	94.6
西　安	13 274	11 344	1 931	12 974	315	676 136	15 803.3
铜　川	350	130	220	488	46	38 660	1 470.3
宝　鸡	4 821	4 421	399	7 513	301	577 120	24 693.4
咸　阳	7 000	4 206	2 795	5 651	180	534 569	12 719.5
渭　南	3 255	3 053	202	13 536	254	478 099	10 759.0
延　安	2 029	2 006	23	5 810	304	156 252	20 500.8
兰　州	4 094	3 001	1 094	6 191	89	551 848	7 338.8
金　昌	1 964	1 964	0	2 141	62	112 124	6 285.1
西　宁	3 144	3 144	0	13 774	83	519 633	5 175.1
银　川	6 448	5 270	1 178	6 187	76	291 143	11 036.0
石嘴山	1 435	1 333	102	8 528	215	715 213	4 771.6
乌鲁木齐	4 820	4 139	681	20 222	154	294 656	14 056.7
克拉玛依	1 680	1 547	133	6 044	39	221 002	16 747.7

重点城市工业废水排放及处理情况（二）

（2011）   单位：吨

城 市 名 称	工业废水中污染物产生量				
	化学需氧量	氨氮	石油类	挥发酚	氰化物
总　　　计	13 522 466.4	854 229.5	202 600.0	43 782.9	4 245.3
北　　京	81 321.5	3 190.8	2 730.4	877.0	6.9
天　　津	157 653.5	6 814.7	1 251.3	3.2	1.4
石 家 庄	283 169.6	19 832.9	1 425.9	1 095.4	25.4
唐　　山	376 932.8	5 628.8	10 959.2	5 071.6	479.5
秦 皇 岛	69 696.2	1 705.4	195.6	…	0.3
邯　　郸	115 838.9	6 706.2	3 113.7	1 762.1	71.0
保　　定	219 353.9	3 944.8	578.1	4.0	1.5
太　　原	32 032.1	995.2	336.8	718.7	46.6
大　　同	49 783.6	4 942.4	290.9	244.3	…
阳　　泉	3 219.4	3 388.6	28.7	0	0
长　　治	113 862.8	8 282.3	2 329.4	3 160.3	95.2
临　　汾	51 157.6	5 773.2	2 248.0	2 602.7	302.7
呼和浩特	53 843.9	4 087.8	117.4	16.7	…
包　　头	42 783.7	23 823.7	2 479.4	960.8	22.3
赤　　峰	51 481.6	551.2	16.5	…	…
沈　　阳	47 909.0	5 747.2	394.4	71.4	2.3
大　　连	100 477.8	9 568.4	1 433.5	15.0	0.6
鞍　　山	50 010.9	987.6	352.4	11.0	16.9
抚　　顺	9 605.1	327.1	598.6	4.4	0.7
本　　溪	58 694.6	1 540.6	2 305.8	1 266.5	124.3
锦　　州	43 918.4	1 139.5	4 441.0	291.5	405.8
长　　春	274 485.3	16 290.4	337.9	1.6	2.2
吉　　林	83 849.0	1 720.3	141.6	188.8	10.9
哈 尔 滨	176 157.2	6 511.5	173.3	3 244.0	0.1
齐齐哈尔	49 593.7	4 086.5	264.1	164.6	0.1
牡 丹 江	147 102.9	968.2	143.4	87.0	2.3
上　　海	320 510.9	12 564.1	9 509.3	1 119.2	243.9
南　　京	196 618.8	13 262.2	6 573.1	758.7	200.6
无　　锡	179 211.8	4 495.8	1 045.2	94.1	4.1
徐　　州	203 825.5	16 240.2	459.5	244.8	15.3
常　　州	117 574.8	4 417.1	1 825.4	17.5	15.2
苏　　州	439 179.6	17 164.1	2 316.3	186.3	48.4
南　　通	172 449.1	5 686.2	1 633.9	26.5	69.9
连 云 港	369 242.3	3 673.5	650.3	198.9	13.2
扬　　州	106 824.0	2 876.9	487.1	23.8	28.9
镇　　江	84 422.7	10 449.7	194.2	65.0	31.9
杭　　州	1 195 762.4	11 995.1	586.1	328.7	2.2

重点城市工业废水排放及处理情况（二）（续表）

（2011）

单位：吨

城 市 名 称	工业废水中污染物产生量				
	化学需氧量	氨氮	石油类	挥发酚	氰化物
宁　波	292 766.4	24 506.8	19 185.1	0.1	179.2
温　州	148 674.1	4 957.5	1 202.0	4.9	289.0
湖　州	82 302.3	2 553.6	258.3	0	…
绍　兴	315 551.8	15 648.0	87.8	…	1.5
合　肥	66 456.4	22 472.0	1 015.3	0.1	3.0
芜　湖	36 391.4	475.8	464.5	0.2	0.3
马 鞍 山	87 366.6	1 550.6	3 249.6	1 532.9	166.8
福　州	186 223.8	2 479.3	1 502.6	187.7	5.7
厦　门	74 131.6	2 828.7	5 284.2	0.6	24.3
泉　州	412 665.2	15 648.6	2 726.2	0.3	51.7
南　昌	95 768.4	9 131.3	385.8	148.4	31.5
九　江	65 777.8	1 544.5	1 421.1	217.2	0.5
济　南	78 927.0	14 205.3	4 991.4	1 486.9	30.3
青　岛	92 464.0	11 699.2	1 168.7	0.1	10.9
淄　博	277 235.8	20 821.1	2 434.9	653.5	20.9
枣　庄	112 827.7	7 239.9	155.0	95.9	3.3
烟　台	140 256.6	3 366.8	319.9	4.7	6.9
潍　坊	410 763.0	9 212.1	1 437.7	520.2	3.3
济　宁	178 824.0	3 299.0	222.7	0.5	6.8
泰　安	54 272.9	1 740.8	837.5	319.5	10.3
日　照	164 162.3	3 170.8	170.9	111.6	…
郑　州	67 282.5	1 484.5	432.5	0.6	7.2
开　封	47 262.4	7 128.0	122.2	0	5.9
洛　阳	37 990.7	1 464.7	50 338.1	10.0	1.0
平 顶 山	81 032.0	6 489.6	1 124.9	644.0	32.7
安　阳	98 937.7	2 009.4	1 516.6	980.8	55.1
焦　作	182 942.3	9 732.8	365.5	1.1	0.3
三 门 峡	41 988.0	8 132.6	170.1	3.8	5.6
武　汉	150 117.4	5 298.1	886.5	660.9	14.9
宜　昌	84 176.6	20 189.4	163.5	68.4	4.4
荆　州	92 969.9	3 969.8	130.6	50.6	22.6
长　沙	35 570.1	859.9	179.9	0	1.2
株　洲	25 934.1	4 926.1	310.6	1.7	8.0
湘　潭	31 304.7	1 653.9	1 706.7	793.8	158.3
岳　阳	138 548.7	27 891.3	474.8	58.8	5.2
常　德	60 524.2	1 292.2	98.6	…	3.4
张 家 界	3 188.1	77.3	11.6	0	0
广　州	304 694.7	5 876.9	1 430.4	103.5	55.3

重点城市工业废水排放及处理情况（二）（续表）

（2011）

单位：吨

城 市名 称	工业废水中污染物产生量				
	化学需氧量	氨氮	石油类	挥发酚	氰化物
韶 关	22 779.5	642.3	141.6	1.0	13.4
深 圳	67 467.9	4 600.0	120.0	0.3	195.1
珠 海	66 389.1	1 348.4	69.6	16.0	0.1
汕 头	65 477.6	1 146.1	2.0	…	0.3
湛 江	38 216.6	1 536.9	24.3	0	0
南 宁	289 885.0	2 476.2	50.8	29.2	1.7
柳 州	99 870.1	12 402.6	1 847.2	1 062.3	11.3
桂 林	30 877.6	2 557.1	53.6	0.4	0.7
北 海	51 174.2	334.1	26.0	…	0.1
海 口	4 964.1	181.4	4.1	0	0.2
重 庆	260 752.3	28 728.3	3 536.0	1 219.5	88.4
成 都	88 311.1	5 088.8	1 044.0	…	58.8
自 贡	7 122.8	789.7	80.9	84.6	…
攀 枝 花	43 466.6	936.8	2 060.9	797.3	12.8
泸 州	42 524.5	33 733.7	80.9	0.1	0.3
德 阳	89 216.4	1 712.7	251.4	…	1.1
绵 阳	55 119.0	1 011.4	642.1	…	…
南 充	11 960.4	653.8	241.6	0	0
宜 宾	105 828.5	1 291.6	90.4	…	1.7
贵 阳	19 135.8	1 251.2	715.8	238.2	95.0
遵 义	55 261.3	12 503.6	64.9	0.8	2.0
昆 明	73 848.0	4 393.7	910.8	907.5	30.2
曲 靖	78 401.2	22 062.5	2 104.2	2 053.4	86.7
玉 溪	36 907.9	4 320.7	274.8	72.3	17.5
拉 萨	5 118.4	128.2	4.7	0	0
西 安	141 915.6	9 455.4	909.0	0.2	15.3
铜 川	2 661.3	12.1	104.6	…	0
宝 鸡	133 723.2	12 993.7	348.5	185.5	15.9
咸 阳	58 414.3	3 726.9	399.4	0.1	0.3
渭 南	96 025.0	2 416.6	1 151.2	1 674.0	31.5
延 安	26 714.2	2 537.5	4 286.5	54.9	0.1
兰 州	35 894.6	6 803.4	1 705.6	190.9	1.4
金 昌	13 145.5	7 671.8	68.7	0.2	0.3
西 宁	26 572.4	2 460.2	496.4	2.4	0
银 川	192 797.0	76 676.7	662.0	13.4	0.6
石 嘴 山	35 630.9	3 177.5	1 311.3	535.6	22.8
乌鲁木齐	47 995.8	41 582.1	2 931.6	1 037.0	20.0
克拉玛依	13 070.8	475.7	1 831.1	16.5	0

重点城市工业废水排放及处理情况（三）

（2011）

单位：吨

城 市 名 称	工业废水中污染物排放量				
	化学需氧量	氨氮	石油类	挥发酚	氰化物
总 计	1 489 270.8	138 056.7	10 154.0	1 374.0	115.5
北 京	7 117.7	433.8	77.5	0.2	0.2
天 津	24 294.2	3 253.1	199.4	1.3	…
石 家 庄	43 683.0	6 459.7	174.7	4.4	1.6
唐 山	20 240.6	1 473.6	281.9	187.4	7.3
秦 皇 岛	16 070.5	1 056.1	62.5	…	0
邯 郸	9 488.4	607.0	98.2	1.5	0.6
保 定	18 254.7	1 323.1	30.0	4.0	…
太 原	3 280.6	237.1	24.6	0.1	0.8
大 同	9 406.2	1 328.1	147.7	220.1	…
阳 泉	284.1	59.2	0.1	0	0
长 治	15 785.9	1 513.2	438.1	275.9	9.0
临 汾	9 390.6	911.7	234.6	12.8	4.4
呼 和 浩 特	9 486.4	564.5	3.0	…	…
包 头	4 876.4	5 213.1	559.0	1.4	1.2
赤 峰	6 212.7	218.0	2.9	0	…
沈 阳	9 544.9	829.5	42.6	51.4	…
大 连	25 254.1	2 528.8	236.2	2.3	…
鞍 山	7 976.6	677.7	81.0	0.5	1.8
抚 顺	1 765.0	180.2	34.8	0.7	0.3
本 溪	9 838.7	527.4	80.8	1.5	1.5
锦 州	7 899.6	279.2	61.2	1.8	…
长 春	12 305.2	1 314.9	27.7	…	…
吉 林	15 596.3	866.5	61.6	0.1	…
哈 尔 滨	8 619.6	1 131.5	27.6	0.4	…
齐 齐 哈 尔	22 863.0	1 881.3	60.3	18.3	…
牡 丹 江	5 855.3	373.4	25.8	1.1	1.1
上 海	27 357.0	2 611.4	773.8	5.7	3.5
南 京	21 871.1	1 195.8	291.3	9.4	0.9
无 锡	13 702.5	569.7	86.6	0.2	0.2
徐 州	16 693.1	923.2	94.3	…	0.4
常 州	9 415.7	512.0	103.5	2.5	1.4
苏 州	47 887.8	3 492.5	131.7	4.5	3.3
南 通	28 075.0	1 666.0	274.4	7.6	2.0
连 云 港	17 155.0	1 766.0	12.8	11.5	…
扬 州	12 053.3	1 003.5	132.5	6.1	2.0
镇 江	6 628.7	431.2	41.3	7.4	2.3
杭 州	40 366.7	1 595.1	51.0	17.4	0.2

重点城市工业废水排放及处理情况（三）（续表）

（2011）

单位：吨

城 市 名 称	工业废水中污染物排放量				
	化学需氧量	氨氮	石油类	挥发酚	氰化物
宁 波	20 098.0	1 051.2	78.8	0.1	1.9
温 州	14 213.3	1 374.5	197.8	1.5	3.8
湖 州	10 063.4	587.6	14.3	0	0
绍 兴	31 830.6	2 516.9	19.6	…	0.1
合 肥	8 468.1	537.4	24.6	0	…
芜 湖	6 792.2	104.3	42.3	0.1	…
马 鞍 山	6 673.5	274.7	70.9	2.5	1.8
福 州	6 820.3	730.1	28.2	…	0.3
厦 门	4 313.1	345.5	15.1	…	0.9
泉 州	28 842.4	1 998.8	123.2	…	0.9
南 昌	11 549.3	1 484.1	70.8	3.3	5.4
九 江	11 793.3	858.2	87.0	2.0	0.1
济 南	5 614.4	412.7	83.9	8.3	3.6
青 岛	7 621.4	756.0	33.4	…	0.1
淄 博	14 339.3	1 200.5	50.5	3.1	0.2
枣 庄	6 802.3	347.7	42.1	0	0.2
烟 台	6 513.9	384.3	36.4	2.4	0.1
潍 坊	19 498.1	2 828.0	60.1	2.1	0.6
济 宁	8 685.7	423.6	77.0	…	0
泰 安	6 190.1	260.1	49.5	1.3	0.1
日 照	6 390.9	437.3	14.0	29.1	…
郑 州	12 041.8	551.2	134.5	0.5	3.9
开 封	11 729.4	1 485.3	86.3	0	0.7
洛 阳	9 411.6	588.7	87.8	3.2	0.9
平 顶 山	12 644.7	974.6	214.2	109.3	4.4
安 阳	10 875.6	548.0	57.5	0.4	3.0
焦 作	15 458.6	1 207.7	39.3	0.3	0
三 门 峡	7 593.7	544.7	48.2	1.0	…
武 汉	17 841.4	2 043.5	147.3	0.7	2.1
宜 昌	18 134.3	3 035.3	58.9	0.9	0.3
荆 州	26 402.9	3 270.3	86.5	4.7	0.4
长 沙	14 885.5	574.5	84.1	0	…
株 洲	6 212.7	1 434.1	61.4	0.4	1.7
湘 潭	4 702.4	614.0	105.9	1.7	2.6
岳 阳	32 819.0	14 099.0	91.1	0.4	2.1
常 德	18 103.0	953.0	31.2	…	…
张 家 界	2 908.0	76.2	10.7	0	0
广 州	22 502.0	1 309.9	111.2	0.2	2.1

重点城市工业废水排放及处理情况（三）（续表）

（2011）

单位：吨

城　市 名　称	工业废水中污染物排放量				
	化学需氧量	氨氮	石油类	挥发酚	氰化物
韶　关	5 649.6	304.6	14.6	0.2	0.9
深　圳	11 388.3	894.4	63.2	0.3	1.8
珠　海	6 376.7	556.9	27.0	0.2	0.1
汕　头	10 305.1	559.2	1.4	…	…
湛　江	11 564.3	533.2	9.2	0	0
南　宁	31 239.2	1 468.0	29.2	11.9	1.7
柳　州	12 350.0	1 630.0	44.7	2.2	1.6
桂　林	8 565.1	556.1	22.6	0.3	…
北　海	9 533.0	112.0	0.5	…	0.1
海　口	737.2	40.3	3.7	0	…
重　庆	58 028.2	3 205.3	521.1	35.0	2.7
成　都	14 027.1	833.8	41.5	…	…
自　贡	3 121.7	519.8	28.2	50.8	…
攀枝花	1 728.6	129.9	5.4	0.1	…
泸　州	11 322.0	287.5	18.7	0.1	0.3
德　阳	7 232.8	486.2	74.0	…	…
绵　阳	5 294.3	157.1	35.3	…	0
南　充	5 535.8	390.6	14.6	0	0
宜　宾	19 368.5	566.5	66.7	…	0.7
贵　阳	5 433.2	228.7	113.3	…	1.6
遵　义	18 203.3	396.6	24.5	0.2	1.9
昆　明	8 148.0	260.0	78.4	0.4	0.4
曲　靖	9 520.9	1 246.3	77.4	0.2	0.7
玉　溪	13 421.9	269.3	52.5	0.5	0.1
拉　萨	774.8	51.4	0.2	0	0
西　安	32 848.6	1 347.0	140.5	…	0.1
铜　川	1 342.7	7.4	100.8	…	0
宝　鸡	14 061.9	1 268.4	123.4	0.6	2.2
咸　阳	14 841.4	1 467.3	74.4	…	0.2
渭　南	21 313.6	1 141.3	91.9	198.2	5.1
延　安	4 623.5	252.3	126.7	0.2	0.1
兰　州	4 568.6	2 431.6	86.6	8.4	…
金　昌	4 470.0	4 804.8	35.0	0.2	0.3
西　宁	15 469.3	520.8	60.3	0.6	0
银　川	18 074.1	2 884.8	41.0	4.5	0.4
石嘴山	4 941.5	1 353.7	14.9	0.2	0.3
乌鲁木齐	8 124.6	1 572.9	57.2	18.3	1.3
克拉玛依	1 744.8	88.9	90.8	1.3	0

重点城市工业废水排放及处理情况（四）

（2011）   单位：吨

城　市 名　称	工业废水中污染物排放量				
	汞	镉	总铬	铅	砷
总　　计	0.433	8.275	126.457	46.252	27.700
北　京	0	0.001	0.446	0.068	0.013
天　津	0.001	0.010	0.285	1.454	0.023
石 家 庄	…	0.001	2.140	0.013	0.001
唐　山	…	0	0.296	0.043	0
秦 皇 岛	0	0	0.021	0	0
邯　郸	0	0	0	0	0
保　定	0	0.006	0.082	0.383	0.005
太　原	0.002	0.013	0.127	0.172	0.064
大　同	…	0.004	0.008	0.237	0.511
阳　泉	…	0	0	0	0
长　治	0.001	0	0.013	0	0
临　汾	0	0.027	0.004	0.068	0.175
呼和浩特	0.002	0	…	0.002	0
包　头	0.001	0	0.002	0.002	0
赤　峰	0.034	0.120	0.035	1.010	1.845
沈　阳	0	…	0.093	0.036	0
大　连	0	0	0.059	0.160	0
鞍　山	0	0	0	0	0
抚　顺	0	0.015	0	0.227	0.077
本　溪	…	0	0	0	0
锦　州	0	0	0.263	0	0.026
长　春	0	0	0.098	0	0
吉　林	…	0.001	0.003	0.001	0.003
哈 尔 滨	0	0	0.152	0.016	0
齐齐哈尔	0	0	0.019	0	0
牡 丹 江	0	0	0	0	0
上　海	…	0.003	2.505	0.104	0.010
南　京	…	0.007	0.421	0.020	0.046
无　锡	0	0	0.599	0.205	0.012
徐　州	0.001	0.097	0.940	0.632	0.206
常　州	0	0.005	0.274	0.031	0.007
苏　州	0	…	2.730	0.158	0
南　通	0.091	0.007	4.349	1.024	0.349
连 云 港	0	0.001	0.002	0.037	0.001
扬　州	0.001	…	1.018	0.253	0.001
镇　江	0	0	0.225	0.049	0.081
杭　州	0	0.002	2.657	0.040	0.012

重点城市工业废水排放及处理情况（四）（续表）

（2011）

单位：吨

城 市 名 称	工业废水中污染物排放量				
	汞	镉	总铬	铅	砷
宁 波	0	0	3.843	0.040	0.002
温 州	0	0.244	5.302	0.122	0.034
湖 州	0	0	0.329	0.031	0
绍 兴	0	0	0.382	0.030	0
合 肥	0	0.002	0.010	0.045	0.080
芜 湖	…	0.003	0.101	0.272	0.170
马 鞍 山	0	0	0.003	…	0
福 州	0.008	…	0.502	0.026	0.002
厦 门	0	…	2.845	0.204	0.134
泉 州	0.002	0.036	9.272	0.209	0.134
南 昌	…	0.008	16.945	0.043	0.019
九 江	0.002	0.178	0.128	0.182	0.602
济 南	0	0	0.241	0	0.042
青 岛	0	0	0.217	0.002	0
淄 博	0	0	0.477	0.017	0.038
枣 庄	0.001	0	0.072	0.001	0.015
烟 台	0.012	0.008	0.193	0.023	0.038
潍 坊	0	0	0.640	0	0
济 宁	0	0	0.004	0	0
泰 安	0	0	0.066	0.006	0.092
日 照	0	0	0.128	0.005	0
郑 州	0	0	0.045	0	0.024
开 封	…	0.480	1.127	1.324	0.114
洛 阳	0.008	1.450	3.684	3.186	1.047
平 顶 山	0.002	…	0.074	0.005	0.040
安 阳	0	…	0.002	0.140	0
焦 作	0.003	0.010	22.768	0.201	0.018
三 门 峡	0.003	0.146	0.010	0.596	0.325
武 汉	0.002	0.005	1.191	0.189	0.219
宜 昌	0.002	0.001	0.110	0.026	0.052
荆 州	0.001	0	2.798	0	0.098
长 沙	0	0.022	0.546	0.147	0.029
株 洲	0.145	0.394	0.071	2.896	0.979
湘 潭	0	0.611	17.292	1.619	0.701
岳 阳	0.001	0.032	0.007	0.323	0.363
常 德	0	0	0.399	…	0.232
张 家 界	0	0.006	0.012	0.021	0.012
广 州	0.001	0.011	5.339	0.667	0.048

重点城市工业废水排放及处理情况（四）（续表）

（2011） 单位：吨

城 市名 称	工业废水中污染物排放量				
	汞	镉	总铬	铅	砷
韶 关	0.016	0.898	0.772	7.744	1.077
深 圳	0.018	0.002	0.371	0.033	0
珠 海	…	0.007	0.758	0.011	…
汕 头	…	0.008	0.106	0.033	0.042
湛 江	0	…	0.018	0.001	0.011
南 宁	0.012	0.001	0.527	0.076	0.079
柳 州	0.016	0.162	1.588	2.105	0.344
桂 林	0.007	0.010	0.131	0.371	0.220
北 海	0	0	0.050	0	0
海 口	0	…	0.131	…	…
重 庆	0	…	0.592	0.075	1.377
成 都	0	…	0.391	0.030	0.097
自 贡	0	0	0	0	0
攀 枝 花	0	0.001	0.019	0.002	0.029
泸 州	0	0	0.077	0	1.600
德 阳	0.003	…	0.060	0.003	0.050
绵 阳	0	…	0.416	0.002	0.042
南 充	0	0	…	0	0
宜 宾	0.005	…	0.014	0.060	0.113
贵 阳	0	0.001	0.059	0.007	0
遵 义	0	…	0.035	0.092	0.007
昆 明	0.002	2.039	0.024	11.659	6.746
曲 靖	0.005	0.130	0.028	0.335	0.493
玉 溪	0.002	0.006	0	0.019	0.198
拉 萨	0	0	…	0	2.721
西 安	…	0.005	0.182	0.061	0
铜 川	0	0	0	0	0
宝 鸡	…	0.292	0.017	1.112	0.436
咸 阳	0	0.003	1.333	…	0
渭 南	0	0.016	0	1.791	0
延 安	…	0.004	0.023	0.007	0.043
兰 州	0	0.016	0.149	0.089	0
金 昌	0	0.630	0	1.357	2.663
西 宁	0.002	0.073	1.025	0.117	0.064
银 川	0	0	0.175	0.010	0.038
石 嘴 山	0.002	0	0	0	0.011
乌鲁木齐	0.010	0.005	0.342	0.010	0.032
克拉玛依	0	0	0	0	0

重点城市工业废气排放及处理情况（一）

（2011）

城市名称	工业废气排放总量（标态）/亿米³	工业废气中污染物产生量/万吨			工业废气中污染物排放量/万吨		
		二氧化硫	氮氧化物	烟（粉）尘	二氧化硫	氮氧化物	烟（粉）尘
总　计	378 126	3 083.1	1 028.6	926.1	537.1	972.1	42 764.5
北　京	4 897	16.1	6.1	9.0	2.9	10.1	361.5
天　津	8 919	57.9	22.2	30.0	6.5	31.0	769.9
石 家 庄	6 606	64.2	19.7	21.5	9.6	21.7	989.5
唐　山	28 030	70.2	33.2	32.0	50.6	33.3	1 945.7
秦 皇 岛	3 201	55.1	7.6	6.0	7.0	6.0	189.6
邯　郸	17 518	52.7	22.0	17.6	22.1	18.0	1 194.5
保　定	2 117	24.2	7.6	6.6	3.8	7.0	211.5
太　原	5 442	35.2	10.8	11.4	4.4	12.7	722.3
大　同	3 638	43.7	14.3	13.0	6.6	13.2	792.1
阳　泉	1 563	21.9	10.1	7.4	3.1	7.4	256.1
长　治	4 647	33.7	14.7	11.7	21.5	12.6	627.7
临　汾	6 655	39.5	10.1	6.7	10.3	7.6	401.5
呼 和 浩 特	3 366	38.5	10.7	16.7	2.5	16.7	431.9
包　头	6 403	92.8	21.0	13.3	9.7	13.3	691.8
赤　峰	1 568	68.4	12.8	6.1	1.3	6.1	364.4
沈　阳	1 923	18.7	9.5	8.1	5.1	8.1	225.7
大　连	2 871	20.5	13.2	11.2	5.1	11.9	453.9
鞍　山	5 942	12.9	11.9	6.8	7.1	6.9	283.6
抚　顺	2 386	14.7	5.9	5.7	3.9	5.8	264.8
本　溪	4 691	7.8	8.2	5.6	7.3	5.6	659.6
锦　州	947	10.3	5.0	3.6	4.4	3.8	142.2
长　春	2 610	11.7	6.9	9.3	11.6	9.9	449.5
吉　林	2 415	12.8	8.4	10.0	12.1	10.1	482.8
哈 尔 滨	1 995	14.1	9.0	10.1	4.8	10.4	400.9
齐 齐 哈 尔	1 133	7.2	6.2	7.8	4.9	8.0	202.0
牡 丹 江	575	4.0	2.6	4.5	4.5	4.5	166.6
上　海	13 704	54.5	21.0	32.7	6.6	35.7	732.9
南　京	6 963	37.4	12.6	15.0	5.6	18.0	623.7
无　锡	5 753	30.9	9.6	14.7	6.4	16.1	503.3
徐　州	3 862	44.0	14.3	16.5	4.7	19.5	665.7
常　州	3 698	20.7	4.4	8.6	3.9	8.6	201.1
苏　州	15 161	58.1	19.2	23.3	5.9	24.1	573.3
南　通	2 743	24.8	7.2	6.6	4.9	6.6	269.4
连 云 港	831	6.6	4.5	3.4	1.9	3.4	97.9
扬　州	1 214	19.4	5.6	7.7	1.3	7.7	145.1
镇　江	2 755	28.2	8.7	8.3	3.7	8.5	232.4
杭　州	4 613	16.6	9.2	8.3	3.9	8.4	452.9

重点城市工业废气排放及处理情况（一）（续表）

（2011）

城 市名 称	工业废气排放总量（标态）/亿米³	工业废气中污染物产生量/万吨			工业废气中污染物排放量/万吨		
		二氧化硫	氮氧化物	烟（粉）尘	二氧化硫	氮氧化物	烟（粉）尘
宁 波	5 910	112.8	15.3	25.3	3.2	29.3	611.4
温 州	1 786	13.5	4.0	4.9	2.5	5.2	91.6
湖 州	2 043	9.6	4.0	6.2	2.8	6.2	367.4
绍 兴	1 480	11.4	6.0	4.3	2.3	4.3	127.3
合 肥	2 775	9.6	4.9	8.8	4.0	8.9	424.0
芜 湖	2 459	8.7	3.6	6.4	7.9	6.4	713.0
马 鞍 山	5 493	12.7	6.6	10.4	3.1	10.4	196.7
福 州	3 593	27.2	9.1	10.2	3.4	13.2	214.0
厦 门	1 104	5.5	1.9	1.8	0.3	3.1	49.2
泉 州	2 902	20.8	9.9	8.0	5.0	8.2	118.1
南 昌	1 175	9.5	3.5	2.7	1.9	2.7	138.9
九 江	2 382	16.0	9.0	6.7	3.5	6.7	250.7
济 南	4 555	25.3	10.9	8.3	10.4	8.6	473.5
青 岛	2 438	25.4	7.6	7.8	2.7	7.8	270.3
淄 博	5 194	60.2	22.9	13.4	6.7	13.7	710.3
枣 庄	2 217	21.7	7.6	6.8	2.6	6.8	628.5
烟 台	2 659	38.4	8.8	8.3	3.4	8.3	476.5
潍 坊	2 724	54.5	13.1	8.3	3.7	8.5	447.8
济 宁	4 001	45.3	14.0	16.6	4.1	16.7	706.9
泰 安	2 237	26.8	7.8	6.1	2.4	6.2	343.6
日 照	5 328	16.0	6.0	6.1	3.2	6.1	287.4
郑 州	3 567	20.2	11.1	14.9	4.2	14.9	614.1
开 封	913	11.4	4.5	3.9	2.5	3.9	271.4
洛 阳	11 074	62.4	17.5	14.8	5.7	15.0	672.2
平 顶 山	2 813	24.4	9.3	8.5	6.4	8.6	782.1
安 阳	3 998	25.4	12.7	6.5	9.1	6.5	339.1
焦 作	1 714	9.5	5.2	5.3	3.0	5.3	312.8
三 门 峡	1 969	24.1	11.4	8.0	3.7	8.0	331.5
武 汉	6 329	27.9	10.9	12.0	2.3	12.1	364.5
宜 昌	2 591	13.1	7.0	3.8	1.8	3.9	261.2
荆 州	861	13.0	5.6	1.2	1.5	1.9	100.3
长 沙	1 022	7.4	2.6	2.2	1.6	3.3	200.5
株 洲	1 104	29.5	4.4	4.0	1.1	4.0	192.8
湘 潭	2 356	15.7	5.7	5.0	2.5	5.0	320.3
岳 阳	1 206	14.9	5.2	6.3	4.3	6.3	283.6
常 德	1 568	13.6	5.2	6.2	2.1	6.2	210.1
张 家 界	162	2.6	2.1	0.6	0.4	0.6	24.0
广 州	3 512	45.0	6.8	7.4	1.5	11.5	311.7

重点城市工业废气排放及处理情况（一）（续表）

（2011）

城市名称	工业废气排放总量（标态）/亿米³	工业废气中污染物产生量/万吨			工业废气中污染物排放量/万吨		
		二氧化硫	氮氧化物	烟（粉）尘	二氧化硫	氮氧化物	烟（粉）尘
韶　关	1 446	11.2	6.2	3.3	0.6	3.3	117.6
深　圳	1 885	4.1	1.0	3.5	0.1	3.7	50.7
珠　海	1 518	10.7	3.2	5.4	1.0	5.5	89.5
汕　头	984	12.2	2.7	3.2	0.4	3.5	86.1
湛　江	953	8.9	3.0	2.9	1.7	2.9	113.1
南　宁	1 248	5.7	3.2	2.4	2.7	2.4	202.2
柳　州	3 427	13.5	5.4	5.3	5.8	5.4	279.3
桂　林	949	10.7	4.1	3.6	1.7	3.6	88.7
北　海	501	3.4	1.1	1.8	0.6	1.8	65.4
海　口	28	0.2	0.2	…	0.1	…	0.2
重　庆	9 121	158.9	53.1	29.4	17.1	29.6	1 800.9
成　都	2 832	11.3	5.3	5.7	2.2	6.4	203.2
自　贡	355	3.8	3.7	0.9	1.4	0.9	31.3
攀枝花	3 137	13.2	9.8	3.5	5.7	3.6	332.0
泸　州	861	17.4	6.1	2.8	1.4	3.0	902.1
德　阳	873	10.6	2.7	1.3	1.4	1.3	108.3
绵　阳	1 098	10.5	3.8	3.5	1.5	3.5	268.3
南　充	188	1.1	0.9	0.2	1.2	0.2	1.6
宜　宾	3 138	30.5	12.3	4.0	2.1	4.3	243.8
贵　阳	1 557	35.4	8.2	3.0	3.1	3.0	206.5
遵　义	1 524	31.1	8.4	4.0	6.3	4.0	274.0
昆　明	4 069	40.6	10.4	7.3	6.2	7.4	310.3
曲　靖	4 914	78.9	21.8	13.6	9.8	13.7	944.3
玉　溪	1 299	12.2	3.5	1.7	3.9	1.7	158.0
拉　萨	63	0.1	0.1	0.2	0.1	0.2	18.8
西　安	996	17.2	9.8	4.9	1.7	4.9	146.8
铜　川	1 360	8.6	2.0	5.5	3.9	5.6	31.9
宝　鸡	1 120	26.9	4.4	6.2	1.5	6.7	194.3
咸　阳	1 849	21.7	7.0	8.6	3.0	9.5	211.3
渭　南	4 800	79.9	30.6	17.0	4.6	19.5	718.3
延　安	235	2.3	1.8	0.4	0.7	0.5	10.0
兰　州	3 183	19.1	9.3	8.0	4.0	8.4	917.2
金　昌	697	148.5	11.2	2.6	1.5	2.7	95.7
西　宁	2 884	12.3	7.2	5.0	4.9	5.0	242.5
银　川	1 752	26.4	7.5	6.7	2.6	7.1	428.6
石嘴山	2 498	36.3	9.6	10.2	8.5	10.9	1 035.9
乌鲁木齐	3 268	21.8	12.9	11.2	6.2	11.2	369.1
克拉玛依	942	5.1	4.8	2.5	1.2	2.5	47.9

重点城市工业废气排放及处理情况（二）

（2011）

城 市 名 称	废气治理设施数/套	废气治理设施处理能力 （标态）/ （万米³/时）	废气治理设施运行费用/万元
总 计	121 476	907 981	9 459 895
北 京	2 962	10 947	82 699
天 津	3 837	23 588	436 181
石 家 庄	2 382	13 863	97 686
唐 山	3 453	89 513	688 925
秦 皇 岛	3 589	7 946	177 709
邯 郸	1 584	24 063	202 167
保 定	1 445	4 556	23 223
太 原	1 348	17 579	121 213
大 同	1 538	6 751	63 550
阳 泉	492	3 520	37 501
长 治	1 004	8 622	105 048
临 汾	892	5 080	92 787
呼和浩特	851	9 649	38 670
包 头	851	9 125	111 292
赤 峰	796	4 078	35 365
沈 阳	1 952	13 286	28 955
大 连	1 948	7 705	44 311
鞍 山	1 525	12 041	104 197
抚 顺	462	4 520	25 745
本 溪	803	7 981	122 234
锦 州	530	3 236	29 575
长 春	616	4 210	42 188
吉 林	728	6 564	43 824
哈 尔 滨	1 211	16 231	32 294
齐齐哈尔	752	8 159	1 031 692
牡 丹 江	303	2 029	10 519
上 海	4 730	30 581	416 226
南 京	1 249	12 735	323 680
无 锡	2 595	11 898	91 167
徐 州	1 034	11 670	106 274
常 州	994	12 412	45 533
苏 州	4 279	27 492	272 913
南 通	1 389	5 128	64 713
连 云 港	464	2 112	28 656
扬 州	471	4 848	52 908
镇 江	669	6 081	51 720
杭 州	2 416	9 473	109 251

重点城市工业废气排放及处理情况（二）（续表）

（2011）

城市名称	废气治理设施数/套	废气治理设施处理能力（标态）/（万米³/时）	废气治理设施运行费用/万元
宁 波	2 679	16 411	234 265.0
温 州	1 984	5 899	45 696.6
湖 州	1 180	7 316	22 753.0
绍 兴	2 253	4 106	35 649.6
合 肥	590	1 981	61 641.0
芜 湖	528	10 335	33 009.1
马 鞍 山	493	12358	125 878.0
福 州	665	9 496	108 056.8
厦 门	918	3 533	25 657.2
泉 州	1 081	3 633	46 129.8
南 昌	589	3 190	33 218.4
九 江	602	3 551	51 000.9
济 南	925	14 680	119 114.5
青 岛	1 197	8 562	62 181.8
淄 博	3 231	14 253	210 423.4
枣 庄	910	5 863	39 145.4
烟 台	1 132	7 937	64 357.5
潍 坊	1 468	8 940	69 302.8
济 宁	1 053	12 236	100 472.9
泰 安	689	5 153	27 249.9
日 照	701	9 530	96 044.9
郑 州	1 184	6 230	59 327.6
开 封	184	591	5 559.8
洛 阳	1 099	8 703	81 011.9
平 顶 山	480	4 231	26 894.3
安 阳	918	6 187	50 279.8
焦 作	941	6 012	25 894.4
三 门 峡	613	5 209	41 386.7
武 汉	853	9 083	111 276.1
宜 昌	1 451	33 516	21 497.0
荆 州	428	1 387	14 913.0
长 沙	458	1 790	7 689.6
株 洲	698	2 407	40 358.3
湘 潭	199	1 818	20 491.0
岳 阳	396	3 788	38 904.0
常 德	502	3 860	21 172.9
张 家 界	49	101	1 006.2
广 州	2 073	10 895	226 475.6

重点城市工业废气排放及处理情况（二）（续表）

（2011）

城 市 名 称	废气治理设施数/套	废气治理设施处理能力 （标态）/ （万米³/时）	废气治理设施运行费用/万元
韶 关	637	4 929	59 092.2
深 圳	1 041	2 040	33 853.1
珠 海	949	3 419	32 045.5
汕 头	573	4 799	33 215.0
湛 江	378	4 514	30 545.8
南 宁	799	4 069	14 234.7
柳 州	912	9 650	74 649.4
桂 林	566	7 214	17 315.9
北 海	119	1 225	9 534.0
海 口	32	42	169.7
重 庆	4 134	22 710	166 041.2
成 都	1 735	4 662	36 375.3
自 贡	160	469	5 789.4
攀 枝 花	690	5 355	70 146.1
泸 州	199	800	32 460.6
德 阳	744	2 452	161 464.3
绵 阳	622	12 059	29 080.6
南 充	21	119	825.0
宜 宾	538	4 663	83 998.4
贵 阳	429	2 969	43 097.0
遵 义	576	4 007	56 815.3
昆 明	1 390	8 564	106 937.1
曲 靖	872	7 697	50 453.2
玉 溪	878	2 294	21 695.7
拉 萨	107	47	424.3
西 安	748	3 675	27 587.6
铜 川	427	1 439	3 357.3
宝 鸡	893	2 388	18 279.5
咸 阳	497	3 273	23 819.0
渭 南	736	7 796	84 227.0
延 安	210	423	4 702.9
兰 州	933	4 730	45 511.8
金 昌	202	1 106	70 892.3
西 宁	525	5 312	51 975.7
银 川	344	5 099	46 786.6
石 嘴 山	569	3 996	57 866.1
乌鲁木齐	586	12 504	47 844.1
克拉玛依	167	1 432	8 811.5

重点城市工业固体废物产生及处置利用情况

（2011）

单位：万吨

城　市名　称	一般工业固体废物产生量	一般工业固体废物综合利用量	一般工业固体废物处置量	一般工业固体废物贮存量	一般工业固体废物倾倒丢弃量	一般工业固体废弃物综合利用率/%
总　　计	143 731	96 079	37 342	12 286	102	66.0
北　京	1 126	749	349	28	0	66.5
天　津	1 752	1 749	9	0	0	99.5
石 家 庄	1 520	1 517	9	124	0	91.9
唐　山	9 935	7 175	2 855	386	…	68.9
秦 皇 岛	1 987	1 150	663	173	0	57.9
邯　郸	4 014	3 607	22	445	0	87.8
保　定	795	728	67	0	0	91.5
太　原	3 154	1 673	1 423	44	1	53.0
大　同	3 901	2 428	544	922	15	62.1
阳　泉	1 766	419	1 344	1	2	23.7
长　治	2 500	1 701	656	253	8	66.6
临　汾	2 408	1 533	317	556	1	63.7
呼 和 浩 特	892	359	480	53	0	40.2
包　头	2 848	966	1 535	480	…	33.9
赤　峰	2 068	355	1 696	17	0	17.2
沈　阳	705	661	137	30	0	93.8
大　连	529	508	21	0	0	96.1
鞍　山	5 005	1 230	1 379	2 396	0	24.6
抚　顺	2 839	1 154	1 242	439	…	40.6
本　溪	6 750	1 051	5 413	278	8	15.6
锦　州	335	233	100	2	0	69.5
长　春	617	613	4	0	0	99.4
吉　林	1 555	607	9	940	0	39.0
哈 尔 滨	564	518	36	104	0	84.2
齐 齐 哈 尔	308	239	12	60	7	73.7
牡 丹 江	337	325	0	11	0	96.6
上　海	2 442	2 358	75	11	…	96.5
南　京	1 759	1 504	115	146	0	85.2
无　锡	1 031	940	67	24	0	91.2
徐　州	1 540	1 573	22	…	0	99.7
常　州	624	611	13	0	0	98.0
苏　州	2 298	2 250	48	…	0	97.9
南　通	465	453	13	1	0	97.5
连 云 港	448	410	8	46	0	88.3
扬　州	322	317	2	3	0	98.2
镇　江	753	739	14	0	0	98.1
杭　州	763	708	56	…	0	92.7

重点城市工业固体废物产生及处置利用情况（续表）

（2011）   单位：万吨

城　市名　称	一般工业固体废物产生量	一般工业固体废物综合利用量	一般工业固体废物处置量	一般工业固体废物贮存量	一般工业固体废物倾倒丢弃量	一般工业固体废弃物综合利用率/%
宁　波	1 422	1 281	112	33	…	89.8
温　州	243	227	16	0	…	93.5
湖　州	195	188	6	…	0	96.3
绍　兴	353	321	31	1	0	91.1
合　肥	1 066	1 001	8	58	0	93.8
芜　湖	399	337	28	18	0	84.5
马鞍山	2 131	1 462	556	113	0	68.6
福　州	694	623	69	5	…	89.8
厦　门	119	112	5	5	0	91.5
泉　州	712	672	25	14	…	94.4
南　昌	185	182	2	…	1	98.3
九　江	942	505	95	322	9	53.5
济　南	1 126	1 117	6	4	0	99.2
青　岛	888	882	10	14	0	97.3
淄　博	1 881	1 822	12	54	0	96.5
枣　庄	750	749	2	4	0	99.1
烟　台	2 358	1 946	369	44	0	82.5
潍　坊	927	850	…	0	0	91.6
济　宁	2 687	2 423	199	82	0	89.6
泰　安	1 119	1 261	…	22	0	98.3
日　照	957	952	4	0	0	99.5
郑　州	1 249	919	303	28	0	73.5
开　封	427	427	0	0	0	100.0
洛　阳	2 806	1 343	1 363	92	0	47.9
平顶山	1 819	1 472	291	90	0	79.5
安　阳	1 289	1 233	52	3	0	95.7
焦　作	782	499	295	27	…	61.7
三门峡	1 790	925	60	827	0	51.0
武　汉	1 380	1 374	50	6	0	96.1
宜　昌	1 167	511	656	…	0	43.8
荆　州	118	118	0	0	0	100.0
长　沙	178	175	3	…	…	98.4
株　洲	434	393	34	45	0	83.5
湘　潭	737	737	1	33	…	95.6
岳　阳	486	452	14	37	0	89.9
常　德	310	292	1	16	0	94.2
张家界	46	41	0	5	…	89.8
广　州	659	626	30	5	0	94.8

重点城市工业固体废物产生及处置利用情况（续表）

（2011）   单位：万吨

城　市名　称	一般工业固体废物产生量	一般工业固体废物综合利用量	一般工业固体废物处置量	一般工业固体废物贮存量	一般工业固体废物倾倒丢弃量	一般工业固体废弃物综合利用率/%
韶　关	799	760	188	24	…	78.1
深　圳	96	92	4	…	…	96.0
珠　海	310	300	8	0	0	97.0
汕　头	120	110	1	10	0	91.3
湛　江	262	249	9	3	…	94.3
南　宁	349	316	13	23	0	89.9
柳　州	1 179	1 103	52	28	0	93.3
桂　林	303	262	29	18	…	86.6
北　海	369	164	205	0	…	44.4
海　口	5	4	…	0	0	90.2
重　庆	3 299	2 585	518	199	24	77.8
成　都	518	512	6	…	0	98.8
自　贡	145	129	17	0	0	89.0
攀枝花	4 897	1 455	3 323	123	0	29.7
泸　州	345	329	12	7	0	94.5
德　阳	420	339	101	…	0	77.1
绵　阳	300	302	7	7	0	96.0
南　充	10	10	…	0	0	98.2
宜　宾	628	526	75	28	0	83.6
贵　阳	1 140	642	473	26	…	56.3
遵　义	640	334	267	21	…	52.2
昆　明	3 670	1 683	1 863	357	14	43.1
曲　靖	2 428	1 466	271	683	0	60.0
玉　溪	2 496	1 049	1 247	203	1	42.0
拉　萨	248	8	16	233	…	3.1
西　安	278	271	5	1	…	97.6
铜　川	214	174	20	20	0	80.8
宝　鸡	504	300	205	…	0	59.4
咸　阳	518	462	56	…	…	89.2
渭　南	2 849	1 435	1 375	38	0	50.4
延　安	71	64	…	…	7	89.5
兰　州	605	561	43	…	0	92.8
金　昌	1 150	183	828	138	0	15.9
西　宁	501	493	11	6	0	96.8
银　川	625	523	72	28	1	83.8
石嘴山	898	471	356	73	0	52.5
乌鲁木齐	1 001	810	188	2	1	80.9
克拉玛依	56	43	13	…	…	77.1

重点城市工业危险废物产生及处置利用情况

（2011）

单位：万吨

城　市 名　称	危险废物 产生量	危险废物 综合利用量	危险废物 处置量	危险废物 贮存量	危险废物 倾倒丢弃量
总　　计	1 703.97	1 081.58	650.19	12.82	…
北　　京	11.92	5.05	6.86	0	0
天　　津	10.27	3.09	7.18	…	0
石　家　庄	24.67	12.45	12.25	0	0
唐　　山	16.54	16.05	0.49	0	0
秦　皇　岛	2.44	0.13	2.31	0	0
邯　　郸	0.80	0.76	0.04	0	0
保　　定	1.72	0.48	1.24	0	0
太　　原	5.23	2.09	3.14	…	0
大　　同	0.24	0.20	0.04	…	0
阳　　泉	0.13	0.01	0.13	0	0
长　　治	3.06	3.05	0	0.01	0
临　　汾	4.97	4.90	0.07	…	0
呼和浩特	5.17	5.02	0.15	0	0
包　　头	3.54	3.03	0.53	0	0
赤　　峰	34.03	8.97	25.06	0	0
沈　　阳	7.40	5.07	2.33	…	0
大　　连	13.84	1.23	12.60	…	0
鞍　　山	1.95	1.64	0.31	…	0
抚　　顺	4.33	3.52	0.76	0.05	0
本　　溪	13.03	12.86	0.17	…	…
锦　　州	7.27	13.64	0.36	0	0
长　　春	1.69	0.11	1.58	0	0
吉　　林	85.70	57.46	28.24	…	0
哈　尔　滨	2.36	1.07	1.30	…	0
齐齐哈尔	0.99	0.02	0.26	0.72	0
牡　丹　江	0.23	0.02	0.21	0	0
上　　海	56.36	30.13	26.01	0.34	0
南　　京	32.18	16.01	15.84	0.59	0
无　　锡	58.52	27.92	30.49	0.11	0
徐　　州	0.59	0.02	0.56	0.01	0
常　　州	10.34	5.85	4.49	0	0
苏　　州	48.29	18.57	29.72	…	0
南　　通	10.81	7.13	3.63	0.06	0
连　云　港	0.68	0.26	0.44	…	0
扬　　州	10.72	9.29	1.63	0	0
镇　　江	4.24	2.26	1.96	0.02	0
杭　　州	11.45	8.68	2.74	0.06	0

重点城市工业危险废物产生及处置利用情况（续表）

（2011）

单位：万吨

城 市 名 称	危险废物 产生量	危险废物 综合利用量	危险废物 处置量	危险废物 贮存量	危险废物 倾倒丢弃量
宁 波	22.76	13.24	9.43	0.17	0
温 州	3.96	1.67	2.29	…	0
湖 州	9.65	0.40	9.28	0.04	0
绍 兴	16.99	2.99	13.93	0.39	0
合 肥	1.66	1.18	0.47	…	0
芜 湖	0.93	0.38	0.55	…	0
马 鞍 山	2.07	1.34	0.70	0.04	0
福 州	2.16	0.96	1.10	0.1	0
厦 门	2.07	0.74	1.32	0.03	0
泉 州	2.33	0.17	2.15	…	0
南 昌	1.53	1.09	0.44	…	…
九 江	2.69	2.12	0.56	0.04	0
济 南	15.39	4.27	13.20	0.02	0
青 岛	3.60	1.67	1.92	0.01	0
淄 博	175.48	163.00	12.48	0	0
枣 庄	1.50	1.47	0.03	…	0
烟 台	127.63	123.90	12.47	0	0
潍 坊	9.23	7.85	1.38	…	0
济 宁	203.37	75.87	127.46	0.04	0
泰 安	3.63	1.33	2.28	0.02	0
日 照	0.19	0.01	0.18	0	0
郑 州	0.89	0.21	0.67	…	0
开 封	0.20	0.20	…	0	0
洛 阳	1.62	1.33	0.29	…	0
平 顶 山	0.29	0.27	0.02	0	0
安 阳	2.10	2.06	0.03	0.02	0
焦 作	3.54	3.20	0.31	0.03	0
三 门 峡	19.75	18.35	1.38	0.02	0
武 汉	5.83	2.38	3.46	…	0
宜 昌	1.16	0.97	0.19	0	0
荆 州	1.01	0.94	0.07	0	0
长 沙	0.11	0.07	0.01	0.03	0
株 洲	17.75	16.40	1.34	0	0
湘 潭	11.92	13.48	0.39	0.05	0
岳 阳	82.40	46.55	35.84	0.01	0
常 德	5.26	4.09	1.17	0	0
张 家 界	0.29	0.29	0	0	0
广 州	30.29	11.83	18.46	0	0

重点城市工业危险废物产生及处置利用情况（续表）

（2011）

单位：万吨

城 市 名 称	危险废物 产生量	危险废物 综合利用量	危险废物 处置量	危险废物 贮存量	危险废物 倾倒丢弃量
韶 关	15.45	10.67	4.79	0.43	0
深 圳	36.97	19.26	17.72	0	0
珠 海	7.59	4.67	2.93	0	0
汕 头	0.80	0.57	0.23	…	0
湛 江	0.21	0	0.21	0	0
南 宁	0.17	0.14	0.03	…	0
柳 州	1.91	1.29	0.58	0.04	0
桂 林	0.15	0.14	0.01	0	0
北 海	0.53	0	0.53	0	0
海 口	0.18	0.06	0.12	…	0
重 庆	46.50	5.64	43.61	0.51	0
成 都	4.74	1.52	3.18	0.05	0
自 贡	0.56	0.06	0.50	…	0
攀 枝 花	82.74	43.47	39.27	0.09	0
泸 州	2.84	2.82	0.02	…	0
德 阳	0.58	0.03	0.54	0	0
绵 阳	7.59	16.37	1.38	0	0
南 充	0.55	0.55	0	…	0
宜 宾	0.41	0.14	0.29	0	0
贵 阳	1.27	1.77	0.30	0	0
遵 义	0.70	0.15	0.55	0	0
昆 明	58.45	54.74	3.71	0	0
曲 靖	13.13	8.93	2.14	2.25	0
玉 溪	1.12	1.06	0.03	0.04	0
拉 萨	0	0	0	0	0
西 安	0.92	0.07	0.86	…	0
铜 川	0	0	0	0	0
宝 鸡	1.11	0.83	0.28	…	0
咸 阳	0.48	0.06	0.42	0	0
渭 南	14.64	14.17	0.47	0	0
延 安	7.22	1.41	5.81	0	0
兰 州	10.10	5.06	5.04	…	0
金 昌	1.76	1.54	0.32	0.04	0
西 宁	25.71	18.92	6.09	5.22	0
银 川	3.03	2.60	0.42	0	0
石 嘴 山	7.28	7.19	0.09	0	0
乌鲁木齐	22.39	19.53	2.86	…	0
克拉玛依	23.27	19.82	2.53	1.04	0

重点城市汇总工业企业概况（一）

（2011）

城　市名　称	汇总工业企业数/个	工业总产值（现价）/亿元	工业炉窑数/台	工业锅炉数/台	35 蒸吨及以上的	20（含）～35蒸吨的	10（含）～20蒸吨的	10 蒸吨以下的
总　　　计	71 223	243 291.8	49 540	54 870	7 516	4 481	7 672	35 201
北　京	914	6 471.6	330	1 751	301	288	304	858
天　津	1 988	10 596.8	718	2 150	412	288	336	1 114
石　家　庄	1 578	2 182.2	855	1 181	147	42	97	895
唐　山	1 560	6 320.5	1 444	629	154	61	162	252
秦　皇　岛	395	1 203.6	101	383	50	25	49	259
邯　郸	623	3 213.8	728	362	105	45	55	157
保　定	1 627	4 697.2	615	1 057	25	26	162	844
太　原	320	2 034.8	563	413	108	50	91	164
大　同	664	1 003.8	320	959	45	40	123	751
阳　泉	332	411.2	476	238	27	13	37	161
长　治	629	1 637.4	530	647	80	52	133	382
临　汾	486	1 486.1	319	329	68	57	48	156
呼和浩特	488	718.8	81	787	124	67	250	346
包　头	143	2 358.0	355	244	56	19	41	128
赤　峰	479	378.1	239	478	54	76	102	246
沈　阳	884	2 543.5	287	1 551	265	191	396	699
大　连	1 013	3 921.4	478	1 223	185	224	204	610
鞍　山	627	2 610.1	2 263	607	98	69	131	309
抚　顺	253	1 116.2	347	353	57	41	75	180
本　溪	429	1 614.9	330	407	54	24	59	270
锦　州	227	1 014.2	316	380	63	60	65	192
长　春	346	3 826.8	232	601	229	65	57	250
吉　林	257	1 707.4	163	380	108	57	35	180
哈　尔　滨	376	1 305.2	97	674	232	107	95	240
齐齐哈尔	484	449.3	429	771	83	64	119	505
牡　丹　江	141	151.9	33	255	36	19	49	151
上　海	2 283	16 561.2	1 503	2 254	158	77	185	1 834
南　京	723	6 027.9	505	435	78	14	30	313
无　锡	1 460	7 530.6	782	927	165	16	78	668
徐　州	614	1 781.9	249	377	88	24	50	215
常　州	1 104	4 051.0	488	576	57	13	27	479
苏　州	1 870	11 002.0	583	1 755	214	96	221	1 224
南　通	1 381	2 208.2	560	706	76	25	66	539
连　云　港	248	890.7	30	210	27	15	48	120
扬　州	569	1 602.9	161	262	38	7	15	202
镇　江	599	1 431.7	142	345	33	9	12	291
杭　州	1 658	4 437.3	578	1 074	164	51	145	714

重点城市汇总工业企业概况（一）（续表）

（2011）

城 市 名 称	汇总工业企业数/个	工业总产值（现价）/亿元	工业炉窑数/台	工业锅炉数/台	35蒸吨及以上的	20（含）～35蒸吨的	10（含）～20蒸吨的	10蒸吨以下的
宁 波	1 427	6 317.9	1 226	912	107	32	85	688
温 州	1 942	1 220.2	668	1 139	8	18	83	1 030
湖 州	1 147	1 123.2	371	576	40	20	39	477
绍 兴	1 073	2 833.2	261	1 021	65	47	118	791
合 肥	853	2 010.3	356	282	43	12	30	197
芜 湖	574	1 237.9	311	194	17	13	32	132
马 鞍 山	585	1 309.0	420	155	36	15	20	84
福 州	487	1 592.9	286	420	18	11	86	305
厦 门	386	2 441.4	76	196	17	5	33	141
泉 州	1 316	1 891.0	814	860	39	36	138	647
南 昌	287	1 639.9	1 030	260	11	25	49	175
九 江	654	1 047.2	367	218	25	7	37	149
济 南	637	2 266.5	607	333	118	29	74	112
青 岛	761	4 627.9	207	606	148	73	81	304
淄 博	1 066	3 610.9	1 106	538	197	62	73	206
枣 庄	223	590.5	131	219	64	6	18	131
烟 台	593	3 630.4	240	593	123	42	82	346
潍 坊	881	5 261.8	0	860	0	284	122	454
济 宁	493	2 093.3	417	489	146	23	30	290
泰 安	290	901.1	121	338	113	13	49	163
日 照	144	1 349.8	124	173	55	10	14	94
郑 州	639	1 513.7	606	510	40	25	68	377
开 封	354	251.8	229	130	9	6	25	90
洛 阳	789	1 782.7	497	275	40	20	42	173
平 顶 山	235	831.3	176	150	37	15	13	85
安 阳	450	1 247.1	380	266	32	17	22	195
焦 作	424	838.5	363	459	55	24	30	350
三 门 峡	403	542.9	588	192	39	9	35	109
武 汉	456	4 290.0	272	360	58	13	47	242
宜 昌	379	1 292.6	352	287	39	14	52	182
荆 州	467	437.1	120	297	11	6	27	253
长 沙	438	1 611.7	108	269	6	9	18	236
株 洲	580	834.1	523	225	13	11	21	180
湘 潭	203	1 552.7	145	146	16	0	30	100
岳 阳	310	1 228.8	143	258	33	9	33	183
常 德	212	881.9	511	176	22	12	28	114
张 家 界	78	15.6	52	29	2	0	1	26
广 州	1 307	7 101.7	313	926	83	29	155	659

重点城市汇总工业企业概况（一）（续表）

（2011）

城 市 名 称	汇总工业企业数/个	工业总产值（现价）/亿元	工业炉窑数/台	工业锅炉数/台	35 蒸吨及以上的	20（含）～35蒸吨的	10（含）～20蒸吨的	10 蒸吨以下的
韶 关	349	604.6	214	144	18	4	23	99
深 圳	1 859	6 853.4	64	418	31	18	38	331
珠 海	522	2 642.5	82	349	21	10	30	288
汕 头	502	359.1	20	526	9	5	69	443
湛 江	434	944.3	129	264	48	40	20	156
南 宁	407	395.9	96	342	46	33	30	233
柳 州	342	1 713.2	302	212	49	22	13	128
桂 林	486	389.2	260	209	4	10	34	161
北 海	82	151.6	37	57	14	6	23	14
海 口	100	302.7	30	81	7	6	4	64
重 庆	3 212	5 349.4	5 305	1 438	109	49	119	1 161
成 都	1 234	1 961.2	1 277	935	28	23	100	784
自 贡	339	363.1	158	124	11	3	7	103
攀 枝 花	223	2 734.7	354	107	22	7	16	62
泸 州	603	413.0	218	347	16	4	9	318
德 阳	394	870.4	516	218	10	17	42	149
绵 阳	566	1 127.1	423	255	8	6	38	203
南 充	525	153.5	180	88	4	13	11	60
宜 宾	289	683.4	184	213	29	30	13	141
贵 阳	97	614.4	1 011	79	33	9	14	23
遵 义	522	720.6	710	293	14	22	35	222
昆 明	583	1 777.9	329	315	44	15	26	230
曲 靖	534	1 179.2	319	224	50	13	39	122
玉 溪	276	1 456.4	206	157	16	16	16	109
拉 萨	46	26.5	12	21	0	0	1	20
西 安	513	1 813.1	1 032	577	84	76	111	306
铜 川	104	228.3	45	67	2	1	5	59
宝 鸡	202	1 162.5	111	261	27	48	84	102
咸 阳	734	1 395.2	381	398	30	19	52	297
渭 南	454	890.2	285	334	46	9	62	217
延 安	191	506.8	215	312	13	12	70	217
兰 州	277	1 496.8	963	413	36	64	166	147
金 昌	39	639.6	81	79	30	10	2	37
西 宁	140	584.1	218	182	24	14	26	118
银 川	152	439.2	83	353	89	77	66	121
石 嘴 山	280	563.4	595	371	32	14	42	283
乌 鲁 木 齐	227	730.5	171	483	195	69	66	153
克 拉 玛 依	69	1 304.1	177	556	78	268	58	152

重点城市汇总工业企业概况（二）

（2011）

单位：万吨

城 市 名 称	工业用水 总量	取水量	重 复 用水量	工业煤炭 消耗量	燃料煤
总　　计	24 761 997	3 844 909	20 917 088	203 080	152 790
北　　京	450 344	20 061	430 283	1 651	1 530
天　　津	1 028 306	40 283	988 023	4 537	4 262
石 家 庄	607 661	38253	569 408	3 893	3 460
唐　　山	1 354 072	55 114	1298 958	7 494	3 663
秦 皇 岛	85 778	10 929	74 848	1 166	1 092
邯　　郸	748 029	27 140	720 889	5 027	2 871
保　　定	352 599	24 862	327 737	1 559	1 504
太　　原	349 280	12 224	337 056	4 000	1 854
大　　同	177 923	13 710	164 213	2 875	2 645
阳　　泉	33 663	4 986	28 678	969	957
长　　治	233 416	14 989	218 427	4 388	2 026
临　　汾	187 304	12 473	174 832	4 051	1 844
呼和浩特	101 847	8 776	93 071	3 939	2 739
包　　头	635 000	27 773	607 226	3 261	2 305
赤　　峰	152 422	7 153	145 269	1 501	1 437
沈　　阳	115 916	12 860	103 056	1 714	1 704
大　　连	163 186	154 084	9 102	2 077	1 997
鞍　　山	391 547	61 917	329 630	1 952	1 027
抚　　顺	276 140	12 434	263 705	1 240	1 102
本　　溪	164 126	17 965	146 162	1 376	454
锦　　州	153 866	8 466	145 399	776	765
长　　春	155 037	14 288	140 748	2 274	2 114
吉　　林	323 906	52 563	271 343	2 077	1 845
哈 尔 滨	225 785	18 257	207 528	1 998	1 791
齐齐哈尔	64 791	19 553	45 238	2 495	1 149
牡 丹 江	18 376	7 570	10 806	805	759
上　　海	833 695	85 828	747 867	5 935	4 390
南　　京	816 659	287 875	528 785	3 377	3 048
无　　锡	163 255	42 309	120 946	2 794	2 694
徐　　州	540 493	27 317	513 176	3 697	3 167
常　　州	173 467	22 389	151 078	1 171	898
苏　　州	948 901	105 467	843 434	6 073	5 919
南　　通	146 287	31 712	114 575	1 830	1 780
连 云 港	99 257	11 517	87 739	495	398
扬　　州	196 975	18 766	178 210	1 272	1 266
镇　　江	117 610	16 160	101 450	1 912	1 785
杭　　州	325 903	67 421	258 482	1 552	1 270

重点城市汇总工业企业概况（二）（续表）

（2011）

单位：万吨

城 市 名 称	工业用水 总量	取水量	重 复 用水量	工业煤炭 消耗量	燃料煤
宁 波	874 926	557 422	317 505	4 702	4 491
温 州	144 560	123 072	21 489	1 123	1 123
湖 州	52 745	19 510	33 235	864	643
绍 兴	100 154	41 317	58 837	1 129	1 045
合 肥	189 432	14 513	174 919	1 369	1 271
芜 湖	100 020	64 689	35 331	1 149	753
马 鞍 山	288 227	12 224	276 003	1 817	880
福 州	50 629	16 698	33 931	2 152	2 045
厦 门	190 021	114 515	75 506	513	513
泉 州	175 974	28 762	147 212	1 232	1 171
南 昌	50 726	12 557	38 169	618	452
九 江	133 952	58 266	75 686	761	554
济 南	299 724	14 655	285 070	1 974	1 332
青 岛	143 779	27 748	116 031	1 476	1 343
淄 博	504 661	32 462	472 199	3 382	2 717
枣 庄	148 140	14 290	133 850	1 807	978
烟 台	71 993	15 089	56 904	1 633	1 506
潍 坊	222 782	45 160	177 622	2 403	1 759
济 宁	494 781	28 440	466 341	3 899	3 382
泰 安	97 532	14 134	83 398	1 446	1 057
日 照	213 194	14 292	198 902	1 302	830
郑 州	234 249	20 345	213 904	2 194	2 129
开 封	149 231	12 025	137 206	767	616
洛 阳	193 764	23 381	170 383	2 801	2 636
平 顶 山	311 402	16 731	294 671	7 640	2 655
安 阳	253 949	16 992	236 957	1 958	668
焦 作	136 915	18 211	118 703	857	641
三 门 峡	176 163	12 499	163 664	1 402	1 143
武 汉	654 584	124 232	530 353	2 879	1 182
宜 昌	139 437	30 084	109 352	973	669
荆 州	23 652	17 266	6 387	381	328
长 沙	7 951	5 544	2 406	557	529
株 洲	64 539	8 436	56 103	591	403
湘 潭	151 911	10 704	141 207	1 145	542
岳 阳	113 342	24 555	88 788	1 090	1 015
常 德	79 707	11 300	68 407	932	740
张 家 界	309	244	65	88	73
广 州	488 186	242 236	245 950	1 929	1 841

重点城市汇总工业企业概况（二）（续表）

（2011）

单位：万吨

城 市 名 称	工业用水 总量	取水量	重 复 用水量	工业煤炭 消耗量	燃料煤
韶 关	176 946	10 513	166 433	1 097	697
深 圳	20 758	14 825	5 934	489	489
珠 海	36 935	9 540	27 395	791	764
汕 头	11 536	7 634	3 902	1 104	1 080
湛 江	50 982	8 201	42 782	579	550
南 宁	118 868	20 410	98 458	378	352
柳 州	370 618	39 341	331 277	1 279	436
桂 林	62 549	7 054	55 495	407	303
北 海	15 335	2 275	13 060	242	238
海 口	3 588	1 468	2 120	2	2
重 庆	645 054	172 196	472 858	4 657	3 626
成 都	159 113	20 548	138 565	850	780
自 贡	38 852	2 957	35 895	218	174
攀 枝 花	147 300	18 601	128 699	1 342	683
泸 州	148 676	7 792	140 884	523	447
德 阳	50 559	9 305	41 255	214	152
绵 阳	88 710	12 665	76 045	545	410
南 充	6 313	3 632	2 681	72	42
宜 宾	33 580	12 540	21 041	770	633
贵 阳	131 461	6 493	124 968	802	475
遵 义	112 342	5 942	106 400	764	732
昆 明	202 019	16 848	185 171	1 707	1 131
曲 靖	364 645	20 003	344 642	3 556	2 657
玉 溪	36 776	6 948	29 828	327	195
拉 萨	2 169	1 366	803	20	19
西 安	36 652	19 383	17 269	827	783
铜 川	18 772	573	18 199	479	424
宝 鸡	67 008	12 657	54 352	853	740
咸 阳	25 831	10 519	15 311	1 230	1 181
渭 南	166 500	11 603	154 897	2 881	2 205
延 安	5 755	4 039	1 716	144	134
兰 州	17 245	11 009	6 236	1 281	1 187
金 昌	73 618	7 603	66 014	485	446
西 宁	110 206	8 060	102 146	750	600
银 川	90 312	11 187	79 125	1 490	1 446
石 嘴 山	93 503	8 499	85 004	3 078	1 369
乌鲁木齐	205 681	14 640	191 041	2 358	1 693
克拉玛依	175 162	5 996	169 166	345	345

重点城市汇总工业企业概况（三）

（2011）

城　市 名　称	燃料油 消耗量/万吨	焦炭消耗量/ 万吨	天然气消耗量/ 亿米^3	其他燃料消耗量/ 万吨标煤
总　　计	914	18 797	1 059	10 966
北　京	21	1	29	94
天　津	15	428	18	262
石 家 庄	4	408	1	77
唐　山	1	3 364	15	997
秦 皇 岛	6	211	2	1
邯　郸	1	1 319	2	583
保　定	…	70	169	26
太　原	1	263	2	113
大　同	…	64	0	31
阳　泉	…	4	2	1
长　治	0	299	0	82
临　汾	…	416	1	83
呼和浩特	…	34	14	13
包　头	…	586	2	218
赤　峰	…	70	0	4
沈　阳	6	1	2	4
大　连	38	6	0	31
鞍　山	11	793	1	1 302
抚　顺	15	167	1	1
本　溪	…	771	0	365
锦　州	11	7	3	36
长　春	2	2	2	13
吉　林	21	174	5	74
哈 尔 滨	2	32	1	4
齐齐哈尔	…	34	0	106
牡 丹 江	0	0	0	3
上　海	69	152	37	645
南　京	57	187	18	941
无　锡	7	258	5	9
徐　州	1	652	9	24
常　州	1	70	9	3
苏　州	48	244	36	76
南　通	4	53	13	93
连 云 港	0	123	0	19
扬　州	13	3	2	18
镇　江	5	99	4	558
杭　州	4	121	17	29

重点城市汇总工业企业概况（三）（续表）

（2011）

城 市 名 称	燃料油消耗量/ 万吨	焦炭消耗量/ 万吨	天然气消耗量/ 亿米³	其他燃料消耗量/ 万吨标煤
宁 波	47	75	28	28
温 州	5	2	0	3
湖 州	7	…	2	44
绍 兴	1	1	0	4
合 肥	1	62	1	48
芜 湖	1	62	37	5
马 鞍 山	…	726	1	398
福 州	13	143	1	95
厦 门	12	0	5	5
泉 州	28	56	11	122
南 昌	1	…	0	89
九 江	3	145	0	36
济 南	…	431	2	27
青 岛	6	135	3	27
淄 博	28	62	24	23
枣 庄	…	7	…	7
烟 台	2	18	3	33
潍 坊	0	0	170	76
济 宁	1	…	61	34
泰 安	…	121	1	11
日 照	8	476	…	…
郑 州	4	3	27	1
开 封	…	…	…	1
洛 阳	4	12	2	36
平 顶 山	…	37	2	15
安 阳	0	656	5	4
焦 作	2	6	1	48
三 门 峡	1	6	…	15
武 汉	17	2	4	26
宜 昌	…	18	4	5
荆 州	…	…	…	65
长 沙	1	…	1	9
株 洲	3	34	3	9
湘 潭	…	4	…	23
岳 阳	1	4	2	26
常 德	…	3	…	20
张 家 界	0	…	0	0
广 州	65	88	9	115

重点城市汇总工业企业概况（三）（续表）

（2011）

城 市 名 称	燃料油 消耗量/ 万吨	焦炭消耗量/ 万吨	天然气消耗量/ 亿米^3	其他燃料消耗量/ 万吨标煤
韶 关	1	247	…	4
深 圳	32	0	26	107
珠 海	23	79	3	6
汕 头	1	0	…	1
湛 江	2	…	…	74
南 宁	5	2	…	44
柳 州	3	426	…	347
桂 林	…	29	…	16
北 海	1	21	2	13
海 口	…	0	8	2
重 庆	3	301	30	258
成 都	…	82	19	38
自 贡	0	1	14	8
攀 枝 花	…	371	…	924
泸 州	1	…	5	12
德 阳	…	3	9	1
绵 阳	0	9	2	66
南 充	…	1	2	1
宜 宾	…	7	1	15
贵 阳	9	10	…	17
遵 义	…	46	3	2
昆 明	9	256	…	13
曲 靖	2	245	0	13
玉 溪	…	373	…	5
拉 萨	11	0	0	…
西 安	2	1	1	…
铜 川	…	0	…	0
宝 鸡	155	9	40	5
咸 阳	2	…	3	24
渭 南	4	252	…	84
延 安	3	…	1	…
兰 州	2	141	12	49
金 昌	7	28	…	19
西 宁	…	588	7	70
银 川	1	2	2	6
石 嘴 山	…	60	3	22
乌鲁木齐	2	322	8	228
克拉玛依	2	0	28	…

重点城市工业污染防治投资情况（一）

（2011）

单位：个

城 市 名 称	汇总工业 企业数	本年施工 项目总数	工业废水 治理项目	工业废气 治理项目	脱硫治 理项目	脱硝治 理项目	工业固体 废物治理 项目	噪声治理 项目	其他治理 项目
总　　　计	3 608	5 002	1 886	2 065	590	63	290	108	653
北　　京	48	66	25	30	14	0	5	2	4
天　　津	92	117	40	51	24	1	4	2	20
石 家 庄	41	51	19	28	11	1	2	0	2
唐　　山	60	123	21	76	20	1	4	1	21
秦 皇 岛	7	13	2	7	3	0	1	0	3
邯　　郸	27	37	4	30	10	1	1	0	2
保　　定	19	29	11	16	7	2	0	0	2
太　　原	37	57	11	23	13	2	16	0	7
大　　同	45	88	13	43	23	0	19	0	13
阳　　泉	16	23	3	10	3	0	3	1	6
长　　治	40	59	9	5	2	1	0	4	41
临　　汾	32	44	4	26	16	3	1	0	13
呼和浩特	38	73	20	43	18	0	1	1	8
包　　头	14	24	3	13	2	0	5	1	2
赤　　峰	15	26	10	11	1	0	2	0	3
沈　　阳	4	9	2	1	1	0	0	0	6
大　　连	30	32	25	3	1	0	0	2	2
鞍　　山	4	8	0	6	3	0	1	0	1
抚　　顺	8	12	5	5	0	0	0	2	0
本　　溪	6	8	1	4	1	0	0	1	2
锦　　州	10	14	4	6	3	0	2	1	1
长　　春	24	25	10	15	11	0	0	0	0
吉　　林	21	30	7	17	5	1	1	0	5
哈 尔 滨	17	19	5	9	5	0	3	0	2
齐齐哈尔	4	5	1	4	2	0	0	0	0
牡 丹 江	5	5	2	2	2	0	0	0	1
上　　海	82	120	33	58	5	2	3	6	20
南　　京	73	102	43	44	9	0	2	4	9
无　　锡	58	82	22	38	6	3	8	3	11
徐　　州	9	13	5	5	1	0	0	0	3
常　　州	66	82	34	36	11	1	2	2	8
苏　　州	138	193	94	64	19	5	14	2	19
南　　通	100	111	72	21	7	1	5	1	12
连 云 港	25	36	17	11	1	0	5	0	3
扬　　州	17	17	8	8	0	1	1	0	0
镇　　江	21	39	23	15	3	0	1	0	0
杭　　州	137	158	58	84	20	3	1	4	11

重点城市工业污染防治投资情况（一）（续表）

（2011）

单位：个

城 市 名 称	汇总工业 企业数	本年施工项 目总数	工业废水 治理项目	工业废气 治理项目	脱硫治 理项目	脱硝治 理项目	工业固 体废物 治理项目	噪声治 理项目	其他治 理项目
宁　波	43	49	17	16	6	1	0	0	16
温　州	35	40	33	7	0	0	0	0	0
湖　州	21	22	11	10	0	0	0	0	1
绍　兴	160	199	99	63	10	0	7	3	27
合　肥	4	4	2	1	1	0	0	1	0
芜　湖	14	16	4	7	2	2	0	0	5
马 鞍 山	8	22	10	11	2	0	0	0	1
福　州	17	17	8	2	0	1	1	0	6
厦　门	40	70	34	20	2	0	4	2	10
泉　州	100	128	51	49	14	2	12	6	10
南　昌	13	24	12	9	4	0	1	0	2
九　江	14	17	14	3	0	0	0	0	0
济　南	35	42	19	17	9	1	0	2	4
青　岛	18	24	6	8	6	0	0	1	9
淄　博	226	271	34	217	26	2	5	1	14
枣　庄	24	27	14	11	8	0	1	0	1
烟　台	49	59	31	14	13	0	6	0	8
潍　坊	99	141	56	69	25	2	4	0	12
济　宁	31	48	17	18	7	2	3	0	10
泰　安	17	21	9	7	4	0	0	0	5
日　照	11	23	4	14	5	1	0	0	5
郑　州	55	69	18	31	19	2	1	3	16
开　封	0	0	0	0	0	0	0	0	0
洛　阳	32	66	23	27	6	0	7	0	9
平 顶 山	15	20	4	5	2	0	1	1	9
安　阳	16	25	1	19	2	1	2	0	3
焦　作	26	40	18	9	5	0	2	2	9
三 门 峡	12	17	6	3	1	1	3	0	5
武　汉	10	21	5	10	3	0	1	0	5
宜　昌	29	48	10	26	4	1	3	1	8
荆　州	13	16	8	7	3	1	0	1	0
长　沙	8	12	6	3	0	0	0	1	2
株　洲	20	31	13	9	5	0	4	4	1
湘　潭	7	10	4	5	0	1	1	0	0
岳　阳	5	5	4	1	0	1	0	0	0
常　德	19	24	14	4	2	0	3	0	3
张 家 界	4	12	5	1	1	0	4	0	2
广　州	50	65	20	33	7	3	3	2	7

重点城市工业污染防治投资情况（一）（续表）

（2011）

单位：个

城市名称	汇总工业企业数	本年施工项目总数	工业废水治理项目	工业废气治理项目	脱硫治理项目	脱硝治理项目	工业固体废物治理项目	噪声治理项目	其他治理项目
韶 关	27	52	16	26	6	0	5	0	5
深 圳	23	28	15	11	2	1	2	0	0
珠 海	22	41	20	20	2	0	1	0	0
汕 头	17	17	16	1	1	0	0	0	0
湛 江	30	42	25	12	2	2	1	2	2
南 宁	17	24	19	3	0	1	1	0	1
柳 州	41	65	27	16	3	0	9	2	11
桂 林	4	5	3	2	1	0	0	0	0
北 海	4	4	1	3	2	0	0	0	0
海 口	12	19	9	10	0	0	0	0	0
重 庆	176	227	115	51	18	0	25	10	26
成 都	25	39	24	7	2	0	1	5	2
自 贡	5	5	2	0	0	0	0	1	2
攀 枝 花	25	33	6	16	7	0	6	0	5
泸 州	24	30	22	6	3	0	1	0	1
德 阳	21	37	13	15	1	1	2	0	7
绵 阳	33	43	17	12	0	0	4	4	6
南 充	6	9	7	0	0	0	0	0	2
宜 宾	14	18	11	5	2	0	0	1	1
贵 阳	4	4	4	0	0	0	0	0	0
遵 义	18	23	15	5	3	0	1	0	2
昆 明	63	93	36	30	8	0	9	1	17
曲 靖	48	71	32	24	5	2	8	2	5
玉 溪	24	41	11	18	0	0	3	2	7
拉 萨	7	8	4	2	0	0	1	0	1
西 安	21	25	12	12	10	0	0	1	0
铜 川	23	40	16	14	1	0	4	1	5
宝 鸡	28	38	15	12	1	2	1	0	10
咸 阳	43	46	22	8	1	0	1	0	15
渭 南	20	38	9	17	3	0	4	0	8
延 安	10	16	8	1	1	0	1	4	2
兰 州	8	12	0	10	0	0	0	0	2
金 昌	11	16	7	8	4	0	1	0	0
西 宁	17	34	5	11	1	0	4	0	14
银 川	12	15	3	9	3	0	0	1	2
石 嘴 山	25	38	4	28	5	0	2	0	4
乌鲁木齐	10	11	5	6	5	0	0	0	0
克拉玛依	0	0	0	0	0	0	0	0	0

重点城市工业污染防治投资情况（二）

（2011）

单位：个

城 市 名 称	本年竣工项目总数	工业废水治理项目	工业废气治理项目	脱硫治理项目	脱硝治理项目	工业固体废物治理项目	噪声治理项目	其他治理项目
总　　计	3 824	1 329	1 670	468	38	225	90	510
北　京	59	21	28	12	0	5	2	3
天　津	99	33	42	18	0	3	1	20
石 家 庄	41	14	23	9	1	2	0	2
唐　山	107	14	71	19	1	4	0	18
秦 皇 岛	8	2	3	3	0	0	0	3
邯　郸	34	4	27	9	1	1	0	2
保　定	18	8	8	4	2	0	0	2
太　原	34	3	20	12	2	8	0	3
大　同	82	11	41	21	0	17	0	13
阳　泉	20	2	10	3	0	2	1	5
长　治	47	8	4	1	1	0	4	31
临　汾	24	2	15	11	0	0	0	7
呼和浩特	38	5	31	16	0	0	0	2
包　头	18	1	11	1	0	4	1	1
赤　峰	24	9	10	0	0	2	0	3
沈　阳	8	1	1	1	0	0	0	6
大　连	15	9	2	0	0	0	2	2
鞍　山	5	0	4	1	0	0	0	1
抚　顺	5	2	3	0	0	0	0	0
本　溪	8	1	4	1	0	0	1	2
锦　州	10	2	5	2	0	1	1	1
长　春	19	9	10	9	0	0	0	0
吉　林	23	6	13	5	1	1	0	3
哈 尔 滨	17	4	9	5	0	2	0	2
齐齐哈尔	4	1	3	1	0	0	0	0
牡 丹 江	4	1	2	2	0	0	0	1
上　海	93	22	47	3	2	1	5	18
南　京	73	28	33	4	0	2	3	7
无　锡	77	21	37	6	3	6	3	10
徐　州	5	2	3	1	0	0	0	0
常　州	73	31	33	9	1	2	1	6
苏　州	158	74	54	18	2	14	2	14
南　通	99	65	18	7	0	4	1	11
连 云 港	27	12	8	1	0	4	0	3
扬　州	13	6	7	0	0	0	0	0
镇　江	38	23	14	3	0	1	0	0
杭　州	137	47	76	18	0	0	4	10

重点城市工业污染防治投资情况（二）（续表）

（2011）

单位：个

城 市 名 称	本年竣工 项目总数	工业废水 治理项目	工业废气 治理项目	脱硫治 理项目	脱硝治 理项目	工业固 体废物 治理项目	噪声治 理项目	其他治 理项目
宁 波	32	7	14	6	0	0	0	11
温 州	33	26	7	0	0	0	0	0
湖 州	12	6	5	0	0	0	0	1
绍 兴	140	58	52	10	0	7	2	21
合 肥	2	0	1	1	0	0	1	0
芜 湖	14	3	6	2	1	0	0	5
马 鞍 山	22	10	11	2	0	0	0	1
福 州	14	8	1	0	0	1	0	4
厦 门	49	20	16	2	0	4	1	8
泉 州	58	15	25	10	0	6	6	6
南 昌	7	4	2	0	0	1	0	0
九 江	11	8	3	0	0	0	0	0
济 南	33	11	17	9	1	0	1	4
青 岛	23	6	7	6	0	0	1	9
淄 博	255	32	205	23	2	5	1	12
枣 庄	19	9	8	6	0	1	0	1
烟 台	47	25	10	9	0	5	0	7
潍 坊	116	43	60	24	2	2	0	11
济 宁	35	11	15	5	2	2	0	7
泰 安	19	9	6	3	0	0	0	4
日 照	14	4	6	3	1	0	0	4
郑 州	39	12	16	10	0	1	2	8
开 封	0	0	0	0	0	0	0	0
洛 阳	61	21	25	4	0	7	0	8
平 顶 山	14	4	2	1	0	1	1	6
安 阳	16	1	10	0	0	2	0	3
焦 作	30	9	9	5	0	2	2	8
三 门 峡	14	5	3	1	1	3	0	3
武 汉	12	4	6	3	0	1	0	1
宜 昌	46	9	26	4	1	2	1	8
荆 州	13	7	5	1	1	0	1	0
长 沙	6	2	3	0	0	0	0	1
株 洲	29	12	8	4	0	4	4	1
湘 潭	9	4	4	0	1	1	0	0
岳 阳	4	4	0	0	0	0	0	0
常 德	19	12	3	2	0	1	0	3
张 家 界	11	5	1	1	0	4	0	1
广 州	43	14	25	6	1	2	1	1

重点城市工业污染防治投资情况（二）（续表）

（2011）

单位：个

城　市名　称	本年竣工项目总数	工业废水治理项目	工业废气治理项目	脱硫治理项目	脱硝治理项目	工业固体废物治理项目	噪声治理项目	其他治理项目
韶　关	46	13	23	5	0	5	0	5
深　圳	18	8	9	2	1	1	0	0
珠　海	22	14	7	0	0	1	0	0
汕　头	6	5	1	1	0	0	0	0
湛　江	33	18	10	0	2	1	2	2
南　宁	16	14	1	0	0	1	0	0
柳　州	45	22	10	0	0	4	2	7
桂　林	4	3	1	0	0	0	0	0
北　海	1	1	0	0	0	0	0	0
海　口	17	8	9	0	0	0	0	0
重　庆	169	73	42	14	0	21	9	24
成　都	29	17	6	2	0	1	4	1
自　贡	5	2	0	0	0	0	1	2
攀枝花	25	5	12	6	0	3	0	5
泸　州	24	18	4	3	0	1	0	1
德　阳	33	11	13	1	1	2	0	7
绵　阳	35	12	10	0	0	3	4	6
南　充	7	5	0	0	0	0	0	2
宜　宾	16	10	4	2	0	0	1	1
贵　阳	3	3	0	0	0	0	0	0
遵　义	5	2	3	2	0	0	0	0
昆　明	64	26	20	5	0	7	1	10
曲　靖	51	24	14	2	2	6	2	5
玉　溪	33	9	14	0	0	2	2	6
拉　萨	4	1	2	0	0	1	0	0
西　安	18	10	8	6	0	0	0	0
铜　川	20	7	6	1	0	4	1	2
宝　鸡	27	10	11	1	1	0	0	6
咸　阳	34	14	6	1	0	1	0	13
渭　南	22	3	12	1	0	1	0	6
延　安	7	1	0	0	0	1	3	2
兰　州	10	0	8	0	0	0	0	2
金　昌	15	6	8	4	0	1	0	0
西　宁	29	5	9	1	0	4	0	11
银　川	11	2	6	1	0	0	1	2
石嘴山	30	3	23	4	0	2	0	2
乌鲁木齐	5	0	5	4	0	0	0	0
克拉玛依	0	0	0	0	0	0	0	0

重点城市工业污染治理项目建设情况（三）

（2011）

城 市 名 称	本年竣工项目新增设计处理能力		
	治理废水/ (万吨/日)	治理废气（标态）/ (万米³/时)	治理固废/ (万吨/日)
总　　计	450	38 238	224
北　京	2	63	5
天　津	10	514	…
石 家 庄	1	547	0
唐　山	9	1 485	…
秦 皇 岛	…	1	…
邯　郸	…	428	0
保　定	2	1	0
太　原	2	26	…
大　同	1	8	…
阳　泉	1	3	0
长　治	5	259	0
临　汾	7	16	0
呼和浩特	…	64	0
包　头	…	319	3
赤　峰	1	8	4
沈　阳	…	17	0
大　连	2	66	0
鞍　山	0	501	2
抚　顺	4	58	0
本　溪	…	312	0
锦　州	…	69	10
长　春	2	186	0
吉　林	…	794	…
哈 尔 滨	…	72	4
齐齐哈尔	0	3	0
牡 丹 江	1	31	0
上　海	1	312	1
南　京	9	169	7
无　锡	1	1 112	2
徐　州	1	147	0
常　州	19	1 886	…
苏　州	11	2 756	1
南　通	5	121	…
连 云 港	1	2	…
扬　州	1	136	…
镇　江	1	2	0
杭　州	3	109	0

重点城市工业污染治理项目建设情况（三）（续表）

（2011）

城 市名 称	本年竣工项目新增设计处理能力		
	治理废水/（万吨/日）	治理废气（标态）/（万米³/时）	治理固废/（万吨/日）
宁　波	2	17	0
温　州	…	14	0
湖　州	1	1	0
绍　兴	15	149	…
合　肥	…	…	0
芜　湖	…	30	0
马 鞍 山	…	109	0
福　州	1	254	0
厦　门	2	70	0
泉　州	9	95	…
南　昌	1	27	…
九　江	33	4	0
济　南	2	119	0
青　岛	0	72	0
淄　博	2	925	…
枣　庄	5	888	…
烟　台	24	240	3
潍　坊	16	412	4
济　宁	34	122	…
泰　安	2	17	0
日　照	1	8	0
郑　州	1	59	0
开　封	0	0	0
洛　阳	2	23	1
平 顶 山	2	43	0
安　阳	…	14	30
焦　作	8	121	…
三 门 峡	1	121	0
武　汉	21	160	…
宜　昌	3	75	…
荆　州	…	85	0
长　沙	…	1	0
株　洲	2	12	0
湘　潭	…	6	1
岳　阳	7	200	0
常　德	7	2	0
张 家 界	…	1	…
广　州	…	138	100

重点城市工业污染治理项目建设情况（三）（续表）

（2011）

城 市 名 称	本年竣工项目新增设计处理能力		
	治理废水/ (万吨/日)	治理废气（标态）/ (万米³/时)	治理固废/ (万吨/日)
韶 关	1	21	2
深 圳	…	23	0
珠 海	…	44	0
汕 头	2	0	0
湛 江	5	595	…
南 宁	2	87	0
柳 州	6	287	…
桂 林	…	4	0
北 海	0	37	0
海 口	3	9	0
重 庆	5	432	…
成 都	6	200	0
自 贡	…	0	0
攀 枝 花	16	254	20
泸 州	6	5	…
德 阳	1	38	1
绵 阳	1	236	…
南 充	1	0	0
宜 宾	2	42	0
贵 阳	…	0	0
遵 义	…	302	…
昆 明	11	69	…
曲 靖	13	16 693	10
玉 溪	1	10	0
拉 萨	30	…	7
西 安	2	167	0
铜 川	…	…	0
宝 鸡	7	18	4
咸 阳	6	402	…
渭 南	5	774	0
延 安	1	0	0
兰 州	0	49	0
金 昌	3	51	0
西 宁	2	23	…
银 川	2	114	0
石 嘴 山	…	14	0
乌鲁木齐	…	5	0
克拉玛依	0	0	0

重点城市农业污染排放情况

（2011）

单位：吨

城 市名 称	农业污染物排放（流失）总量			
	化学需氧量	总氮	总磷	氨氮
总　　　计	5 145 227	1 778 631	222 365	334 622
北　京	81 692	32 840	4 471	4 813
天　津	114 674	36 713	4 689	5 961
石 家 庄	180 746	56 197	6 281	7 084
唐　山	136 619	45 474	6 633	7 271
秦 皇 岛	37 637	15 289	2 152	2 345
邯　郸	81 152	30 855	3 506	4 671
保　定	105 099	48 270	5 389	5 566
太　原	9 234	4 072	452	611
大　同	15 535	7 420	785	798
阳　泉	3 319	1 223	208	420
长　治	13 400	6 267	739	1 173
临　汾	21 632	11 384	1 150	1 608
呼和浩特	114 408	30 035	3 074	1 319
包　头	63 979	3 190	355	1 027
赤　峰	91 602	24 773	2 648	2 138
沈　阳	215 699	47 845	6 088	6 987
大　连	103 124	17 983	3 204	3 957
鞍　山	51 738	11 367	1 855	2 056
抚　顺	23 007	3 861	652	836
本　溪	17 073	5 115	775	825
锦　州	82 359	21 351	2 754	4 513
长　春	131 223	29 752	4 166	4 931
吉　林	97 966	21 955	2 465	2 931
哈 尔 滨	220 891	55 977	6 159	6 775
齐齐哈尔	186 399	43 747	4 081	4 700
牡 丹 江	24 606	5 672	647	1 017
上　海	33 932	15 292	1 971	3 500
南　京	18 053	8 434	946	1 967
无　锡	8 568	5 372	593	1 209
徐　州	52 288	25 545	2 614	3 420
常　州	18 328	6 763	850	1 529
苏　州	19 509	8 701	1 109	2 217
南　通	51 640	24 127	2 380	6 985
连 云 港	50 953	14 964	1 694	2 774
扬　州	11 032	7 982	679	1 836
镇　江	7 375	4 630	501	1 085
杭　州	30 712	12 662	1 744	3 822

重点城市农业污染排放情况（续表）

（2011）   单位：吨

城 市 名 称	农业污染物排放（流失）总量			
	化学需氧量	总氮	总磷	氨氮
宁　波	17 783	9 960	1 235	2 890
温　州	15 347	7 326	928	1 959
湖　州	22 413	7 659	931	2 092
绍　兴	16 094	7 943	944	2 393
合　肥	69 191	15 357	2 286	3 528
芜　湖	10 028	2 787	4 225	1 460
马 鞍 山	5 416	16 947	1 298	824
福　州	37 164	14 484	2 140	6 380
厦　门	14 668	4 745	637	1 951
泉　州	15 565	7 362	905	2 214
南　昌	35 552	13 383	1 780	3 387
九　江	16 752	4 634	697	2 370
济　南	73 538	27 691	3 216	3 007
青　岛	121 625	36 965	5 228	5 804
淄　博	36 302	15 825	1 937	1 858
枣　庄	29 046	14 302	1 765	1 864
烟　台	114 898	31 521	4 561	5 862
潍　坊	138 801	40 995	5 811	9 322
济　宁	103 288	40 489	4 689	7 675
泰　安	83 336	30 267	3 870	4 000
日　照	27 170	7 969	1 211	1 680
郑　州	53 158	19 429	2 461	2 605
开　封	54 740	22 494	2 369	3 448
洛　阳	41 988	16 025	1 824	2 508
平 顶 山	32 635	17 105	2 195	2 879
安　阳	42 378	18 099	2 260	3 944
焦　作	40 170	14 997	1 835	2 138
三 门 峡	12 553	4 981	529	816
武　汉	40 462	13 808	1 780	3 255
宜　昌	34 134	18 170	2 333	4 850
荆　州	96 425	26 692	3 393	5 836
长　沙	48 116	15 810	2 075	4 689
株　洲	28 785	11 423	1 350	3 639
湘　潭	40 369	12 720	1 348	4 198
岳　阳	68 419	26 039	3 297	7 710
常　德	60 920	26 031	3 184	5 916
张 家 界	5 142	3 202	352	600
广　州	51 160	14 647	2 439	4 164

重点城市农业污染排放情况（续表）

（2011）

单位：吨

城 市 名 称	农业污染物排放（流失）总量			
	化学需氧量	总氮	总磷	氨氮
韶 关	28 743	6 946	936	2 979
深 圳	6 057	969	160	306
珠 海	12 400	3 331	489	1 135
汕 头	13 791	4 289	626	1 468
湛 江	46 839	22 236	3 272	6 026
南 宁	37 954	18 430	2 407	4 495
柳 州	17 437	8 908	995	2 099
桂 林	33 026	13 401	1 831	3 317
北 海	10 489	5 489	768	873
海 口	11 797	5 068	662	1 126
重 庆	127 357	55 088	6 917	13 612
成 都	76 785	31 804	3 943	8 347
自 贡	18 751	5 790	674	1 700
攀 枝 花	2 997	1 549	196	327
泸 州	24 368	9 057	996	2 729
德 阳	26 076	10 268	1 266	2 303
绵 阳	32 342	16 898	2 048	4 047
南 充	53 102	24 073	3 043	5 500
宜 宾	26 270	11 325	1 187	3 163
贵 阳	7 427	2 833	411	568
遵 义	8 739	9 453	908	1 910
昆 明	13 518	7 438	1 307	1 720
曲 靖	15 407	12 927	1 292	2 314
玉 溪	5 158	2 808	439	322
拉 萨	1 733	924	476	123
西 安	26 063	14 642	1 473	1 298
铜 川	8 262	859	98	324
宝 鸡	21 458	9 903	790	1 104
咸 阳	34 753	19 864	2 030	1 451
渭 南	32 186	18 214	1 670	2 355
延 安	7 842	4 853	475	814
兰 州	7 607	3 806	407	507
金 昌	7 665	2 142	155	185
西 宁	9 513	2 957	339	364
银 川	31 382	8 491	935	648
石 嘴 山	7 958	2 788	318	213
乌 鲁 木 齐	7 734	4 548	531	341
克 拉 玛 依	1 504	882	118	120

重点城市城镇生活污染排放及处理情况（一）

（2011）

年　份 地　区	城镇人口/ 万人	生活煤炭消费量/ 万吨	生活天然气消费量/万米³	生活用水总量/ 万吨	居民家庭 用水总量	公共服务 用水总量
总　　计	39 148	8 052	2 179 138	2 934 743	2 397 309	537 434
北　京	1 741	463	507 831	151 935	85 447	66 488
天　津	1 091	204	61 042	52 580	35 980	16 600
石 家 庄	540	83	9 749	34 835	26 115	8 720
唐　山	397	74	5 000	23 171	20 679	2 492
秦 皇 岛	145	29	2 142	8 977	8 070	907
邯　郸	420	151	6 300	24 684	22 538	2 146
保　定	455	111	10 205	27 615	20 711	6 904
太　原	353	370	33 700	25 641	18 956	6 685
大　同	167	69	11 201	9 709	7 322	2 387
阳　泉	85	46	22 500	5 369	4 353	1 016
长　治	146	226	373	6 971	5 598	1 373
临　汾	184	175	42 200	15 306	7 025	8 282
呼 和 浩 特	186	43	4 350	13 063	9 024	4 039
包　头	217	176	28 400	10 841	9 107	1 734
赤　峰	183	106	1	9 122	9 022	100
沈　阳	643	202	11 563	43 903	36 000	7 903
大　连	517	120	0	35 321	32 299	3 022
鞍　山	250	69	471	19 740	14 622	5 118
抚　顺	156	40	0	9 709	6 472	3 237
本　溪	131	57	14	5 632	3 374	2 258
锦　州	156	51	3 740	9 368	7 495	1 874
长　春	385	80	7 202	24 030	7 056	16 974
吉　林	284	70	1 201	16 706	15 247	1 459
哈 尔 滨	681	681	11 229	42 518	39 535	2 984
齐 齐 哈 尔	240	280	18 090	12 565	10 000	2 565
牡 丹 江	167	101	443	7 255	5 182	2 073
上　海	2096	165	205 800	188 100	173 052	15 048
南　京	647	7	67 200	59 121	36 526	22 595
无　锡	465	4	3	28 818	21 071	7 747
徐　州	475	16	3 303	35 905	33 477	2 428
常　州	303	5	27 345	25 201	22 966	2 235
苏　州	752	6	35 071	40 650	30 470	10 180
南　通	420	13	1 368	33 700	31 555	2 145
连 云 港	233	39	8 608	18 897	17 450	1 447
扬　州	260	23	8 738	18 490	13 248	5 242
镇　江	199	16	3 023	16 097	14 936	1 160
杭　州	644	4	17 642	58 527	53 594	4 933

重点城市城镇生活污染排放及处理情况（一）（续表）

（2011）

年 份 地 区	城镇人口/ 万人	生活煤炭 消费量/万吨	生活天然气 消费量/万米^3	生活用 水总量/万吨	居民家庭 用水总量	公共服务 用水总量
宁 波	530	37	6 105	29 736	22 439	7 297
温 州	609	5	546	50 007	45 784	4 223
湖 州	155	12	1 458	14 663	12 876	1 787
绍 兴	292	9	12 127	26 528	14 958	11 570
合 肥	425	23	10 065	35 990	32 923	3 067
芜 湖	210	10	3 604	13 354	12 366	989
马 鞍 山	121	12	4 706	8 505	7 484	1 021
福 州	441	7	4 059	35 751	33 175	2 577
厦 门	319	3	2 110	19 142	11 369	7 773
泉 州	487	22	417	39 449	36 606	2 843
南 昌	323	6	21	38 906	35 015	3 891
九 江	221	25	12	15 236	11 321	3 915
济 南	448	140	22 000	30 577	27 797	2 780
青 岛	586	122	19 025	39 472	35 198	4 274
淄 博	292	52	8 510	20 794	17 188	3 605
枣 庄	180	55	4 000	10 521	8 221	2 300
烟 台	391	102	7 008	24 982	22 697	2 284
潍 坊	439	152	3 613	25 232	20 186	5 046
济 宁	357	135	3 178	20 975	19 151	1 824
泰 安	284	106	7 824	6 775	4 561	2 214
日 照	135	58	…	8 081	7 180	901
郑 州	529	75	27 222	42 495	28 729	13 766
开 封	155	114	5 277	13 002	11 981	1 021
洛 阳	303	138	1 200	23 339	21 812	1 528
平 顶 山	180	4	2 100	17 237	10 342	6 895
安 阳	185	48	7 191	13 186	12 395	791
焦 作	172	58	482	13 443	12 563	879
三 门 峡	103	308	118	6 425	5 735	690
武 汉	751	150	9 600	62 525	58 137	4 388
宜 昌	197	43	3 223	12 298	10 060	2 238
荆 州	275	39	3 094	18 871	17 666	1 205
长 沙	509	20	16 900	42 414	39 464	2 950
株 洲	211	18	20 448	18 484	15 403	3 081
湘 潭	144	11	3 630	12 613	8 829	3 784
岳 阳	264	41	3 285	20 113	18 092	2 021
常 德	228	48	5 508	16 781	16 781	0
张 家 界	61	8	304	4 462	3 659	804
广 州	1073	17	20 000	129 935	93 502	36 433

重点城市城镇生活污染排放及处理情况（一）（续表）

（2011）

年　份 地　区	城镇人口/ 万人	生活煤炭 消费量/万吨	生活天然气 消费量/万米³	生活用 水总量/万吨	居民家庭 用水总量	公共服务 用水总量
韶　关	150	7	0	10 046	10 046	0
深　圳	1 047	4	20 000	110 033	109 933	100
珠　海	138	3	0	14 993	9 592	5 401
汕　头	376	6	400	21 752	19 261	2 491
湛　江	263	11	5 000	25 437	22 833	2 604
南　宁	367	51	5 326	27 069	19 109	7 960
柳　州	215	21	714	11 625	9 139	2 486
桂　林	194	25	1 200	15 318	13 653	1 665
北　海	78	6	1 628	6 062	5 519	543
海　口	157	0	1 778	14 113	10 260	3 854
重　庆	1 606	150	182 608	114 536	101 874	12 662
成　都	938	29	153 300	83 714	69 396	15 318
自　贡	114	15	9 057	7 170	6 378	792
攀枝花	75	5	0	4 057	2 737	1 320
泸　州	169	42	12 341	10 500	8 972	1 528
德　阳	154	8	9 781	11 732	10 660	1 071
绵　阳	196	21	58 903	12 918	11 626	1 292
南　充	235	8	17 865	13 785	10 552	3 233
宜　宾	175	23	10 133	10 451	8 648	1 803
贵　阳	255	98	9 120	29 200	16 812	12 388
遵　义	264	180	0	13 693	13 164	529
昆　明	428	30	0	45 958	37 117	8 840
曲　靖	222	60	271	14 087	12 386	1 700
玉　溪	93	3	0	5 428	4 764	664
拉　萨	37	2	4	2 065	1 852	213
西　安	597	132	68 725	42 183	29 907	12 276
铜　川	50	11	2 507	1 219	853	366
宝　鸡	153	9	3 380	8 704	7 920	784
咸　阳	218	124	2 796	11 062	9 292	1 770
渭　南	160	21	5 856	7 968	7 304	664
延　安	93	49	6 612	5 778	4 993	785
兰　州	284	87	30 160	15 966	13 506	2 460
金　昌	30	24	2 000	2 119	1 929	190
西　宁	146	50	26 028	8 248	5 321	2 927
银　川	146	32	52 285	12 175	10 105	2 071
石嘴山	51	27	3 600	4 174	1 483	2 691
乌鲁木齐	249	68	48 200	22 777	16 952	5 825
克拉玛依	28	2	542	4 258	3 244	1 015

重点城市城镇生活污染排放及处理情况（二）

（2011）

城市名称	城镇生活污水排放量/万吨	城镇生活化学需氧量产生量/吨	城镇生活氨氮产生量/吨	城镇生活化学需氧量排放量/吨	城镇生活氨氮排放量/吨
总　计	2 534 878	9 823 087.1	1 252 174.9	4 294 436.1	745 652.0
北　京	136 741	501 931.0	61 629.0	99 097.0	15 513.0
天　津	47 322	274 010.3	37 356.4	96 422.4	17 128.0
石 家 庄	28 631	118 334.0	16 172.3	11 707.0	3 139.8
唐　山	19 696	89 823.3	11 940.6	14 064.7	5 581.2
秦 皇 岛	7 195	37 415.8	4 829.8	6 540.3	1 867.8
邯　郸	22 683	90 511.9	12 370.2	46 006.3	8 843.0
保　定	23 473	101 221.6	12 422.6	40 640.8	6 811.2
太　原	16 751	82 464.6	10 952.3	13 822.9	4 705.7
大　同	7 301	39 763.7	5 294.4	17 468.0	3 103.2
阳　泉	4 326	19 986.8	2 537.6	7 670.6	1 382.3
长　治	5 577	34 023.4	4 518.7	15 301.7	2 985.6
临　汾	9 076	45 714.5	5 485.8	27 483.0	4 403.5
呼和浩特	11 103	42 674.8	4 736.2	17 266.7	2 845.4
包　头	5 639	50 500.0	5 187.3	29 494.8	4 587.2
赤　峰	9 022	42 772.2	5 346.5	16 273.0	5 127.0
沈　阳	33 806	150 249.2	20 424.5	47 107.2	15 487.8
大　连	30 033	120 886.1	16 433.0	56 674.3	8 929.6
鞍　山	12 323	58 418.7	7 941.3	46 160.6	7 529.3
抚　顺	7 095	36 324.8	4 937.9	11 951.3	4 197.7
本　溪	4 279	30 563.0	4 160.0	16 658.4	2 128.3
锦　州	7 963	36 401.9	4 948.4	20 015.1	4 148.3
长　春	20 425	88 503.5	11 512.0	31 448.6	7 231.2
吉　林	14 210	65 344.9	8 505.2	33 520.9	5 842.6
哈 尔 滨	36 054	149 595.7	19 916.4	101 901.5	15 830.5
齐齐哈尔	11 360	52 597.4	6 981.6	40 989.7	6 168.8
牡 丹 江	3 628	36 661.6	4 840.9	19 375.4	3 854.9
上　海	169 290	581 430.4	74 208.9	181 866.3	43 962.8
南　京	57 345	169 484.4	21 796.5	73 659.5	14 685.1
无　锡	24 495	117 780.8	15 149.4	21 834.5	2 484.9
徐　州	30 528	119 684.3	15 611.1	77 084.2	10 230.4
常　州	22 131	76 817.8	9 850.5	17 797.0	4 105.0
苏　州	60 988	189 327.2	24 694.7	30 530.3	11 230.9
南　通	28 798	107 181.6	13 786.5	39 130.5	9 365.4
连 云 港	16 003	58 733.9	7 660.9	45 553.0	6 597.0
扬　州	17 841	65 481.0	8 541.0	34 638.7	4 982.5
镇　江	13 631	50 728.0	6 524.8	31 508.9	4 155.0
杭　州	48 247	157 572.3	20 746.7	39 293.2	8 739.5

重点城市城镇生活污染排放及处理情况（二）（续表）

（2011）

城市名称	城镇生活污水排放量/万吨	城镇生活化学需氧量产生量/吨	城镇生活氨氮产生量/吨	城镇生活化学需氧量排放量/吨	城镇生活氨氮排放量/吨
宁　波	35 782	119 918.6	18 181.2	32 924.0	10 427.5
温　州	43 117	159 381.2	20 891.7	109 879.0	16 959.2
湖　州	10 995	40 126.5	4 227.4	19 140.0	3 967.6
绍　兴	20 663	76 684.0	9 982.0	40 853.0	5 536.0
合　肥	34 167	100 947.5	11 958.4	52 044.3	7 643.3
芜　湖	12 312	49 770.3	5 895.9	35 078.8	5 284.9
马 鞍 山	7 969	28 695.4	3 399.3	17 575.6	2 809.7
福　州	30 437	107 897.9	14 010.6	64 266.0	9 094.0
厦　门	17 232	78 011.5	10 129.9	28 985.0	6 354.0
泉　州	33 585	119 059.2	15 459.9	87 931.4	13 018.5
南　昌	31 125	83 489.0	9 224.0	44 436.0	6 357.0
九　江	12 906	194 546.1	21 577.9	41 690.0	5 308.0
济　南	23 382	106 283.3	14 225.6	41 354.0	6 764.0
青　岛	29 918	138 990.6	18 603.4	28 809.0	6 641.0
淄　博	15 447	61 203.4	9 397.9	18 310.6	3 332.9
枣　庄	8 417	42 776.2	5 725.4	20 029.0	3 714.0
烟　台	21 270	92 786.1	12 421.0	33 064.2	6 665.9
潍　坊	22 709	104 983.1	12 921.0	25 866.3	6 079.3
济　宁	18 630	84 681.5	11 334.3	34 754.0	6 317.7
泰　安	14 487	64 031.3	8 503.5	34 603.0	5 879.0
日　照	7 046	32 029.5	4 287.0	15 411.6	2 830.8
郑　州	32 837	115 894.8	14 873.0	36 525.8	9 640.0
开　封	11 752	36 941.0	4 740.0	19 462.3	3 110.9
洛　阳	19 086	65 437.2	8 397.8	20 633.7	4 536.1
平 顶 山	13 790	41 369.1	5 309.0	18 357.0	3 921.1
安　阳	11 208	39 558.0	5 076.6	19 583.3	3 690.0
焦　作	11 433	37 689.9	4 836.9	10 751.9	2 702.4
三 门 峡	5 836	22 208.1	3 332.6	10 436.1	1 972.7
武　汉	53 201	224 870.1	24 324.4	103 334.2	13 618.8
宜　昌	12 246	46 103.3	5 762.9	24 892.1	4 632.8
荆　州	16 060	66 247.5	7 508.1	51 647.1	7 003.1
长　沙	38 172	122 625.3	14 863.7	63 148.4	8 955.0
株　洲	14 787	50 829.9	6 161.2	32 640.3	4 700.1
湘　潭	10 511	34 682.4	4 204.2	21 020.0	3 372.3
岳　阳	16 071	63 515.7	7 698.9	53 199.3	6 778.0
常　德	11 699	54 831.6	6 646.2	47 219.0	5 665.1
张 家 界	3 570	14 725.8	1 784.6	12 905.0	1 543.1
广　州	116 942	301 585.8	38 000.2	121 252.3	19 289.6

重点城市城镇生活污染排放及处理情况（二）（续表）

（2011）

城 市名 称	城镇生活污水排放量/万吨	城镇生活化学需氧量产生量/吨	城镇生活氨氮产生量/吨	城镇生活化学需氧量排放量/吨	城镇生活氨氮排放量/吨
韶 关	9 148	34 598.7	4 449.9	27 967.2	3 955.4
深 圳	110 033	294 282.8	37 081.8	77 510.8	12 693.6
珠 海	13 494	37 686.5	4 825.0	17 636.2	3 111.0
汕 头	19 577	86 429.8	11 122.6	59 440.5	9 213.6
湛 江	18 267	60 622.6	7 799.9	43 535.4	5 915.0
南 宁	21 659	90 009.0	11 444.0	67 144.0	9 076.0
柳 州	13 636	54 960.0	6 992.9	34 820.4	5 430.6
桂 林	13 020	47 383.6	6 017.9	28 833.2	4 228.0
北 海	5 147	18 804.0	2 390.8	11 992.0	1 227.0
海 口	11 291	36 759.3	4 537.5	6 336.5	3 428.0
重 庆	97 356	398 609.2	50 998.5	230 769.5	38 069.0
成 都	71 554	240 495.2	29 469.1	112 569.1	15 140.1
自 贡	6 086	29 292.6	3 589.4	20 363.1	2 824.7
攀 枝 花	3 651	19 183.7	2 350.7	14 120.2	1 846.5
泸 州	9 141	43 384.3	5 316.1	39 274.4	4 839.1
德 阳	9 190	39 401.0	4 827.6	23 946.8	3 448.7
绵 阳	11 482	50 190.3	6 150.1	34 753.6	4 743.1
南 充	12 010	60 213.7	7 378.3	46 101.2	5 799.7
宜 宾	8 539	43 131.9	5 285.2	32 483.1	4 251.5
贵 阳	12 519	57 410.0	7 428.3	28 867.9	4 132.6
遵 义	10 954	59 166.8	7 550.2	50 315.2	6 593.5
昆 明	39 064	117 081.4	13 037.4	8 526.4	4 979.7
曲 靖	10 524	54 904.1	6 622.9	48 276.1	6 067.8
玉 溪	4 460	20 592.7	2 484.0	14 523.7	2 161.8
拉 萨	1 759	7 993.5	1 025.8	7 993.5	1 025.8
西 安	27 458	128 907.1	17 216.0	63 156.0	11 041.2
铜 川	1 168	29 477.5	3 701.6	9 354.1	1 240.3
宝 鸡	7 410	34 418.4	4 801.1	16 393.8	3 209.1
咸 阳	9 568	42 201.1	6 097.1	15 252.9	3 467.5
渭 南	6 761	33 504.6	4 569.8	27 950.4	4 176.7
延 安	4 730	22 815.1	2 829.2	17 393.6	2 330.1
兰 州	12 000	61 262.2	8 099.3	44 483.2	7 198.1
金 昌	1 403	6 359.5	840.7	1 970.9	484.7
西 宁	7 024	32 183.3	4 150.7	16 591.7	3 506.0
银 川	11 249	32 728.2	4 745.3	3 600.1	2 914.8
石 嘴 山	3 548	11 628.0	1 462.9	3 488.4	980.1
乌 鲁 木 齐	19 360	62 807.8	7 150.7	14 165.6	4 817.1
克 拉 玛 依	3 832	5 883.9	505.7	564.8	58.4

重点城市城镇生活污染排放及处理情况（三）

（2011）

单位：吨

城 市 名 称	城镇生活二氧化硫排放量	城镇生活氮氧化物排放量	城镇生活烟尘排放量
总　　计	957 708	188 255	564 338
北　京	36 539	12 145	31 967
天　津	8 959	4 447	4 071
石 家 庄	10 233	3 191	1 783
唐　山	4 680	1 550	1 090
秦 皇 岛	2 370	708	1 969
邯　郸	15 545	2 436	1 711
保　定	14 191	1 774	2 677
太　原	32 943	8 694	34 055
大　同	7 791	1 503	6 221
阳　泉	3 377	687	342
长　治	2 708	4 530	11 882
临　汾	13 517	4 380	2 622
呼 和 浩 特	5 117	645	4 838
包　头	17 984	3 753	32 437
赤　峰	17 937	1 662	8 512
沈　阳	15 737	2 200	9 745
大　连	14 298	2 403	9 612
鞍　山	8 880	1 492	6 931
抚　顺	4 224	600	4 800
本　溪	5 612	1 139	2 149
锦　州	4 889	925	3 900
长　春	7 344	1 600	8 000
吉　林	6 426	1 410	6 300
哈 尔 滨	32 130	13 646	86 542
齐 齐 哈 尔	18 658	7 449	66 275
牡 丹 江	6 868	4 668	4 327
上　海	29 860	9 079	14 850
南　京	827	683	463
无　锡	512	300	387
徐　州	124	336	1 395
常　州	82	595	97
苏　州	791	397	117
南　通	2 596	348	1 017
连 云 港	6 945	904	775
扬　州	2 908	582	454
镇　江	2 649	497	2 100
杭　州	739	569	87

重点城市城镇生活污染排放及处理情况（三）（续表）

（2011）

单位：吨

城　市 名　称	城镇生活二氧化硫排放量	城镇生活氮氧化物排放量	城镇生活烟尘排放量
宁　波	2 343	648	461
温　州	598	400	466
湖　州	2 176	328	239
绍　兴	1 662	268	239
合　肥	2 347	110	58
芜　湖	2 282	200	498
马鞍山	2 073	333	165
福　州	1 226	236	547
厦　门	486	54	54
泉　州	3 199	559	548
南　昌	1 159	124	496
九　江	5 871	1 439	756
济　南	11 306	3 629	8 355
青　岛	27 059	2 601	9 795
淄　博	4 973	1 223	4 144
枣　庄	11 684	1 350	5 335
烟　台	13 073	2 354	9 693
潍　坊	24 294	3 227	13 646
济　宁	12 106	2 841	9 796
泰　安	18 013	2 487	8 477
日　照	10 753	1 250	5 508
郑　州	5 900	1 222	5 000
开　封	14 878	1 997	1 141
洛　阳	24 190	2 070	1 794
平顶山	477	384	400
安　阳	7 861	1 015	1 521
焦　作	4 922	1 158	4 922
三门峡	22 503	1 474	2 770
武　汉	5 720	1 416	1 001
宜　昌	5 458	853	853
荆　州	5 569	596	596
长　沙	2 328	225	2 946
株　洲	2 592	524	810
湘　潭	271	310	229
岳　阳	7 014	851	825
常　德	5 760	1 012	1 338
张家界	1 920	172	800
广　州	1 813	1 906	654

重点城市城镇生活污染排放及处理情况（三）（续表）

城　市 名　称	城镇生活二氧化硫排放量	城镇生活氮氧化物排放量	城镇生活烟尘排放量
韶　关	453	456	239
深　圳	488	285	82
珠　海	280	118	13
汕　头	455	117	55
湛　江	1 999	363	208
南　宁	8 747	1 123	4 631
柳　州	2 993	435	966
桂　林	5 065	531	249
北　海	1 240	141	122
海　口	21	60	2
重　庆	55 585	4 817	2 463
成　都	4 331	1 815	589
自　贡	5 256	382	309
攀 枝 花	556	109	109
泸　州	7 515	872	1 467
德　阳	1 362	291	132
绵　阳	8 680	1 083	700
南　充	4 290	294	151
宜　宾	11 022	694	1 852
贵　阳	33 515	2 393	5 380
遵　义	48 692	3 317	9 240
昆　明	5 261	942	320
曲　靖	8 205	1 086	1 207
玉　溪	964	208	43
拉　萨	468	41	206
西　安	16 923	7 675	11 899
铜　川	3 626	633	1 400
宝　鸡	815	165	346
咸　阳	21 173	2 593	10 952
渭　南	4 867	475	321
延　安	5 044	1 104	5 606
兰　州	11 836	1 979	1 741
金　昌	4 021	589	1 905
西　宁	5 965	1 208	4 010
银　川	5 428	929	2 538
石 嘴 山	6 307	559	2 650
乌鲁木齐	9 207	1 540	6 700
克拉玛依	205	62	160

重点城市机动车污染排放情况

（2011）　　　　　　　　　　　　　　　　　　　　　单位：万吨

城　市 名　称	机动车污染物排放总量			
	总颗粒物	氮氧化物	一氧化碳	碳氢化合物
总　　计	32.0	321.6	1 769.3	220.6
北　　京	0.4	8.6	89.4	9.9
天　　津	0.6	5.4	43.5	5.0
石 家 庄	0.8	6.7	24.6	3.2
唐　　山	0.9	7.9	34.8	4.4
秦 皇 岛	0.5	3.9	18.0	2.4
邯　　郸	1.2	7.6	25.0	4.0
保　　定	0.1	7.6	45.4	5.2
太　　原	0.4	3.1	16.0	1.9
大　　同	0.3	3.0	14.9	1.8
阳　　泉	…	2.4	22.4	2.4
长　　治	0.2	1.7	6.5	0.9
临　　汾	0.6	5.1	26.6	3.5
呼和浩特	0.4	2.7	8.6	1.3
包　　头	0.5	3.3	10.9	1.6
赤　　峰	0.4	2.1	6.6	1.0
沈　　阳	0.5	4.8	25.8	3.3
大　　连	0.7	5.0	21.5	2.9
鞍　　山	0.2	2.5	13.1	1.6
抚　　顺	0.2	1.8	8.1	1.1
本　　溪	0.1	1.1	4.4	0.6
锦　　州	0.3	2.5	7.3	1.2
长　　春	0.6	5.9	38.1	4.8
吉　　林	0.4	4.1	21.4	2.8
哈 尔 滨	0.5	4.0	24.4	3.1
齐齐哈尔	0.4	3.0	13.1	1.8
牡 丹 江	0.2	1.7	11.3	1.4
上　　海	0.8	9.9	46.1	6.7
南　　京	0.3	3.3	15.8	1.9
无　　锡	0.2	3.0	15.1	1.7
徐　　州	0.5	5.8	17.6	2.5
常　　州	0.1	1.9	11.6	1.4
苏　　州	0.5	6.3	52.6	6.2
南　　通	0.1	1.9	17.8	2.3
连 云 港	0.1	1.3	6.5	0.8
扬　　州	0.2	1.7	9.2	1.1
镇　　江	0.1	1.1	6.5	0.8
杭　　州	0.4	4.2	24.2	2.8

重点城市机动车污染排放情况（续表）

（2011）

单位：万吨

城 市 名 称	机动车污染物排放总量			
	总颗粒物	氮氧化物	一氧化碳	碳氢化合物
宁 波	0.3	2.5	16.4	1.9
温 州	0.1	1.6	13.3	1.5
湖 州	0.1	0.8	8.4	1.1
绍 兴	0.1	1.3	9.1	1.0
合 肥	0.3	2.4	10.9	1.3
芜 湖	0.1	1.2	6.6	0.7
马 鞍 山	0.1	0.6	2.7	0.3
福 州	0.2	2.2	12.4	1.5
厦 门	0.2	2.1	9.0	1.1
泉 州	0.2	1.8	18.7	2.1
南 昌	0.4	3.5	16.1	1.9
九 江	0.2	1.9	6.8	1.0
济 南	0.3	3.0	16.7	2.0
青 岛	0.6	5.0	25.0	3.3
淄 博	0.3	2.3	14.7	1.8
枣 庄	0.2	1.9	9.6	1.2
烟 台	0.4	3.7	21.9	2.9
潍 坊	0.6	5.9	30.6	4.1
济 宁	0.5	3.7	14.3	2.1
泰 安	0.2	1.7	9.8	1.3
日 照	0.1	1.2	6.3	0.8
郑 州	0.6	6.0	28.7	3.6
开 封	0.2	1.7	5.8	0.9
洛 阳	0.4	3.2	12.7	1.6
平 顶 山	0.2	1.8	8.0	1.0
安 阳	0.3	3.1	12.6	1.8
焦 作	0.3	3.6	16.4	2.1
三 门 峡	0.1	1.3	8.3	1.0
武 汉	0.5	4.6	20.3	2.4
宜 昌	0.2	1.9	10.3	1.4
荆 州	0.1	1.2	9.9	1.2
长 沙	0.2	2.5	15.0	1.7
株 洲	0.1	1.0	4.9	0.6
湘 潭	…	0.6	3.7	0.4
岳 阳	0.1	0.8	4.5	0.5
常 德	0.1	1.1	6.3	0.8
张 家 界	…	0.5	2.4	0.3
广 州	0.6	6.9	33.7	3.9

重点城市机动车污染排放情况（续表）

（2011）

单位：万吨

城　市 名　称	机动车污染物排放总量			
	总颗粒物	氮氧化物	一氧化碳	碳氢化合物
韶　关	0.1	0.6	4.6	0.5
深　圳	0.9	8.0	44.2	4.8
珠　海	0.3	1.9	11.0	1.3
汕　头	0.3	1.9	14.5	1.8
湛　江	0.1	1.2	6.6	0.8
南　宁	0.4	3.8	21.9	2.6
柳　州	0.1	1.1	6.0	0.7
桂　林	0.1	1.2	8.2	0.9
北　海	…	0.4	3.4	0.4
海　口	0.1	1.0	5.8	0.6
重　庆	0.7	10.4	62.9	7.9
成　都	0.3	4.6	30.0	3.4
自　贡	…	0.7	3.9	0.5
攀 枝 花	0.1	0.6	2.0	0.3
泸　州	…	0.9	8.4	0.9
德　阳	0.1	1.0	6.2	0.8
绵　阳	0.1	1.1	5.9	0.7
南　充	0.2	1.4	5.5	0.7
宜　宾	0.1	0.8	4.6	0.6
贵　阳	0.2	2.2	10.3	1.3
遵　义	0.1	1.2	7.6	0.9
昆　明	0.1	4.0	22.0	3.1
曲　靖	0.3	2.6	14.2	2.0
玉　溪	0.2	1.7	12.6	1.9
拉　萨	0.1	1.1	12.4	1.2
西　安	0.2	4.2	24.0	2.8
铜　川	…	0.6	3.8	0.5
宝　鸡	0.1	1.5	15.5	1.7
咸　阳	0.1	1.6	10.5	1.0
渭　南	0.1	2.2	8.2	1.1
延　安	1.0	1.5	6.4	1.2
兰　州	0.1	2.0	20.1	2.4
金　昌	…	0.4	4.1	0.5
西　宁	0.1	1.6	10.8	1.6
银　川	0.4	3.3	13.5	1.8
石 嘴 山	0.1	0.6	2.6	0.4
乌鲁木齐	0.5	5.7	20.2	2.7
克拉玛依	0.2	1.8	7.0	0.9

重点城市污水处理情况（一）

（2011）

城　市 名　称	污水处理厂数/座	污水处理厂 设计处理能力/ （万吨/日）	本年运行费用/万 元	污水处理厂累计完 成投资/万元	新增固定 资产/万元
总　　计	2 288	9 428.3	2 084 882.1	18 373 133.5	1 115 640.3
北　京	116	410.5	106 592.9	826 604.6	10 395.8
天　津	52	234.8	47 655.9	367 651.3	3 692.8
石 家 庄	27	180.6	28 727.6	358 982.9	19 720.1
唐　山	20	113.2	19 857.6	245 677.7	22 664.0
秦 皇 岛	17	55.6	10 649.8	90 438.2	9 149.0
邯　郸	25	79.4	14 935.1	135 014.4	26 047.5
保　定	38	119.1	26 007.8	294 924.7	44 838.8
太　原	14	58.8	15 253.7	110 484.1	3 823.1
大　同	15	24.3	6 641.7	67 007.7	1 370.0
阳　泉	6	16.2	3 862.6	35 960.0	504.0
长　治	19	21.9	6 298.5	59 924.9	4 370.5
临　汾	18	28.6	5 049.7	75 134.0	775.0
呼和浩特	9	30.3	7 724.5	99 786.6	4 500.0
包　头	12	55.6	10 055.7	179 845.0	7 038.9
赤　峰	12	44.5	7 149.2	91 276.6	695.9
沈　阳	35	159.7	34 781.4	229 399.4	32 258.0
大　连	19	96.3	20 603.7	127 646.4	12 540.0
鞍　山	4	30.0	7 890.7	63 148.6	84.7
抚　顺	15	34.5	3 010.9	44 754.9	1 080.0
本　溪	5	31.5	3 411.5	48 519.1	1 239.0
锦　州	8	33.0	8 458.4	73 088.0	687.7
长　春	12	85.8	17 116.4	29 849.5	260.0
吉　林	8	65.0	14 590.0	104 810.4	680.0
哈 尔 滨	11	108.2	21 684.6	146 849.2	360.6
齐齐哈尔	4	23.0	3 129.9	49 712.0	48.6
牡 丹 江	6	20.0	5 193.7	55 359.0	80.0
上　海	55	669.7	136 142.7	1 233 708.5	26 472.9
南　京	57	176.7	29 354.0	458 669.4	4 705.5
无　锡	76	220.3	71 668.4	501 762.6	55 926.2
徐　州	32	91.2	16 159.6	169 126.2	10 155.0
常　州	43	101.9	31 174.7	369 258.1	1 732.4
苏　州	125	314.5	91 675.1	841 031.1	78 527.6
南　通	84	107.8	22 597.3	272 840.3	21 561.0
连 云 港	17	30.4	6 279.3	75 878.0	8 635.0
扬　州	36	70.5	10 918.4	101 564.0	17 255.3
镇　江	34	48.9	9 322.7	141 616.6	368.7
杭　州	35	324.4	85 097.9	530 256.8	18 648.0

重点城市污水处理情况（一）（续表）

（2011）

城 市 名 称	污水处理厂数/座	污水处理厂 设计处理能力/ （万吨/日）	本年运行费用/ 万元	污水处理厂累计完 成投资/万元	新增固定资产/ 万元
宁 波	31	148.5	32 687.1	309 636.8	1 613.0
温 州	26	81.7	23 602.2	172 939.9	3 039.8
湖 州	28	60.5	18 474.2	157 828.1	2 855.4
绍 兴	5	126.4	58 716.4	367 291.5	18 888.0
合 肥	13	88.8	23 570.8	162 833.8	22 220.7
芜 湖	6	37.0	7 354.0	74 173.7	2 163.1
马 鞍 山	7	30.0	6 487.9	72 183.0	11 301.8
福 州	15	78.5	11 825.7	116 798.1	801.1
厦 门	8	83.0	10 437.3	20 180.9	0
泉 州	34	88.6	19 860.0	135 890.8	2 307.3
南 昌	12	105.6	15 611.6	107 211.4	12 664.8
九 江	18	30.6	7 260.4	66 849.5	10 984.9
济 南	31	88.1	24 564.7	142 028.9	15 982.0
青 岛	33	151.7	50 587.1	377 240.3	49 221.7
淄 博	13	76.5	18 575.0	110 123.8	4 070.9
枣 庄	9	40.3	9 477.3	61 565.0	7 800.0
烟 台	22	74.4	18 600.7	173 117.5	14 700.0
潍 坊	28	122.4	38 930.4	136 013.4	0
济 宁	19	80.9	17 425.9	192 802.7	2 740.4
泰 安	9	42.0	9 480.5	70 990.0	102.4
日 照	11	25.0	6 510.5	69 908.0	12 014.0
郑 州	13	127.3	20 845.3	134 086.5	8 101.0
开 封	6	34.5	4 074.6	85 538.0	13.6
洛 阳	15	60.6	5 694.4	65 960.0	409.1
平 顶 山	7	36.0	6 639.8	60 586.4	1 716.2
安 阳	7	31.3	4 940.5	37 729.8	3 934.9
焦 作	14	46.9	10 677.2	86 285.9	6 718.0
三 门 峡	6	17.0	2 680.5	17 118.0	302.0
武 汉	17	188.0	33 791.4	478 698.7	148 841.4
宜 昌	24	49.8	9 604.9	75 006.1	3 187.0
荆 州	9	39.0	6 601.6	46 474.9	0
长 沙	14	151.0	23 383.7	418 500.7	5 385.6
株 洲	9	39.7	8 104.8	94 392.6	4 167.2
湘 潭	6	27.0	7 045.0	35 445.6	302.0
岳 阳	11	33.0	5 839.2	89 166.3	80.0
常 德	10	37.0	6 395.1	90 466.0	3 316.0
张 家 界	5	9.5	1 988.8	23 644.0	2.8
广 州	24	383.9	82 371.2	580 595.8	7 021.3

重点城市污水处理情况（一）（续表）

（2011）

城市名称	污水处理厂数/座	污水处理厂设计处理能力/（万吨/日）	本年运行费用/万元	污水处理厂累计完成投资/万元	新增固定资产/万元
韶 关	11	16.1	4 928.3	26 103.2	321.1
深 圳	36	396.3	44 363.6	624 870.3	4 241.9
珠 海	10	50.8	14 926.0	123 306.8	9 800.4
汕 头	8	58.8	12 425.8	141 081.3	6.9
湛 江	6	42.0	4 139.7	54 902.0	4 284.9
南 宁	12	92.4	12 284.9	212 172.5	48 348.1
柳 州	10	60.5	12 048.1	151 720.6	39 942.2
桂 林	21	45.5	10 017.5	81 195.4	1 199.1
北 海	2	15.0	3 448.6	48 338.0	2 077.0
海 口	6	84.5	14 236.2	86 006.4	1 144.0
重 庆	117	261.0	93 180.7	675 037.1	65 046.7
成 都	116	213.4	53 147.0	410 465.7	21 000.2
自 贡	4	9.5	2 595.9	17 851.3	6.0
攀 枝 花	6	10.0	2 358.5	26 780.0	2 469.0
泸 州	9	12.1	2 069.0	63 908.5	11.7
德 阳	8	23.5	5 481.0	38 483.5	1 997.5
绵 阳	4	26.7	6 974.0	37 415.3	500.8
南 充	9	25.1	5 985.0	40 298.0	831.0
宜 宾	9	13.6	4 213.5	31 077.0	135.0
贵 阳	14	64.6	11 617.5	138 453.8	10 861.0
遵 义	21	23.0	3 644.1	59 772.1	456.0
昆 明	17	115.1	35 666.7	265 142.5	3 680.8
曲 靖	5	17.0	3 790.6	32 796.7	7 369.6
玉 溪	7	13.8	2 463.3	26 468.4	1 305.5
拉 萨	1	0	0	0	0
西 安	31	118.8	20 156.2	105 324.2	65.0
铜 川	3	6.7	958.0	27 635.0	4 575.7
宝 鸡	11	26.1	6 029.0	32 825.0	8 183.0
咸 阳	15	39.4	5 035.5	55 421.7	4 717.6
渭 南	12	25.5	7 212.3	50 130.0	15.0
延 安	15	10.4	1 881.6	133 985.2	1 356.9
兰 州	4	44.5	6 655.8	53 427.5	1 118.6
金 昌	1	8.0	1 314.3	16 016.6	0
西 宁	10	28.4	5 732.8	66 095.6	0
银 川	7	36.5	5 521.3	57 475.0	0
石 嘴 山	4	17.5	3 386.0	17 656.0	18.0
乌鲁木齐	7	70.5	15 612.9	95 062.9	26 693.1
克拉玛依	3	18.0	1 940.3	23 570.0	0

重点城市污水处理情况（二）

（2011）　　　　　　　　　　　　　　　　　　　单位：万吨

城 市 名 称	污水实际 处理量	污水再生 利用量	工业用水量	市政用水量	景观用水量
总　　计	2 825 346.9	63 645.6	38 351.2	7 240.7	18 053.7
北　京	119 624.1	8 926.3	4 179.7	1 687.8	3 058.9
天　津	60 344.1	210.1	151.7	58.4	…
石 家 庄	52 401.9	1 099.6	1 099.6	0	0
唐　山	27 933.6	5 558.9	4 630.5	0	928.4
秦 皇 岛	13 538.2	365.0	0	365.0	0
邯　郸	18 782.8	4 414.3	1 587.5	51.1	2 775.7
保　定	30 175.6	394.9	233.8	118.1	42.9
太　原	17 293.6	2 835.7	2 672.0	159.7	4.0
大　同	5 454.9	1 265.1	1 265.1	0	0
阳　泉	4 414.8	555.0	196.7	0.3	358.0
长　治	6 245.1	653.6	653.6	0	0
临　汾	5 094.1	0	0	0	0
呼 和 浩 特	8 770.7	255.1	0	55.9	199.2
包　头	14 228.6	6 236.7	5 882.5	354.2	0
赤　峰	9 098.2	83.8	83.8	0	0
沈　阳	41 505.6	3 564.9	1 866.0	0	1 698.9
大　连	27 680.4	821.9	25.7	6.4	789.8
鞍　山	6 657.0	0	0	0	0
抚　顺	8 707.0	0	0	0	0
本　溪	9 410.2	0	0	0	0
锦　州	9 054.8	480.5	480.5	0	0
长　春	21 522.1	0	0	0	0
吉　林	18 250.4	0	0	0	0
哈 尔 滨	21 078.1	0	0	0	0
齐 齐 哈 尔	3 928.7	0	0	0	0
牡 丹 江	6 492.1	0	0	0	0
上　海	204 982.2	619.5	274.0	178.8	166.7
南　京	61 465.8	42.6	27.6	15.0	0
无　锡	55 534.2	674.5	630.8	0	43.7
徐　州	23 748.2	621.5	529.3	56.2	36.0
常　州	29 226.2	185.0	2.5	182.5	0
苏　州	75 989.0	513.6	463.1	35.9	14.6
南　通	29 593.1	0	0	0	0
连 云 港	7 847.2	0	0	0	0
扬　州	18 197.7	169.2	169.2	0	0
镇　江	11 636.3	83.5	47.5	10.5	25.6
杭　州	84 006.3	2 234.6	2 177.4	57.2	0

重点城市污水处理情况（二）（续表）

单位：万吨

城 市 名 称	污水实际 处理量	污水再生 利用量	工业用水量	市政用水量	景观用水量
宁 波	39 839.9	3 059.6	2 356.3	575.0	128.3
温 州	24 399.3	14.8	14.8	0	0
湖 州	14 406.3	664.0	664.0	...	0
绍 兴	38 768.9	2.0	2.0	0	0
合 肥	31 864.3	0	0	0	0
芜 湖	10 332.2	0	0	0	0
马 鞍 山	7 650.4	0	0	0	0
福 州	25 741.3	0	0	0	0
厦 门	20 903.3	43.8	0	35.5	8.3
泉 州	24 476.9	154.8	154.8	0	0
南 昌	29 899.3	0	0	0	0
九 江	10 007.4	312.0	107.0	0.3	204.7
济 南	26 724.3	160.2	102.2	58.1	0
青 岛	40 496.9	0	0	0	0
淄 博	22 388.5	0	0	0	0
枣 庄	10 840.0	59.1	0	0	59.1
烟 台	23 037.3	120.0	0	120.0	0
潍 坊	38 136.2	0	0	0	0
济 宁	233 636.7	1 952.4	1 678.4	208.1	65.9
泰 安	11 732.0	559.9	559.9	0	0
日 照	7 745.3	167.0	0	167.0	0
郑 州	39 486.5	0	0	0	0
开 封	7 893.9	0	0	0	0
洛 阳	16 882.6	1 606.0	1 606.0	0	0
平 顶 山	12 446.5	1.0	1.0	0	0
安 阳	7 997.4	0	0	0	0
焦 作	12 752.2	49.4	49.4	0	0
三 门 峡	4 738.0	0	0	0	0
武 汉	59 036.3	546.5	0	546.5	0
宜 昌	12 041.5	0	0	0	0
荆 州	9 656.6	0	0	0	0
长 沙	38 641.4	0	0	0	0
株 洲	11 749.9	0	0	0	0
湘 潭	8 350.3	0	0	0	0
岳 阳	8 016.4	0.6	0.4	0	0.2
常 德	7 903.0	47.0	47.0	0	0
张 家 界	2 436.3	0	0	0	0
广 州	113 305.0	380.8	334.2	11.0	35.6

重点城市污水处理情况（二）（续表）

（2011）

单位：万吨

城　市名　称	污水实际处理量	污水再生利用量	工业用水量	市政用水量	景观用水量
韶　关	8 068.8	0	0	0	0
深　圳	104 462.9	3 019.1	0	105.0	2 914.0
珠　海	16 499.2	109.5	0	109.5	0
汕　头	17 601.6	86.8	0	86.8	0
湛　江	10 180.9	0	0	0	0
南　宁	20 386.6	10.2	0	1.0	9.2
柳　州	14 439.0	0	0	0	0
桂　林	10 968.2	0	0	0	0
北　海	4 248.6	6.0	6.0	0	0
海　口	26 538.2	3.1	0	0	3.1
重　庆	76 059.5	0	0	0	0
成　都	68 275.0	2 044.6	0	529.9	1 514.8
自　贡	3 936.5	0	0	0	0
攀 枝 花	2 059.3	0	0	0	0
泸　州	2 204.7	0	0	0	0
德　阳	7 065.2	1.8	0	0	1.8
绵　阳	7 674.8	0	0	0	0
南　充	6 576.6	0	0	0	0
宜　宾	4 296.1	0	0	0	0
贵　阳	20 451.9	2 739.8	0	27.8	2 712.0
遵　义	5 403.0	159.6	0	159.5	0.2
昆　明	39 018.8	0	0	0	0
曲　靖	4 025.4	0	0	0	0
玉　溪	3 171.7	0	0	0	0
拉　萨	0	0	0	0	0
西　安	30 264.9	861.8	569.6	292.2	0
铜　川	962.1	0	0	0	0
宝　鸡	7 368.0	440.0	409.9	30.2	0
咸　阳	8 781.0	661.0	134.0	478.0	49.0
渭　南	3 965.9	10.4	2.4	0	8.0
延　安	1 826.0	3.9	0	3.9	0
兰　州	10 760.2	128.5	128.5	0	0
金　昌	1 241.0	0	0	0	0
西　宁	5 781.3	0	0	0	0
银　川	12 203.2	0	0	0	0
石 嘴 山	2 764.8	0	0	0	0
乌鲁木齐	15 799.0	290.8	93.6	0	197.2
克拉玛依	3 744.2	302.5	0	302.5	0

重点城市污水处理情况（三）

（2011）

单位：万吨

城 市 名 称	污泥产生量	污泥处置量	土地利用量	填埋处置量	建筑材料利用量	焚烧处置量	污泥倾倒丢弃量
总　　　计	1 741.3	1 741.2	276.8	916.9	176.3	371.2	0.04
北　京	113.3	113.3	79.2	8.4	5.7	20.0	0
天　津	21.7	21.7	1.3	19.3	0.9	0.3	0
石 家 庄	35.6	35.6	6.9	28.1	0	0.5	0
唐　山	26.8	26.8	2.7	23.2	0.9	0	0
秦 皇 岛	7.8	7.8	5.6	2.3	0	0	0
邯　郸	10.5	10.5	0.2	9.7	0.1	0.4	0
保　定	25.8	25.8	0.5	24.7	0	0.6	0
太　原	11.6	11.6	10.5	1.0	0	0	0
大　同	3.1	3.1	0	3.1	0	…	0
阳　泉	2.5	2.5	2.3	0.3	0	0	0
长　治	3.3	3.3	0.8	2.4	0	0.1	0
临　汾	2.7	2.7	1.8	0.5	0	0.4	0
呼和浩特	3.1	3.1	0.1	3.0	…	0	0
包　头	12.6	12.6	0	12.6	0	0	0
赤　峰	1.4	1.4	0	1.4	0	0	0
沈　阳	28.5	28.5	0.2	28.4	0	0	0
大　连	13.6	13.6	0	6.3	6.0	1.2	0
鞍　山	5.1	5.1	0	5.1	0	0	0
抚　顺	4.5	4.5	0	4.5	0	0	0
本　溪	3.3	3.3	0	0.1	3.2	0	0
锦　州	3.8	3.8	0	3.8	0	0	0
长　春	9.9	9.9	0.2	9.7	0	0	0
吉　林	9.6	9.6	0	9.6	0	0	0
哈 尔 滨	18.9	18.9	0	18.9	0	0	0
齐齐哈尔	1.0	1.0	0	1.0	0	0	0
牡 丹 江	0.9	0.9	0	0.9	0	0	0
上　海	115.7	115.7	4.9	101.0	4.4	5.4	0
南　京	16.7	16.7	1.3	7.3	0.3	7.8	0
无　锡	67.0	67.0	0.6	15.0	21.3	30.1	0
徐　州	12.2	12.2	1.4	1.6	2.1	7.1	0
常　州	32.8	32.8	0.3	1.1	6.6	24.7	0
苏　州	71.9	71.9	8.5	17.0	0.7	45.6	0
南　通	18.2	18.2	0.8	0.5	4.5	12.4	0
连 云 港	0.8	0.8	…	0.1	0.2	0.6	0
扬　州	7.7	7.7	0.4	1.5	0.1	5.7	0
镇　江	5.8	5.8	0	0.1	0	5.7	0
杭　州	160.0	160.0	29.2	7.3	56.7	66.8	0

重点城市污水处理情况（三）（续表）

（2011）   单位：万吨

城 市 名 称	污泥产生量	污泥处置量	土地利用量	填埋处置量	建筑材料 利用量	焚烧处置量	污泥倾倒 丢弃量
宁 波	27.2	27.2	2.0	6.8	3.9	14.5	0
温 州	18.2	18.2	1.0	13.9	2.5	0.9	0
湖 州	9.4	9.4	…	3.2	2.3	3.8	0
绍 兴	64.8	64.8	0	38.7	4.0	22.1	0
合 肥	18.2	18.2	8.8	0.6	0	8.8	0
芜 湖	5.2	5.2	0	1.0	0	4.2	0
马 鞍 山	3.7	3.7	0	3.7	0	0	0
福 州	10.6	10.6	3.4	7.2	0	0	0
厦 门	13.4	13.4	3.2	6.8	1.3	2.2	0
泉 州	8.8	8.8	0	2	2	4.7	0
南 昌	8.8	8.8	0	3.6	1.8	3.4	0
九 江	4.1	4.1	0.1	3.9	0	…	0
济 南	16.3	16.3	15.5	0.2	0	0.6	0
青 岛	33.0	33.0	11.4	9.0	2.9	9.6	0
淄 博	26.0	26.0	0	21.6	0.6	3.8	0
枣 庄	5.6	5.6	…	0.4	0	5.2	0
烟 台	18.8	18.8	1.2	16.4	0	1.2	0
潍 坊	22.9	22.9	5.4	16.1	1.3	0.1	0
济 宁	12.2	12.2	0.7	4.1	2.3	5.2	0
泰 安	5.9	5.9	3.3	0.8	0.2	1.6	0
日 照	4.0	4.0	3.3	0.6	0	0.1	0
郑 州	36.6	36.6	2.0	34.6	0	0	0
开 封	2.7	2.7	0	2.7	0	0	0
洛 阳	10.2	10.2	0	10.2	0	0	0
平 顶 山	6.1	6.1	0	6.1	0	0	0
安 阳	6.2	6.2	0	6.2	0	0	0
焦 作	5.1	5.1	1.5	3.6	0	0	0
三 门 峡	2.2	2.2	0	2.2	0	0	0
武 汉	24.3	24.3	…	24.3	0	0	0
宜 昌	1.5	1.5	…	0.7	0	0.8	0
荆 州	1.0	1.0	0.1	0.9	0	0	0
长 沙	16.8	16.8	0.2	16.6	0	0	0
株 洲	5.6	5.6	0	5.6	0	0	0
湘 潭	4.3	4.3	0	1.5	2.8	0	0
岳 阳	1.3	1.3	…	1.3	0	0	0
常 德	1.6	1.6	0	1.2	0.4	0	0
张 家 界	0.8	0.8	0	0.8	0	0	0
广 州	65.8	65.8	11.6	14.7	11.6	27.9	0

重点城市污水处理情况（三）（续表）

（2011）

单位：万吨

城　市名　称	污泥产生量	污泥处置量	土地利用量	填埋处置量	建筑材料利用量	焚烧处置量	污泥倾倒丢弃量
韶　关	0.9	0.9	…	0.8	0	0	0
深　圳	77.0	77.0	1.6	75.4	0	0	0
珠　海	8.5	8.5	0	3.5	0	4.9	0
汕　头	8.0	8.0	0	8.0	0	0	0
湛　江	4.1	4.1	4.1	0	0	0	0
南　宁	4.7	4.7	4.7	…	…	0	0
柳　州	4.9	4.9	0	0.3	4.6	0	0
桂　林	1.6	1.6	1.5	0.1	0.1	0	0
北　海	2.3	2.3	0.2	2.1	0	0	0
海　口	4.8	4.8	4.8	…	0	0	0
重　庆	36.4	36.4	5.0	17.0	13.2	1.2	…
成　都	23.2	23.0	2.5	19.5	0.9	0.1	0
自　贡	2.0	2.0	0	2.0	0	0	0
攀枝花	1.2	1.2	0	1.2	0	0	0
泸　州	1.4	1.4	0	1.4	0	…	0
德　阳	0.7	0.7	0	0.5	0.1	0.1	0
绵　阳	3.9	3.9	0	0.2	3.6	0	0
南　充	3.3	3.3	…	3.3	0	0	0
宜　宾	1.7	1.7	0.1	1.5	0.1	…	0
贵　阳	8.7	8.7	0	8.7	0.1	0	0
遵　义	1.7	1.7	…	1.7	0	0	0
昆　明	48.1	48.1	0.3	47.8	0	0	0
曲　靖	2.1	2.1	1.7	0.4	0	0	0
玉　溪	1.2	1.2	1.2	…	0	0	0
拉　萨	0	0	0	0	0	0	0
西　安	26.4	26.4	3.6	18.8	0	4.0	0
铜　川	0.7	0.7	0	0.7	0	0	0
宝　鸡	4.6	4.6	0.2	4.5	…	0	0
咸　阳	7.2	7.2	3.4	3.8	0	0	0
渭　南	2.9	2.9	0	2.9	0	0	0
延　安	1.8	1.8	0	1.8	0	0	0
兰　州	7.9	7.9	…	7.9	0	0	0
金　昌	0.5	0.5	0.5	0	0	0	0
西　宁	4.1	4.1	0	4.1	0	0	0
银　川	6.5	6.5	5.6	0.3	0	0.7	0
石嘴山	4.7	4.7	0	0.6	0	4.1	0
乌鲁木齐	2.1	2.1	0.9	1.1	0	0	0.04
克拉玛依	0.6	0.6	0	0.6	0	0	0

重点城市污水处理情况（四）

（2011）单位：吨

城 市 名 称	污染物去除量						
	化学需氧量	氨氮	油类	总氮	总磷	挥发酚	氰化物
总　　计	7 613 695.4	643 053.7	40 492.4	535 106.3	96 732.2	647.9	999.4
北　　京	387 310.7	51 995.8	6 482.5	42 280.0	4 473.5	21.2	1.4
天　　津	231 509.0	22 592.8	291.4	39 507.0	2 533.0	0.1	0.1
石 家 庄	183 442.4	18 454.4	75.8	3 424.1	1 447.5	0	…
唐　　山	116 875.3	7 427.3	4.9	5 121.9	851.4	0	0
秦 皇 岛	43 270.4	3 461.3	0	759.1	84.0	0	0
邯　　郸	50 415.4	4 983.7	40.0	3 649.4	538.5	0	0
保　　定	89 876.4	7 372.6	49.8	5 146.2	4 817.7	0	0
太　　原	73 499.3	4 668.1	3 615.8	4 660.1	933.3	2.8	…
大　　同	17 345.4	1 910.7	0	2 200.2	527.7	0	0
阳　　泉	12 475.6	1 163.8	0	362.3	17.8	0	0
长　　治	19 172.1	1 576.8	5.4	1 535.7	194.6	27.5	0.1
临　　汾	15 467.8	1 034.9	0.5	638.3	76.2	0	0
呼和浩特	25 202.6	1 905.9	0	0	0	0	0
包　　头	29 033.2	2 038.7	967.1	4 154.1	472.3	0	8.9
赤　　峰	34 162.0	1 399.4	0	0	0	0	0
沈　　阳	107 256.4	5 058.5	90.4	4 645.7	1 088.7	0	0
大　　连	65 173.0	6 763.7	160.7	4 128.2	684.4	0	0
鞍　　山	14 707.4	412.3	0	575.7	84.1	0	0
抚　　顺	24 373.6	740.2	0	1 162.3	210.3	0	0
本　　溪	14 163.5	2 029.3	0	2 912.2	266.1	0	0
锦　　州	18 323.7	1 096.2	122.2	1 187.9	4 638.0	0	0
长　　春	56 525.5	4 242.4	0.0	0	0	0	0
吉　　林	68 098.8	2 879.9	528.6	137.0	217.4	74.0	8.4
哈 尔 滨	52 807.9	4 568.4	268.6	5 143.1	551.9	5.4	0
齐齐哈尔	11 753.3	817.9	0	824.6	163.8	0	0
牡 丹 江	17 286.1	986.0	3.0	877.4	202.4	0	0
上　　海	519 251.6	33 778.6	2 115.3	41 740.9	6 632.2	149.4	2.6
南　　京	117 797.8	11 934.1	236.4	7 412.3	1 184.8	0.2	…
无　　锡	166 711.4	15 219.6	2 121.5	15 382.2	1 553.1	0	0.1
徐　　州	43 873.5	4 938.0	0	2 209.0	577.0	0	0
常　　州	83 209.1	7 049.6	117.0	2 499.0	647.8	29.8	0
苏　　州	236 565.9	18 559.4	374.0	12 331.3	2 412.1	1.0	0
南　　通	82 265.9	6 329.9	0	0	51.9	0	…
连 云 港	12 977.2	1 339.9	7.5	1 019.6	119.0	2.0	0
扬　　州	42 280.0	3 721.4	164.0	2 352.6	386.9	0	0
镇　　江	21 530.3	2 575.0	6.0	1 378.7	221.4	0	0
杭　　州	414 611.0	17 091.9	47.2	2 325.9	1 770.7	0	0

重点城市污水处理情况（四）（续表）

（2011）

单位：吨

城 市 名 称	污染物去除量						
	化学需氧量	氨氮	油类	总氮	总磷	挥发酚	氰化物
宁　波	107 546.5	8 812.0	395.1	7 137.5	1 053.9	0	122.7
温　州	73 578.6	5 125.3	213.5	1 759.3	522.0	0	168.1
湖　州	35 953.1	1 989.4	323.2	1 348.8	237.6	0	0
绍　兴	301 030.2	16 224.5	5 129.1	15 112.8	1 798.3	0	0
合　肥	53 342.8	5 065.5	406.1	4 897.5	880.4	34.7	0
芜　湖	15 744.2	694.0	13.8	1 333.4	188.6	0	0
马鞍山	12 404.0	668.4	0	918.8	108.5	0	0
福　州	45 155.6	4 988.2	351.6	5 588.8	459.3	0	0
厦　门	53 894.5	5 337.5	72.3	4 838.5	668.3	0	0.6
泉　州	111 043.3	3 396.0	377.4	2 863.9	282.6	0	669.5
南　昌	39 055.4	2 866.9	0	588.8	453.6	0	0
九　江	14 385.4	1 168.5	260.0	829.6	68.0	0	0
济　南	71 178.4	8 343.8	0	4 989.3	1 040.2	0	0
青　岛	146 599.3	12 120.9	81.1	13 447.3	2 139.6	0	13.0
淄　博	72 007.2	6 233.5	114.3	431.9	904.2	0	0
枣　庄	25 855.0	2 089.9	0	1 438.0	149.1	0	0
烟　台	64 312.6	7 199.1	10.4	3 640.4	801.8	0	0
潍　坊	129 231.6	9 056.4	110.5	5 494.7	541.9	0	0
济　宁	625 989.5	73 033.6	0	60 104.6	9 437.0	0	0
泰　安	32 013.6	2 917.1	0	2 082.2	286.8	0	0
日　照	20 671.4	1 740.3	2.3	1 194.9	152.3	0	1.0
郑　州	98 456.3	10 184.7	4.0	6 316.0	1 852.9	0.2	0
开　封	18 065.2	1 668.7	0	710.3	80.6	0	0
洛　阳	43 158.0	4 109.3	16.0	4 792.1	852.2	0	0
平顶山	28 680.7	1 547.2	0	2 205.9	117.1	0	0
安　阳	20 912.0	1 605.7	0	812.1	107.5	0	0
焦　作	56 741.3	2 572.9	0	1 129.4	316.2	0	0
三门峡	11 772.0	1 087.9	0	1 589.8	255.0	0	0
武　汉	70 820.6	6 491.7	10.3	3 087.5	603.0	0	0
宜　昌	21 208.7	1 129.9	13.3	2 014.2	211.7	0	0
荆　州	13 747.0	834.0	299.2	27.9	6.5	0	0
长　沙	68 692.4	5 965.0	0	2 538.5	479.0	0	0
株　洲	18 542.9	1 553.5	…	1 816.7	170.5	0	0
湘　潭	14 849.4	905.3	1.1	1 067.9	115.1	0	0
岳　阳	13 145.5	1 080.5	0	937.2	97.4	0	0
常　德	8 314.5	1 041.4	61.6	628.6	60.2	0	0
张家界	1 820.8	241.6	0	0	5.3	0	0
广　州	228 393.8	19 094.5	340.8	18 334.9	3 094.0	0.6	0

重点城市污水处理情况（四）（续表）

（2011）

单位：吨

城 市 名 称	污染物去除量						
	化学需氧量	氨氮	油类	总氮	总磷	挥发酚	氰化物
韶　关	5 837.5	480.8	20.5	457.6	51.0	0	0
深　圳	243 263.0	23 782.8	21.3	19 757.8	4 002.9	53.2	0
珠　海	21 972.1	1 842.3	218.9	5 927.8	341.6	0	0
汕　头	29 259.9	2 436.5	26.2	2 143.8	181.0	0	0
湛　江	17 478.0	1 771.3	14.3	387.5	249.7	0	0
南　宁	36 271.2	3 707.0	561.9	3 397.4	1 238.7	0	0
柳　州	21 187.8	1 927.4	70.7	1 786.6	181.5	0	0
桂　林	18 879.9	1 595.9	0	1 808.4	378.3	0	0
北　海	7 252.5	1 026.6	62.6	1 284.9	71.8	0	0
海　口	44 953.3	1 218.0	0	2 397.7	1 008.2	0	0
重　庆	184 673.6	13 233.0	987.4	18 810.1	2 395.5	0	0
成　都	137 572.3	14 389.8	790.6	13 326.7	1 318.0	47.8	1.7
自　贡	9 065.3	760.9	0	180.3	735.3	55.4	0
攀 枝 花	5 047.2	501.5	371.7	658.7	59.9	0	0
泸　州	4 110.0	477.0	0	420.5	55.4	0	0
德　阳	17 084.7	1 459.5	10.1	996.4	85.4	1.1	0
绵　阳	16 949.0	1 316.7	0.8	1 011.2	111.7	0	0
南　充	14 723.3	1 522.2	334.2	389.3	44.5	0	0
宜　宾	10 412.5	987.3	46.5	869.5	176.0	0	0
贵　阳	29 072.2	3 729.2	55.4	1 358.5	301.4	0.2	…
遵　义	9 357.4	1 284.1	69.5	283.2	44.2	0	0
昆　明	100 624.8	8 016.1	616.4	9 382.0	2 626.6	0	0
曲　靖	5 838.4	773.1	0	570.1	122.6	0	0
玉　溪	5 298.3	343.5	5.0	530.0	58.1	0	0
拉　萨	0	0	0	0	0	0	0
西　安	89 471.4	9 502.0	609.0	10 634.8	1 388.1	0	0.2
铜　川	2 847.2	221.6	0	188.8	41.0	0	0
宝　鸡	19 915.6	1 710.3	0	273.8	1 261.9	120.9	0
咸　阳	26 939.0	2 677.1	0	3 462.6	788.0	0	0
渭　南	9 261.0	683.6	54.2	855.8	62.4	0	0
延　安	5 464.9	481.8	12.6	23.5	6.3	…	0
兰　州	31 906.1	3 946.9	5 815.5	16 804.8	5 415.5	16.6	1.1
金　昌	3 164.6	402.1	12.3	342.9	61.5	1.1	0
西　宁	17 132.7	873.7	0	0	0	0	0
银　川	27 505.1	1 442.9	182.5	683.2	461.8	2.5	0
石 嘴 山	4 946.6	380.1	0	48.9	41.4	0	0
乌鲁木齐	59 080.8	7 274.4	20.0	1 058.4	40.2	0.3	0
克拉玛依	10 962.5	571.6	2 999.7	859.0	95.1	0	0

重点城市生活垃圾处理情况（一）

（2011）

城 市 名 称	生活垃圾 处理厂数/座	实际处理量/ 万吨	本年运行费用/ 万元	生活垃圾处理厂累 计完成投资/万元	新增固定资产/ 万元
总　　计	770	176 131.4	374 875.8	3 676 137.0	334 516.5
北　　京	19	597.6	68 433.2	293 802.6	3 401.2
天　　津	8	130.2	6 905.6	50 024.9	0
石 家 庄	16	54 070.3	2 638.4	47 692.2	175.0
唐　　山	16	205.1	1 974.1	40 542.0	287.0
秦 皇 岛	4	10.8	824.0	9 535.7	0
邯　　郸	16	9 094.4	2 949.9	69 186.6	9 026.0
保　　定	8	129.4	2 476.0	28 465.8	32.0
太　　原	5	159.1	1 630.0	22 702.0	213.5
大　　同	4	17.6	245.0	6 293.0	0
阳　　泉	2	23.0	606.6	8 344.0	0
长　　治	9	35.0	1 454.5	28 453.1	222.3
临　　汾	17	73 058.9	1 039.3	19 096.0	151.4
呼和浩特	5	53.9	1 331.0	54 575.0	0
包　　头	3	76.4	1 969.3	21 426.0	0
赤　　峰	11	46.4	1 122.3	16 777.9	361.4
沈　　阳	3	248.1	5 500.0	28 776.0	500.0
大　　连	2	156.6	1 615.0	35 935.0	12 096.3
鞍　　山	1	55.0	427.0	7 300.0	88.0
抚　　顺	1	50.0	305.0	305.0	0
本　　溪	2	38.3	550.0	13 501.5	212.4
锦　　州	2	14.9	1 063.1	13 900.0	100.0
长　　春	6	122.1	3 841.1	45 504.5	0
吉　　林	6	54.1	1 158.1	22 257.5	1 379.2
哈 尔 滨	3	112.0	1 545.0	18 000.0	173.0
齐齐哈尔	2	26.0	365.0	10 070.0	0
牡 丹 江	6	60.9	727.9	22 139.0	0
上　　海	25	644.9	44 712.3	223 284.1	26 758.6
南　　京	6	209.9	2 176.5	23 810.8	100.0
无　　锡	3	51.2	4 595.0	45 127.9	2 550.0
徐　　州	3	26.4	1 253.0	12 651.6	5 509.8
常　　州	4	36.8	1 720.0	29 548.0	809.1
苏　　州	6	124.4	9 269.0	139 890.0	3 950.0
南　　通	2	37.3	710.0	11 144.0	0
连 云 港	7	60.3	5 072.6	38 739.0	250.0
扬　　州	5	45.9	1 343.0	22 145.6	200.0
镇　　江	4	41.1	1 044.0	26 610.0	649.0
杭　　州	11	169.1	8 632.5	97 743.4	4 559.0

重点城市生活垃圾处理情况（一）（续表）

（2011）

城 市 名 称	生活垃圾处理厂数/座	实际处理量/万吨	本年运行费用/万元	生活垃圾处理厂累计完成投资/万元	新增固定资产/万元
宁 波	8	224.5	22 430.5	176 120.7	4 837.3
温 州	15	83.7	2 811.0	18 427.0	1 410.0
湖 州	1	14.1	135.0	540.0	0
绍 兴	8	68.7	2 268.2	42 300.0	5 399.0
合 肥	25	110.9	2 708.7	4 541.6	7.5
芜 湖	17	37.2	8 130.2	32 941.2	6 026.0
马 鞍 山	18	35.4	572.1	13 237.0	693.0
福 州	4	44.9	1 590.0	51 710.0	6 165.0
厦 门	1	84.0	2 054.0	140 000.0	0
泉 州	29	78.4	7 887.5	57 684.6	2 000.0
南 昌	1	85.0	1 050.0	11 000.0	0
九 江	12	93.4	674.9	9 316.7	172.0
济 南	5	109.7	4 030.0	19 249.5	813.3
青 岛	6	168.9	3 628.0	37 200.0	1 703.0
淄 博	1	10.1	240.0	2 000.0	400.0
枣 庄	1	12.8	680.0	5 000.0	0
烟 台	8	91.2	3 281.9	39 952.4	260.0
潍 坊	8	119.0	1 812.5	46 299.0	614.0
济 宁	4	59.8	1 069.0	30 713.0	2 354.7
泰 安	2	16.7	205.0	1 261.0	0
日 照	4	27.7	1 320.0	25 030.0	2 051.6
郑 州	5	101.8	1 152.0	32 770.0	4 600.0
开 封	3	11.1	384.0	7 943.0	0
洛 阳	11	97.0	1 708.3	31 141.0	1 700.0
平 顶 山	7	63.0	1 100.0	29 569.0	1 840.0
安 阳	3	39.3	628.0	14 624.9	20.0
焦 作	7	54.5	979.3	29 048.0	2 370.0
三 门 峡	6	40.2	541.4	14 953.0	3 519.0
武 汉	7	349.9	3 693.8	124 648.2	0
宜 昌	18	63.9	1 604.8	37 210.1	460.0
荆 州	9	24.0	1 702.8	11 004.0	718.0
长 沙	4	181.5	10 655.0	39 341.0	11 800.0
株 洲	5	49.2	1 351.7	15 197.5	270.0
湘 潭	7	47.5	1 364.3	20 846.0	876.0
岳 阳	10	69.4	1 419.2	33 245.1	4 392.0
常 德	6	172.2	965.0	12 520.0	1 500.0
张 家 界	1	12.2	83.0	3 800.0	0
广 州	6	900.2	11 445.8	72 154.0	5 245.0

重点城市生活垃圾处理情况（一）（续表）

（2011）

城 市 名 称	生活垃圾处理厂数/座	实际处理量/万吨	本年运行费用/万元	生活垃圾处理厂累计完成投资/万元	新增固定资产/万元
韶 关	9	76.5	1 727.1	14 625.0	1 080.8
深 圳	10	2 390.8	13 461.9	40 456.0	2 416.4
珠 海	1	45.8	1 403.2	34 088.0	3 500.0
汕 头	4	375.0	1 519.7	11 020.0	58.5
湛 江	4	39.3	929.7	14 363.0	0
南 宁	4	65.1	1 452.4	4 114.0	432.4
柳 州	1	38.4	828.0	16 107.7	80.0
桂 林	6	42.9	1 097.9	2 167.1	30.0
北 海	1	14.7	825.0	27 000.0	8 586.0
海 口	0	0	0	0	0
重 庆	43	569.8	15 649.1	225 896.3	160 851.5
成 都	9	265.3	6 747.8	31 681.3	2 787.0
自 贡	2	25.7	830.0	8 048.7	0
攀 枝 花	3	18.1	459.0	7 265.0	0
泸 州	4	35.8	839.0	11 613.3	137.0
德 阳	5	37.0	1 769.3	18 099.0	4 293.5
绵 阳	6	47.4	1 219.0	14 905.0	5.0
南 充	5	41.1	1 699.6	15 673.0	0
宜 宾	5	30.1	1 678.8	17 492.0	120.0
贵 阳	4	107.3	1 291.3	24 025.5	0
遵 义	8	52.0	1 644.0	33 319.0	15.0
昆 明	6	21.2	1 312.0	20 956.0	312.0
曲 靖	2	8.4	410.0	7 030.0	0
玉 溪	11	29.6	1 365.4	7 325.7	113.7
拉 萨	1	18.0	151.0	5 187.8	30.0
西 安	3	325.8	1 570.0	20 030.4	0
铜 川	4	18.2	146.0	5 483.0	20.0
宝 鸡	8	41.4	948.7	16 495.0	86.0
咸 阳	8	17 303.2	868.7	18 791.2	129.8
渭 南	4	9 621.6	374.7	5 330.0	600.0
延 安	11	32.5	1 019.7	20 119.8	461.9
兰 州	11	164.3	1 935.9	14 076.5	0
金 昌	2	18.2	241.0	2 206.4	0
西 宁	7	77.3	1 123.7	15 632.2	0
银 川	3	42.6	780.9	10 537.0	0
石 嘴 山	3	10.1	392.9	5 476.4	70.0
乌鲁木齐	3	119.2	2 066.6	23 137.7	249.5
克拉玛依	3	15.5	515.9	6 524.9	80.0

重点城市生活垃圾处理情况（二）

（2011）

单位：吨

城 市 名 称	渗滤液中污染物排放量					
	化学需氧量	氨氮	油类	总磷	挥发酚	氰化物
总　　计	90 424.7	9 103.7	62.7	112.9	7.9	0.6
北　京	5 274.6	563.1	4.4	4.9	0.7	…
天　津	434.7	35.7	0.2	…	…	…
石 家 庄	291.0	28.6	0.3	0.3	0.1	…
唐　山	1 045.4	122.1	1.2	2.6	0.2	…
秦 皇 岛	2.2	1.1	0.1	0.1	0	…
邯　郸	3 966.3	179.0	2.8	5.2	0.6	…
保　定	1 077.5	106.7	1.1	1.5	0.2	…
太　原	169.1	18.0	0.1	0.2	…	…
大　同	142.9	12.8	0.2	0.3	…	…
阳　泉	240.0	24.0	0.4	0	…	…
长　治	334.6	17.3	0.4	0.5	…	0
临　汾	285.3	24.3	2.0	2.2	0.2	…
呼 和 浩 特	355.6	66.7	0.1	0.1	0	…
包　头	241.8	16.1	0	0	0	0
赤　峰	351.7	25.0	0.3	0.4	0.1	…
沈　阳	38.4	3.5	0.4	0.6	…	…
大　连	2.7	0.3	0	1.0	0	0
鞍　山	1 379.1	473.4	0.3	0.3	0.1	…
抚　顺	1 000.0	100.0	1.5	1.6	…	…
本　溪	152.3	13.7	0.1	0.1	0.1	…
锦　州	0.6	0.1	0	0	0	0
长　春	5 067.4	572.7	3.5	6.6	0.2	…
吉　林	944.1	94.4	1.2	1.7	0.2	…
哈 尔 滨	20.7	3.5	0.2	0.2	0.1	…
齐 齐 哈 尔	274.0	31.3	0.0	0.4	0	…
牡 丹 江	794.9	89.8	0.3	0.4	0.1	…
上　海	5 826.7	295.9	3.2	5.0	0.1	…
南　京	598.7	88.8	0.7	1.1	0.3	…
无　锡	60.1	62.4	0.2	0.1	…	…
徐　州	591.6	7.5	0	0	0	0
常　州	10.4	0.4	…	…	0	0
苏　州	58.3	39.4	2.9	0.3	…	…
南　通	259.0	31.0	0.1	0.5	…	…
连 云 港	1 415.0	119.8	0.9	1.9	…	…
扬　州	887.1	129.8	0.2	0.6	0.1	…
镇　江	155.2	41.4	0	0	0	0
杭　州	247.3	26.1	0.2	0.2	0	…

重点城市生活垃圾处理情况（二）（续表）

（2011）

单位：吨

城 市 名 称	渗滤液中污染物排放量					
	化学需氧量	氨氮	油类	总磷	挥发酚	氰化物
宁　波	1 065.6	98.0	…	…	0	…
温　州	2 173.3	170.0	1.4	3.4	0.4	…
湖　州	41.0	7.2	0.2	0.1	0	…
绍　兴	802.1	67.7	0.8	2.3	…	…
合　肥	968.2	99.4	0.5	0.6	…	…
芜　湖	195.0	16.0	0.1	0.2	…	…
马 鞍 山	487.7	41.2	0.2	0.8	0	…
福　州	754.9	58.5	0.8	1.4	0.1	…
厦　门	89.0	13.4	0	0	0	0
泉　州	203.5	17.1	0.6	0.4	0.1	…
南　昌	710.0	62.0	1.0	1.0	…	…
九　江	1 033.2	93.2	1.3	2.4	0.1	…
济　南	259.1	27.4	0.4	0.6	0.1	…
青　岛	283.4	30.3	0.2	0.3	…	…
淄　博	0.2	…	…	…	…	0
枣　庄	0.6	…	0	0	0	0
烟　台	43.6	4.9	0.3	0.3	0.1	…
潍　坊	739.2	79.8	0.3	0.5	…	0.2
济　宁	217.2	38.4	0.4	0.5	…	…
泰　安	77.3	4.4	0.1	0.1	0	…
日　照	250.9	24.7	0.4	0.4	0.1	…
郑　州	981.2	156.3	1.2	3.7	0.2	…
开　封	303.7	40.6	0.3	0.4	0.1	…
洛　阳	843.4	87.9	0.5	0.4	0.1	…
平 顶 山	169.2	19.0	0.1	0.1	…	…
安　阳	377.6	25.2	0.3	0.5	…	…
焦　作	197.7	23.5	0.1	0.2	…	…
三 门 峡	195.8	18.5	…	0.1	…	…
武　汉	3 956.0	308.7	0.1	…	…	…
宜　昌	1 087.7	234.0	0.9	2.3	0.2	…
荆　州	794.3	149.8	0.5	1.3	0.2	…
长　沙	491.0	83.1	0.8	1.0	…	…
株　洲	529.0	41.5	0.4	0.9	0.1	…
湘　潭	1 557.3	139.3	0.7	2.5	0.1	…
岳　阳	2 331.0	180.6	1.2	2.9	0.3	…
常　德	1 731.1	131.9	0.8	2.3	0.2	…
张 家 界	5.6	1.5	0	0	0	0
广　州	368.2	25.7	1.0	0.8	…	…

重点城市生活垃圾处理情况（二）（续表）

（2011）

单位：吨

城 市名 称	渗滤液中污染物排放量					
	化学需氧量	氨氮	油类	总磷	挥发酚	氰化物
韶　关	1 174.8	94.6	0.2	0.7	…	…
深　圳	1 591.9	335.1	0.1	0.9	…	…
珠　海	236.6	15.8	0	0.5	0	0
汕　头	973.5	96.9	…	0.1	0	…
湛　江	752.4	94.2	0.1	0.2	0	…
南　宁	541.3	46.6	0.7	1.1	…	…
柳　州	27.2	2.7	0.1	0.2	…	…
桂　林	702.2	58.9	0.4	0.6	…	…
北　海	9.4	0.9	0	…	0	0
海　口	0	0	0	0	0	0
重　庆	601.5	155.5	1.2	1.7	…	…
成　都	6 310.0	147.0	1.9	1.2	…	…
自　贡	357.5	44.7	0.4	0.6	…	…
攀枝花	142.6	16.3	…	0.1	0	…
泸　州	77.1	9.3	…	0.1	…	…
德　阳	514.4	2.0	…	2.2	…	0
绵　阳	205.9	16.9	0.3	0.4	0	…
南　充	943.6	122.9	0.4	1.2	…	…
宜　宾	625.2	55.7	…	…	…	…
贵　阳	2 591.7	298.0	0.1	0.1	…	…
遵　义	928.0	119.5	0.8	1.5	…	…
昆　明	1 935.8	213.0	4.3	11.0	1.1	…
曲　靖	119.1	19.2	…	…	…	…
玉　溪	977.6	105.2	0.2	0.5	…	…
拉　萨	3.2	0.2	0	…	…	0
西　安	3 433.0	600.0	0	0	0	0
铜　川	2.5	0.8	…	…	0	…
宝　鸡	247.9	39.7	0.2	0.4	…	…
咸　阳	40.8	3.0	…	0.1	…	…
渭　南	0	0	0	0	0	0
延　安	0	0	0	0	0	0
兰　州	430.3	25.0	1.0	0.8	…	…
金　昌	46.7	4.7	0.1	0.1	0	…
西　宁	771.4	70.8	0.5	0.9	0.1	…
银　川	164.6	15.4	0.2	0.2	0	…
石 嘴 山	10.9	0.9	0.1	…	0	0
乌鲁木齐	2 536.5	177.6	2.5	9.6	0.6	…
克拉玛依	86.8	8.7	0.1	0.2	…	…

重点城市生活垃圾处理情况（三）

（2011）

单位：千克

城 市 名 称	渗滤液中污染物排放量				
	汞	镉	总铬	铅	砷
总　　计	58.6	393.7	884.7	1 409.4	473.5
北　　京	1.2	9.5	27.2	103.8	12.5
天　　津	…	…	0.1	0.4	0.2
石 家 庄	0.3	0.8	1.3	3.3	0.8
唐　　山	0.6	3.5	7.1	16.9	4.4
秦 皇 岛	0.1	0.3	0.7	1.0	0.2
邯　　郸	1.8	14.0	27.5	69.3	17.6
保　　定	0.4	3.5	5.9	14.7	4.9
太　　原	0.1	0.3	0.7	1.9	0.6
大　　同	0.1	0.7	1.4	3.0	0.9
阳　　泉	0.2	1.2	2.4	4.5	0.9
长　　治	…	0	0	0	0
临　　汾	0.1	0.6	1.7	2.8	0.7
呼和浩特	0.1	0.4	0.8	1.8	0.5
包　　头	…	0	0	0	0
赤　　峰	0.2	1.2	2.5	5.7	1.6
沈　　阳	…	1.0	1.5	2.1	0.5
大　　连	0.1	…	25.2	0.5	0.6
鞍　　山	2.9	2.9	1.5	7.7	0.4
抚　　顺	1.0	5.0	10.0	19.0	4.0
本　　溪	…	0.2	0.6	1.3	0.3
锦　　州	…	0	0	0	0
长　　春	…	0.1	4.2	16.5	0.2
吉　　林	1.0	6.7	14.1	27.3	4.6
哈 尔 滨	0.3	1.1	2.3	6.0	2.3
齐齐哈尔	0.1	0.5	1.0	2.9	1.0
牡 丹 江	1.2	1.6	2.7	5.7	1.5
上　　海	3.3	15.6	43.7	71.6	27.6
南　　京	0.7	3.4	7.4	19.6	6.1
无　　锡	…	0.2	0.2	0.1	…
徐　　州	…	0	0	0	0
常　　州	0.1	0.4	0.8	2.7	1.0
苏　　州	0.1	0.7	5.5	15.1	1.5
南　　通	0.2	0.8	1.9	4.6	1.4
连 云 港	1.0	5.7	17.6	14.8	5.3
扬　　州	0.6	2.0	5.4	10.9	3.3
镇　　江	1.0	5.1	21.9	30.7	10.7
杭　　州	0.3	0.5	2.2	3.0	0.7

重点城市生活垃圾处理情况（三）（续表）

（2011）   单位：千克

城 市 名 称	渗滤液中污染物排放量				
	汞	镉	总铬	铅	砷
宁 波	0.1	…	0.3	0.3	0.3
温 州	1.2	6.1	16.6	24.4	7.6
湖 州	0.1	0.5	1.4	2.9	0.7
绍 兴	0.7	4.1	13.6	16.0	4.5
合 肥	0.6	5.3	16.7	26.5	12.2
芜 湖	0.1	0.6	1.9	1.9	0.5
马 鞍 山	0.4	1.5	20.6	21.6	1.5
福 州	0.6	2.3	13.2	11.2	4.3
厦 门	…	0	0	0	0
泉 州	0.3	78.5	6.6	12.4	3.0
南 昌	0.3	1.4	10.3	27.5	5.2
九 江	0.8	4.6	13.9	19.0	5.2
济 南	0.3	1.5	3.1	6.6	1.5
青 岛	0.1	0.5	0	2.9	1.0
淄 博	…	…	…	0.1	…
枣 庄	…	…	…	…	…
烟 台	0.3	1.2	2.4	3.5	2.1
潍 坊	15.3	76.9	153.0	310.1	77.3
济 宁	0.4	3.3	6.7	11.4	2.2
泰 安	…	0.1	0.1	…	…
日 照	0.2	1.3	2.5	4.3	1.4
郑 州	0.2	6.3	7.9	21.2	6.4
开 封	0.1	0.3	0.5	3.7	0.5
洛 阳	0.3	1.6	2.0	6.1	1.8
平 顶 山	0.1	0.2	0.3	0.8	0.3
安 阳	0.1	0.011	0.1	0.3	…
焦 作	0.7	0.9	0.7	2.7	0.5
三 门 峡	0.1	0.6	1.2	3.1	1.0
武 汉	…	…	2.5	0.6	0
宜 昌	1.7	3.3	8.3	18.6	5.6
荆 州	0.5	1.0	2.8	4.5	1.8
长 沙	0.2	0.8	8.0	1.9	1.5
株 洲	0.2	2.6	5.0	6.6	2.1
湘 潭	0.6	3.3	4.9	12.9	3.1
岳 阳	0.6	8.5	12.7	15.9	4.7
常 德	0.6	2.4	6.9	10.4	3.3
张 家 界	…	0.2	0.4	1.0	0.4
广 州	0.2	1.0	12.6	15.2	3.5

重点城市生活垃圾处理情况（三）（续表）

（2011）

单位：千克

城 市 名 称	渗滤液中污染物排放量				
	汞	镉	总铬	铅	砷
韶 关	0.2	1.4	7.5	3.9	1.9
深 圳	0.5	2.7	38.7	11.7	17.2
珠 海	…	0.5	0.4	0.6	0.4
汕 头	…	0.2	2.1	3.1	0.9
湛 江	0.1	0.1	2.7	1.3	0.7
南 宁	0.2	2.5	8.0	8.9	3.4
柳 州	0.1	0.1	2.4	1.2	0.6
桂 林	0.5	3.3	2.5	6.1	0.9
北 海	…	0.3	0	0.3	…
海 口	0	0	0	0	0
重 庆	1.2	5.9	13.3	37.8	11.3
成 都	0.2	1.2	9.2	7.8	1.7
自 贡	0.4	1.4	3.8	7.6	2.0
攀 枝 花	…	0.2	0.5	1.0	0.3
泸 州	0.8	6.0	73.9	89.2	73.8
德 阳	0.2	…	1.3	0.4	0.2
绵 阳	0.4	1.4	3.3	6.9	1.2
南 充	0.6	2.2	6.0	12.1	3.6
宜 宾	0.2	0.7	1.5	4.2	1.1
贵 阳	0.1	0.3	1.0	1.0	0.5
遵 义	0.8	4.0	10.5	16.1	3.1
昆 明	0.9	3.3	10.4	19.2	5.4
曲 靖	…	0.1	0.2	0.2	…
玉 溪	0.2	1.0	4.4	4.0	1.0
拉 萨	…	0.1	0.1	0.2	0.1
西 安	…	0	0	0	0
铜 川	…	…	0.1	…	…
宝 鸡	0.2	0.6	1.4	3.7	1.2
咸 阳	0.1	0.2	0.4	0.7	0.3
渭 南	…	0.4	0.1	0.3	0.3
延 安	…	0	0	0	0
兰 州	0.4	2.7	5.4	10.8	3.8
金 昌	…	0.2	0.5	1.2	0.3
西 宁	0.3	2.4	4.8	12.0	3.0
银 川	0.1	0.5	1.0	2.3	0.7
石 嘴 山	…	…	0.1	0.2	…
乌鲁木齐	2.9	35.1	58.0	12.4	45.9
克拉玛依	0.1	0.4	0.8	2.0	0.5

重点城市生活垃圾处理情况（四）

（2011）

单位：吨

城 市 名 称	焚烧废气中污染物排放量		
	二氧化硫	氮氧化物	烟尘
总　　计	1 139	1 765	1 366
北　京	45	301	26
天　津	0	0	0
石 家 庄	0	0	0
唐　山	0	0	0
秦 皇 岛	0	0	0
邯　郸	0	0	0
保　定	0	0	0
太　原	0	0	0
大　同	0	0	0
阳　泉	0	0	0
长　治	20	86	20
临　汾	0	0	0
呼 和 浩 特	0	0	0
包　头	0	0	0
赤　峰	0	0	0
沈　阳	0	0	0
大　连	0	0	0
鞍　山	0	0	0
抚　顺	0	0	0
本　溪	0	0	0
锦　州	38	8	33
长　春	0	0	0
吉　林	0	0	0
哈 尔 滨	0	0	0
齐 齐 哈 尔	0	0	0
牡 丹 江	0	0	0
上　海	…	…	…
南　京	0	0	0
无　锡	0	0	0
徐　州	0	0	0
常　州	0	0	0
苏　州	162	193	35
南　通	0	0	0
连 云 港	0	0	0
扬　州	0	0	0
镇　江	0	0	0
杭　州	31	74	114

重点城市生活垃圾处理情况（四）（续表）

（2011）   单位：吨

城　市 名　称	焚烧废气中污染物排放量		
	二氧化硫	氮氧化物	烟尘
宁　波	0	0	0
温　州	0	0	0
湖　州	0	0	0
绍　兴	53	35	25
合　肥	0	0	0
芜　湖	395	296	874
马　鞍　山	0	0	0
福　州	0	0	0
厦　门	0	0	0
泉　州	104	234	91
南　昌	0	0	0
九　江	0	0	0
济　南	27	61	14
青　岛	0	0	0
淄　博	0	0	0
枣　庄	0	0	0
烟　台	0	0	0
潍　坊	0	0	0
济　宁	0	0	0
泰　安	0	0	0
日　照	0	0	0
郑　州	0	0	0
开　封	0	0	0
洛　阳	0	0	0
平　顶　山	0	0	0
安　阳	0	0	0
焦　作	24	136	22
三　门　峡	0	0	0
武　汉	…	…	…
宜　昌	0	0	0
荆　州	0	0	0
长　沙	0	0	0
株　洲	0	0	0
湘　潭	0	0	0
岳　阳	0	0	0
常　德	0	0	0
张　家　界	0	0	0
广　州	1	5	…

重点城市生活垃圾处理情况（四）（续表）

城　市 名　称	焚烧废气中污染物排放量		
	二氧化硫	氮氧化物	烟尘
韶　关	0	0	0
深　圳	0	0	0
珠　海	0	0	0
汕　头	0	0	0
湛　江	0	0	0
南　宁	0	0	0
柳　州	0	0	0
桂　林	0	0	0
北　海	0	0	0
海　口	0	0	0
重　庆	0	0	0
成　都	155	166	62
自　贡	0	0	0
攀　枝　花	…	…	…
泸　州	0	0	0
德　阳	79	170	43
绵　阳	0	0	0
南　充	0	0	0
宜　宾	0	0	0
贵　阳	0	0	0
遵　义	0	0	0
昆　明	6	…	3
曲　靖	0	0	0
玉　溪	0	0	0
拉　萨	0	0	0
西　安	0	0	0
铜　川	0	0	0
宝　鸡	0	0	0
咸　阳	0	0	0
渭　南	0	0	0
延　安	0	0	0
兰　州	0	0	0
金　昌	0	0	0
西　宁	0	0	0
银　川	0	0	0
石　嘴　山	0	0	0
乌鲁木齐	0	0	0
克拉玛依	…	…	5

重点城市危险（医疗）废物集中处置情况（一）

（2011）

城　市 名　称	危险废物集中 处置厂数/个	医疗废物集中 处置厂数/个	本年运行费用/ 万元	危险（医疗）废 物集中处置厂累计 完成投资/ 万元	新增固定资产/ 万元
总　　计	445	108	384 464.6	1 351 992.7	116 409.3
北　京	2	1	25 574.2	72 726.0	1 391.0
天　津	13	3	19 298.5	27 480.0	443.1
石 家 庄	4	2	1 118.4	2 650.0	520.0
唐　山	2	2	667.5	3 128.0	0
秦 皇 岛	5	0	925.0	930.0	1 000.0
邯　郸	0	2	266.3	1 905.0	0.0
保　定	2	2	521.0	5 340.0	60.0
太　原	3	2	763.6	4 155.8	20.0
大　同	0	1	274.0	1 419.9	119.8
阳　泉	0	1	243.3	1 500.0	0
长　治	0	0	0	0	0
临　汾	3	2	294.6	2 156.2	2 006.8
呼 和 浩 特	0	1	160.7	381.2	0
包　头	1	2	278.0	21 957.0	0
赤　峰	0	1	330.0	1 228.0	0
沈　阳	3	1	3 040.3	28 376.0	900.0
大　连	18	1	19 302.7	30 747.5	13 643.3
鞍　山	0	1	231.0	1 395.0	0
抚　顺	3	0	1 421.0	3 358.0	500.0
本　溪	2	0	202.0	1 850.0	15.0
锦　州	5	0	0	59 308.0	0
长　春	1	1	2 927.0	4 840.0	148.0
吉　林	1	1	400.0	3 496.0	0
哈 尔 滨	5	1	2 070.2	37 972.0	6 300.0
齐 齐 哈 尔	0	1	184.0	2 069.5	0
牡 丹 江	0	1	71.0	300.0	0
上　海	47	2	57 042.1	160 987.7	23 325.2
南　京	4	1	5 354.0	24 217.0	1 246.9
无　锡	28	2	20 928.2	70 638.3	2 755.4
徐　州	2	0	1 274.0	1 200.0	0
常　州	32	0	8 834.0	26 834.4	189.0
苏　州	27	0	20 376.3	73 658.0	4 510.0
南　通	10	1	2 562.9	11 987.2	60.0
连 云 港	1	1	488.0	6 000.0	4 500.0
扬　州	11	1	1 591.5	4 246.7	907.3
镇　江	5	1	2 309.0	10 899.0	1 015.0
杭　州	12	1	11 829.5	44 944.1	3 947.9

重点城市危险（医疗）废物集中处置情况（一）（续表）

（2011）

城　市 名　称	危险废物集中 处置厂数/个	医疗废物集中 处置厂数/个	本年运行费用/ 万元	危险（医疗）废物 集中处置厂累计完 成投资/万元	新增固定资产/ 万元
宁　波	14	0	6 798.7	18 914.5	392.5
温　州	5	1	8 076.0	9 026.0	1 859.0
湖　州	4	1	2 175.0	8 018.0	0
绍　兴	8	1	4 383.0	16 620.0	1 590.0
合　肥	1	2	2 801.5	12 611.4	2 485.5
芜　湖	1	1	403.5	1 130.0	450.0
马 鞍 山	0	1	90.0	578.0	578.0
福　州	5	1	1 948.0	14 590.0	990.0
厦　门	0	1	1 100.0	2 679.5	0
泉　州	0	1	680.0	0	0
南　昌	2	1	1 045.0	3 770.0	0
九　江	3	0	0	16 664.2	0
济　南	1	1	2 020.3	5 089.6	1 566.2
青　岛	10	1	4 844.1	26 780.4	388.0
淄　博	6	1	3 709.7	10 867.6	847.8
枣　庄	1	1	3 200.0	2 800.0	1 200.0
烟　台	4	1	1 766.4	10 480.0	350.0
潍　坊	7	1	992.0	1 690.0	40.0
济　宁	1	1	542.0	1 700.0	52.0
泰　安	1	0	93.0	150.0	20.0
日　照	0	1	170.0	830.0	0
郑　州	0	1	2 753.4	3 410.8	255.6
开　封	0	1	480.0	1 486.0	81.0
洛　阳	0	1	770.0	650.0	0
平 顶 山	0	1	410.0	1 780.0	0
安　阳	0	1	238.0	1 478.0	246.0
焦　作	0	1	360.0	980.0	0
三 门 峡	0	1	96.0	880.0	0
武　汉	8	1	4 148.6	23 232.1	1 588.9
宜　昌	3	1	112.9	1 393.0	10.0
荆　州	1	1	474.2	1 336.2	0.0
长　沙	0	0	0	0	0
株　洲	0	1	495.0	1 751.0	210.0
湘　潭	0	0	0	0	0
岳　阳	0	1	400.0	2 190.0	0
常　德	1	1	487.0	1 980.0	150.0
张 家 界	0	1	90.2	1 191.0	0
广　州	19	2	9 180.9	39 697.0	7 335.0

重点城市危险（医疗）废物集中处置情况（一）（续表）

（2011）

城 市 名 称	危险废物集中 处置厂数/个	医疗废物集中 处置厂数/个	本年运行费用/ 万元	危险（医疗）废 物集中处置厂累 计完成投资/ 万元	新增固定资产/ 万元
韶 关	10	1	414.0	0	0
深 圳	7	1	64 356.0	69 964.2	2 626.7
珠 海	6	0	4 822.2	12 016.5	746.5
汕 头	1	1	862.9	1 702.1	4.2
湛 江	2	1	766.6	0	0
南 宁	3	0	8.0	325.4	0
柳 州	5	1	1 107.1	3 693.0	70.0
桂 林	0	1	285.0	685.0	0
北 海	0	1	25.0	1 324.3	0
海 口	0	1	307.0	2 720.0	365.2
重 庆	6	4	7 710.1	48 987.0	2 103.5
成 都	4	1	4 310.7	10 200.7	2 465.1
自 贡	0	1	0	0	0
攀 枝 花	0	1	200.0	300.0	20.0
泸 州	0	2	672.0	650.0	15.0
德 阳	1	1	12.0	2 000.0	0.0
绵 阳	6	1	4 520.0	13 310.0	60.0
南 充	2	0	0	0	0
宜 宾	0	1	430.0	857.4	12.0
贵 阳	1	1	469.0	14 835.1	0
遵 义	2	1	465.0	5 178.3	280.0
昆 明	0	1	3 618.0	12 100.0	0
曲 靖	0	0	0	0	0
玉 溪	0	0	0	0	0
拉 萨	0	0	0	0	0
西 安	1	1	2 306.0	5 640.5	2 120.0
铜 川	0	0	0	0	0
宝 鸡	0	1	130.0	3 113.0	0.0
咸 阳	0	0	0	0	0
渭 南	0	1	300.0	14.1	0
延 安	11	1	2 386.0	14 366.1	356.5
兰 州	5	1	1 152.0	12 265.3	11 265.3
金 昌	0	1	89.0	698.0	0
西 宁	3	1	928.1	5 041.6	544.0
银 川	1	1	827.7	11 542.0	66.2
石 嘴 山	2	0	0	27.0	0
乌鲁木齐	4	1	497.2	93 730.6	1 080.0
克拉玛依	4	1	0	0	0

重点城市危险（医疗）废物集中处置情况（二）

（2011）

城　市 名　称	危险废物设计 处置能力/ （吨/日）	危险废物实际 处置量/吨	工业危险 废物处置量	医疗废物处置量	其他危险 废物处置量	危险废物综 合利用量/吨
总　　　计	204 485	1 692 836	1 279 274	311 238	102 324	2 493 867
北　　京	622	71 096	60 547	10 549	0	33 852
天　　津	1 671	78 322	52 789	16 983	8 550	112 571
石 家 庄	72	8 775	8 557	217	0	1 143
唐　　山	14	2 379	83	2 296	0	826
秦 皇 岛	75	11 146	9 271	1 743	132	6 611
邯　　郸	11	1 826	0	1 826	0	0
保　　定	53	6 681	3 970	2 711	0	727
太　　原	109	6 991	83	6 906	2	682
大　　同	25	1 062	0	1 062	0	1 874
阳　　泉	5	372	0	372	0	0
长　　治	0	0	0	0	0	0
临　　汾	11	469	0	469	0	0
呼和浩特	0	2 113	0	2 113	0	0
包　　头	157	748	2	746	0	0
赤　　峰	5	488	0	488	0	0
沈　　阳	170	11 530	8 300	3 230	0	5 348
大　　连	1 122	10 643	6 631	4 012	0	103 525
鞍　　山	8	992	0	992	0	0
抚　　顺	81	0	0	0	0	11 883
本　　溪	15	1 645	965	680	0	0
锦　　州	434	0	0	0	0	93 927
长　　春	58	15 645	8 592	5 065	1 988	373
吉　　林	155	15 520	13 877	1 643	0	0
哈 尔 滨	194	4 070	0	3 835	235	356
齐齐哈尔	8	716	0	716	0	0
牡 丹 江	5	779	0	779	0	0
上　　海	2 245	169 472	140 577	22 975	5 920	145 595
南　　京	136	24 800	19 622	5 178	0	8
无　　锡	4 154	102 887	98 539	4 348	0	467 410
徐　　州	30	4 942	2 013	2 929	0	717
常　　州	1 515	54 175	19 917	1 539	32 720	208 499
苏　　州	2 107	111 643	80 459	5 227	25 957	225 063
南　　通	60 228	81 841	67 690	2 085	12 067	10 651
连 云 港	9 000	5 561	4 100	1 461	0	0
扬　　州	157	6 630	5 400	1 230	0	5 461
镇　　江	8 182	30 130	28 130	2 000	0	25 744
杭　　州	2 022	104 374	92 677	11 697	0	53 276

重点城市危险（医疗）废物集中处置情况（二）（续表）

（2011）

城市名称	危险废物设计处置能力/（吨/日）	危险废物实际处置量/吨	工业危险废物处置量	医疗废物处置量	其他危险废物处置量	危险废物综合利用量/吨
宁　波	587	96 363	83 791	8 464	4 108	3 576
温　州	241	10 476	8 768	1 708	0	28 591
湖　州	183	18 729	13 419	5 176	134	3 334
绍　兴	186	16 496	13 496	3 000	0	20 586
合　肥	80	11 051	6 927	4 124	0	0
芜　湖	32	5 199	4 140	1 059	0	2 370
马 鞍 山	2	456	0	456	0	0
福　州	2 392	96	96	0	0	3 824
厦　门	25	8 431	5 109	3 322	0	0
泉　州	8	2 344	0	2 344	0	0
南　昌	19	4 390	660	3 730	0	730
九　江	227	1 514	1 514	0	0	0
济　南	24	5 839	1 237	4 602	0	3 087
青　岛	775	23 428	16 570	4 578	2 280	11 674
淄　博	775	3 959	511	3 447	0	96 885
枣　庄	8	2 370	590	1 780	0	0
烟　台	230	13 917	9 779	4 138	0	9 291
潍　坊	5 072	8 077	5 843	2 114	120	8 378
济　宁	507	3 935	1 532	2 403	0	0
泰　安	0	0	0	0	0	428
日　照	5	799	0	799	0	0
郑　州	30	6 982	0	6 982	0	0
开　封	5	1 248	0	1 248	0	0
洛　阳	24	3 807	0	3 807	0	0
平 顶 山	5	1 366	0	1 366	0	0
安　阳	5	1 605	0	1 605	0	0
焦　作	5	1 297	0	1 297	0	0
三 门 峡	5	548	0	548	0	0
武　汉	235	38 430	27 863	10 567	0	11 489
宜　昌	5 032	2 240	2 040	120	80	2 080
荆　州	3 008	923	0	765	158	0
长　沙	0	0	0	0	0	0
株　洲	8	2 300	600	1 700	0	0
湘　潭	0	0	0	0	0	0
岳　阳	10	1 600	0	1 600	0	0
常　德	41	2 412	47	2 365	0	290
张 家 界	3	451	0	451	0	0
广　州	11 423	20 376	8 986	11 390	0	112 825

重点城市危险（医疗）废物集中处置情况（二）（续表）

（2011）

城 市名 称	危险废物设计处置能力/（吨/日）	危险废物实际处置量/吨	工业危险废物处置量	医疗废物处置量	其他危险废物处置量	危险废物综合利用量/吨
韶　关	7	2 674	0	2 674	0	0
深　圳	2 251	14 383	6 358	7 934	91	141 923
珠　海	16 065	18 767	18 183	0	584	32 253
汕　头	15	2 087	0	2 087	0	2 581
湛　江	60	4 148	498	3 650	0	592
南　宁	167	0	0	0	0	1 820
柳　州	73	7 777	5 700	2 077	0	1 077
桂　林	10	2 374	0	2 374	0	0
北　海	5	190	0	190	0	0
海　口	14	2 628	0	2 628	0	0
重　庆	1 374	27 761	18 661	7 143	1 957	255 893
成　都	182	16 154	3 212	11 113	1 829	5 553
自　贡	0	0	0	0	0	0
攀 枝 花	6	650	0	650	0	0
泸　州	82	2 259	0	2 259	0	0
德　阳	21	54	0	54	0	0
绵　阳	686	1 114	24	1 018	72	165 171
南　充	0	0	0	0	0	0
宜　宾	5	1 600	0	1 600	0	0
贵　阳	41 200	1 822	31	1 766	25	0
遵　义	111	2 130	400	1 730	0	3 365
昆　明	30	8 688	0	8 688	0	0
曲　靖	0	0	0	0	0	0
玉　溪	0	0	0	0	0	0
拉　萨	0	0	0	0	0	0
西　安	20	10 028	0	10 028	0	425
铜　川	0	0	0	0	0	0
宝　鸡	5	444	0	444	0	0
咸　阳	0	0	0	0	0	0
渭　南	10	8	0	8	0	0
延　安	4 024	140 052	136 401	351	3 300	9 265
兰　州	529	131 587	131 039	543	5	0
金　昌	5	540	0	540	0	0
西　宁	169	2 404	0	2 404	0	20 811
银　川	56	1 064	214	840	10	73
石 嘴 山	3 080	0	0	0	0	1 238
乌鲁木齐	4 742	6 019	2 898	3 121	0	6 920
克拉玛依	3 408	9 520	9 350	170	0	9 350

重点城市危险（医疗）废物集中处置情况（三）

（2011）

城 市 名 称	渗滤液中污染物排放量					
	化学需氧量/吨	氨氮/吨	油类/吨	总磷/吨	挥发酚/千克	氰化物/千克
总　　计	1 129.8	36.8	11.1	0.1	6.5	39.2
北　京	2.7	2.5	0.1	…	2.0	4.0
天　津	6.8	0.1	0	0	0	0
石 家 庄	…	…	0	0	0	0
唐　山	0	0	0	0	0	0
秦 皇 岛	0	0	0	0	0	0
邯　郸	0	0	0	0	0	0
保　定	0	0	0	0	0	0
太　原	0	0	0	0	0	0
大　同	0	0	0	0	0	0
阳　泉	0	0	0	0	0	0
长　治	0	0	0	0	0	0
临　汾	0	0	0	0	0	0
呼和浩特	0	0	0	0	0	0
包　头	0	0	0	0	0	0
赤　峰	0	0	0	0	0	0
沈　阳	50.0	9.8	10.0	…	…	…
大　连	646.1	0.1	0	0	0	0
鞍　山	0.2	…	0	0	0	0
抚　顺	0	0	0	0	0	0
本　溪	0	0	0	0	0	0
锦　州	0	0	0	0	0	0
长　春	0	0	0	0	0	0
吉　林	…	…	…	…	…	…
哈 尔 滨	0.3	…	0	0	0	0
齐齐哈尔	0	0	0	0	0	0
牡 丹 江	0	0	0	0	0	0
上　海	0	0	0	0	0	0
南　京	…	0	0	0	0	0
无　锡	5.5	5.9	…	0	…	…
徐　州	0	0	0	0	0	0
常　州	0.1	0.1	0	0	0	0
苏　州	6.2	0.4	0	0	0	…
南　通	1.7	0.1	0.1	0	0	…
连 云 港	0	0	0	0	0	0
扬　州	…	…	…	0	…	…
镇　江	0	0	0	0	0	0
杭　州	117.2	1.2	0	0	0	0

重点城市危险（医疗）废物集中处置情况（三）（续表）

（2011）

城 市 名 称	渗滤液中污染物排放量					
	化学需氧量/吨	氨氮/吨	油类/吨	总磷/吨	挥发酚/千克	氰化物/千克
宁 波	5.5	0.6	…	…	0	0
温 州	1.4	0.3	…	…	0	0
湖 州	0	0	0	0	0	0
绍 兴	5.0	0.9	…	0	0	0
合 肥	0.6	0.1	0	0	0	0
芜 湖	0	0	0	0	0	0
马 鞍 山	0	0	0	0	0	0
福 州	7.7	…	0	0	0	4.7
厦 门	0	0	0	0	0	0
泉 州	0	0	0	0	0	0
南 昌	0	0	0	0	0	0
九 江	0.1	…	…	…	0.3	0.3
济 南	0	0	0	0	0	0
青 岛	0.2	0.1	0	0	0	0
淄 博	0	0	0	0	0	0
枣 庄	0	0	0	0	0	0
烟 台	1.1	0.3	0.4	0	0	0
潍 坊	0	0	0	0	0	0
济 宁	0	0	0	0	0	0
泰 安	0	0	0	0	0	0
日 照	0	0	0	0	0	0
郑 州	0	0	0	0	0	0
开 封	0	0	0	0	0	0
洛 阳	0	0	0	0	0	0
平 顶 山	0	0	0	0	0	0
安 阳	0	0	0	0	0	0
焦 作	0	0	0	0	0	0
三 门 峡	0	0	0	0	0	0
武 汉	19.0	1.9	0	0	0	0
宜 昌	0	0	0	0	0	0
荆 州	0	0	0	0	0	0
长 沙	0	0	0	0	0	0
株 洲	0	0	0	0	0	0
湘 潭	0	0	0	0	0	0
岳 阳	…	…	0	0	0	0
常 德	0.9	0.1	…	0	0	0
张 家 界	0	0	0	0	0	0
广 州	8.0	0.7	…	…	0	0

重点城市危险（医疗）废物集中处置情况（三）（续表）
（2011）

城 市 名 称	渗滤液中污染物排放量					
	化学需氧量/吨	氨氮/吨	油类/吨	总磷/吨	挥发酚/千克	氰化物/千克
韶　关	0	0	0	0	0	0
深　圳	227.7	4.0	…	…	4.1	0
珠　海	0.4	…	0	0	0	0
汕　头	0.2	0	0	0	0	0
湛　江	0	0	0	0	0	0
南　宁	0	0	0	0	0	0
柳　州	0	0	0	0	0	0
桂　林	0	0	0	0	0	0
北　海	0	0	0	0	0	0
海　口	0	0	0	0	0	0
重　庆	15.2	7.5	0.5	…	…	30.0
成　都	…	0	…	0	0	0
自　贡	0	0	0	0	0	0
攀枝花	0	0	0	0	0	0
泸　州	0	0	0	0	0	0
德　阳	0	0	0	0	0	0
绵　阳	0	0	0	0	0	0
南　充	…	…	0	0	0	0
宜　宾	0	0	0	0	0	0
贵　阳	0	0	0	0	0	0
遵　义	0	0	0	0	0	0
昆　明	0	0	0	0	0	0
曲　靖	0	0	0	0	0	0
玉　溪	0	0	0	0	0	0
拉　萨	0	0	0	0	0	0
西　安	0	0	0	0	0	0
铜　川	0	0	0	0	0	0
宝　鸡	0	0	0	0	0	0
咸　阳	0	0	0	0	0	0
渭　南	0	0	0	0	0	0
延　安	0	0	0	0	0	0
兰　州	0	0	0	0	0	0
金　昌	0	0	0	0	0	0
西　宁	0	0	0	0	0	0
银　川	…	…	…	0	…	0
石嘴山	0	0	0	0	0	0
乌鲁木齐	0	0	0	0	0	0
克拉玛依	0	0	0	0	0	0

重点城市危险（医疗）废物集中处置情况（四）

（2011）

单位：千克

城市名称	渗滤液中污染物排放量				
	汞	镉	总铬	铅	砷
总　计	7.9	9.8	210.5	107.2	39.1
北　京	0.5	2.0	36.0	14.0	3.0
天　津	0	0	0	4.9	…
石 家 庄	0	0	0	0	0
唐　山	0	0	0	0	0
秦 皇 岛	0	0	0	0	0
邯　郸	0	0	0	0	0
保　定	0	0	0	0	0
太　原	0	0	0	0	0
大　同	0	0	0	0	0
阳　泉	0	0	0	0	0
长　治	0	0	0	0	0
临　汾	0	0	0	0	0
呼 和 浩 特	0	0	0	0	0
包　头	0	0	0	0	0
赤　峰	0	0	0	0	0
沈　阳	…	0.8	0.7	0	…
大　连	…	0.1	1.6	0.6	0.1
鞍　山	0	0	0	0	0
抚　顺	0	0	0	0	0
本　溪	0	0	0	0	0
锦　州	0	0	0	0	0
长　春	0	0	0	0	0
吉　林	…	…	0.4	0.2	…
哈 尔 滨	0	0	0	0	0
齐 齐 哈 尔	0	0	0	0	0
牡 丹 江	0	0	0	0	0
上　海	0	0	0	0	0
南　京	0	0	0	0	0
无　锡	…	…	…	0	…
徐　州	0	0	0	0	0
常　州	0	0	0	0	0
苏　州	…	…	27.4	0.3	0.3
南　通	0	0	0	…	…
连 云 港	0	0	0	0	0
扬　州	0	…	…	0	…
镇　江	0	0	0	0	0
杭　州	0	0	0	2.0	0

重点城市危险（医疗）废物集中处置情况（四）（续表）

（2011）

单位：千克

城　市 名　称	渗滤液中污染物排放量				
	汞	镉	总铬	铅	砷
宁　波	…	0	0.4	1.5	0.3
温　州	0	0	0.5	0	1.2
湖　州	0	0	0	0	0
绍　兴	0	0.1	0	3.1	1.9
合　肥	0	0	0	0	0
芜　湖	0	0	0	0	0
马 鞍 山	0	0	0	0	0
福　州	4.9	0.3	2.3	2.4	1.9
厦　门	0	0	0	0	0
泉　州	0	0	0	0	0
南　昌	0	0	0	0	0
九　江	…	…	1.0	0.7	0.3
济　南	0	0	0	0	0
青　岛	0	0	0	0	0
淄　博	0	0	0	0	0
枣　庄	0	0	0	0	0
烟　台	0	0	0	0	0
潍　坊	0	0	0	0	0
济　宁	0	0	0	0	0
泰　安	0	0	0	0	0
日　照	0	0	0	0	0
郑　州	0	0	0	0	0
开　封	0	0	0	0	0
洛　阳	0	0	0	0	0
平 顶 山	0	0	0	0	0
安　阳	0	0	0	0	0
焦　作	0	0	0	0	0
三 门 峡	0	0	0	0	0
武　汉	0	0	0	0	0
宜　昌	0	0	0	0	0
荆　州	0	0	0	0	0
长　沙	0	0	0	0	0
株　洲	0	0	0	0	0
湘　潭	0	0	0	0	0
岳　阳	0	0	0	0	0
常　德	0	0	0	0	0
张 家 界	0	0	0	0	0
广　州	0	0	1.3	0	0

重点城市危险（医疗）废物集中处置情况（四）（续表）

（2011）　　　　　　　　　　　　　　　　　　　　单位：千克

城　市 名　称	渗滤液中污染物排放量				
	汞	镉	总铬	铅	砷
韶　　关	0	0	0	0	0
深　　圳	0	0	0.1	0	0
珠　　海	…	0.4	3.4	2.4	…
汕　　头	0	…	0.1	0.1	…
湛　　江	0	0	0	0	0
南　　宁	0	0	0	0	0
柳　　州	0	0	0	0	0
桂　　林	0	0	0	0	0
北　　海	0	0	0	0	0
海　　口	0	0	0	0	0
重　　庆	2.4	6.0	135.2	75.0	30.0
成　　都	0	0	0	0	0
自　　贡	0	0	0	0	0
攀 枝 花	0	0	0	0	0
泸　　州	0	0	0	0	0
德　　阳	0	0	0	0	0
绵　　阳	0	0	0	0	0
南　　充	0	0	0	0	0
宜　　宾	0	0	0	0	0
贵　　阳	0	0	0	0	0
遵　　义	0	0	0	0	0
昆　　明	0	0	0	0	0
曲　　靖	0	0	0	0	0
玉　　溪	0	0	0	0	0
拉　　萨	0	0	0	0	0
西　　安	0	0	0	0	0
铜　　川	0	0	0	0	0
宝　　鸡	0	0	0	0	0
咸　　阳	0	0	0	0	0
渭　　南	0	0	0	0	0
延　　安	0	0	0	0	0
兰　　州	0	0	0	0	0
金　　昌	0	0	0	0	0
西　　宁	0	0	0	0	0
银　　川	0	0	0	0	0
石 嘴 山	0	0	0	0	0
乌鲁木齐	0	0	0	0	0
克拉玛依	0	0	0	0	0

重点城市危险（医疗）废物集中处置情况（五）

（2011）

单位：吨

城 市名 称	焚烧废气中污染物排放量		
	二氧化硫	氮氧化物	烟尘
总　　计	779.2	1 019.6	804.1
北　京	0.5	16.8	1.8
天　津	44.5	45.2	24.7
石 家 庄	9.0	5.6	6.7
唐　山	0.2	2.1	0.2
秦 皇 岛	0.3	1.0	0.1
邯　郸	0.7	0.1	…
保　定	4.6	5.4	1.2
太　原	1.3	0.8	347.9
大　同	0.1	1.3	0.1
阳　泉	0	0	0
长　治	0	0	0
临　汾	0.7	1.7	0.8
呼和浩特	0.6	2.1	2.5
包　头	0	0	0
赤　峰	0.1	0.9	0.1
沈　阳	0	0	0
大　连	5.4	4.1	2.1
鞍　山	0	0	0
抚　顺	0	0	0
本　溪	1.0	1.5	0.2
锦　州	0	0	0
长　春	0	1.4	1.9
吉　林	0.5	1.5	17.8
哈 尔 滨	4.0	8.9	0.4
齐齐哈尔	0.5	0.8	0.7
牡 丹 江	0.2	0.7	1.7
上　海	149.9	331.6	39.1
南　京	4.9	3.7	13.0
无　锡	20.5	31.6	10.1
徐　州	0.5	4.8	0.8
常　州	0	0	19.2
苏　州	23.6	144.6	29.3
南　通	9.7	45.1	46.0
连 云 港	3.6	1.3	1.1
扬　州	0.4	…	0.3
镇　江	15.5	…	6.9
杭　州	28.8	22.4	2.9

重点城市危险（医疗）废物集中处置情况（五）（续表）

（2011）   单位：吨

城 市 名 称	焚烧废气中污染物排放量		
	二氧化硫	氮氧化物	烟尘
宁 波	14.4	13.8	48.4
温 州	24.3	13.3	9.4
湖 州	278.0	21.3	2.8
绍 兴	25.2	12.7	12.8
合 肥	1.1	7.8	2.9
芜 湖	0	0	0
马 鞍 山	1.0	0	1.9
福 州	0	0	0
厦 门	2.6	9.7	0.8
泉 州	0.5	1.9	0.2
南 昌	1.2	3.8	0.3
九 江	5.4	10.9	0.1
济 南	0.5	8.3	0.5
青 岛	1.1	1.3	1.0
淄 博	0.7	2.4	5.0
枣 庄	0.7	2.8	0
烟 台	4.4	12.7	2.7
潍 坊	0	0	0
济 宁	0.9	2.6	1.4
泰 安	0	0	0
日 照	1.0	1.0	0.1
郑 州	1.4	1.2	0.4
开 封	0.2	2.4	0.4
洛 阳	4.6	38.0	3.3
平 顶 山	0.1	2.3	0.7
安 阳	0.5	1.6	0.2
焦 作	0	0	0
三 门 峡	0	0	0
武 汉	0	0	0
宜 昌	3.0	0	0
荆 州	0	0	0
长 沙	0	0	0
株 洲	…	1.2	0.1
湘 潭	0	0	0
岳 阳	0.5	1.5	0.8
常 德	23.0	52.0	26.0
张 家 界	0.1	0.2	0
广 州	13.8	20.1	6.8

重点城市危险（医疗）废物集中处置情况（五）（续表）

（2011）   单位：吨

城 市名 称	焚烧废气中污染物排放量		
	二氧化硫	氮氧化物	烟尘
韶 关	9.1	0	0.2
深 圳	2.5	5.9	2.0
珠 海	1.3	1.4	0.2
汕 头	1.1	3.7	0.8
湛 江	3.0	5.3	0.7
南 宁	1.1	0.1	0.1
柳 州	1.5	7.7	3.6
桂 林	0.7	2.0	16.0
北 海	0	0	0
海 口	0.8	3.4	1.4
重 庆	3.5	14.4	9.7
成 都	4.0	13.1	0.3
自 贡	0	0	0
攀枝花	0.2	0.6	0.1
泸 州	0.9	1.4	1.8
德 阳	0	0.1	0
绵 阳	0.3	0.4	0.4
南 充	1.5	0.1	25.0
宜 宾	0	0	0
贵 阳	0	0	0
遵 义	0.4	1.4	0.5
昆 明	0.6	9.1	0.9
曲 靖	0	0	0
玉 溪	0	0	0
拉 萨	0	0	0
西 安	3.1	9.0	1.1
铜 川	0	0	0
宝 鸡	0.1	0.4	0
咸 阳	0	0	0
渭 南	0	0	0
延 安	0	0	0
兰 州	0	0	0
金 昌	0	0	0
西 宁	0.8	2.3	0.3
银 川	0.2	0.8	1.7
石嘴山	0	0	0
乌鲁木齐	0.7	3.3	27.5
克拉玛依	…	0.1	1.5

各工业行业环境统计

GEGONGYE HANGYE HUANJING TONGJI

ANNUAL STATISTIC REPORT ON ENVIRONMENT IN CHINA

2011

按行业分重点调查工业废水排放及处理情况（一）

（2011）

行业名称	汇总工业企业数/个	工业废水排放量/万吨	直接排入环境的	排入污水处理厂的	工业废水处理量/万吨	废水治理设施数/套	废水治理设施处理能力/（万吨/日）	本年运行费用/万元
行业总计	153 027	2 129 036.0	1 689 831.5	439 204.5	5 805 510.7	915 06	31 405.51	7 321 459.5
煤炭开采和洗选业	6 587	143 493.4	137 857.0	5 636.4	185 121.2	46 03	1 212.31	288 373.6
石油和天然气开采业	247	8 171.9	7 078.6	1 093.3	87 902.8	759	369.08	229 310.5
黑色金属矿采选业	4 662	22 642.8	22 502.9	139.9	287 995.0	22 54	1 361.89	136 894.0
有色金属矿采选业	3 609	51 181.0	50 942.7	238.3	162 009.4	46 94	707.01	411 492.7
非金属矿采选业	1 082	6 190.7	6 079.9	110.8	10 894.3	446	70.43	13 225.9
开采辅助活动	61	1 053.8	1 024.0	29.8	1 040.6	25	3.44	4 117.3
其他采矿业	33	246.8	246.8	0	316.4	9	1.24	297.0
农副食品加工业	11 979	138 116.4	118 962.0	19 154.5	127 643.5	54 54	928.33	175 065.1
食品制造业	3 813	51 950.4	33 536.1	18 414.3	45 946.0	25 10	301.88	124 814.3
酒、饮料和精制茶制造业	4 390	71 663.7	52 869.7	18 794.0	62 391.5	20 15	370.44	193 868.7
烟草制品业	154	2 090.3	1 010.9	1 079.4	2 657.5	125	18.66	5 060.7
纺织业	8 929	240 801.5	116 466.4	124 335.1	205 858.7	92 39	1 077.67	547 962.5
纺织服装、服饰业	1 537	19 877.6	12 585.6	7 292.0	16 136.0	11 33	108.09	220 181.3
皮革、毛皮、羽毛及其制品和制鞋业	2 342	25 785.5	17 630.3	8 155.2	19 759.9	13 32	125.54	57 293.4
木材加工和木、竹、藤、棕、草制品业	2 707	3 522.2	3 152.6	369.6	2 303.8	577	19.81	6 231.4
家具制造业	449	735.5	403.4	332.1	502.4	182	2.34	1 723.2
造纸和纸制品业	5 871	382 264.6	341 237.0	41 027.6	550 236.6	51 22	2 709.07	610 290.0
印刷和记录媒介复制业	727	1 303.2	620.1	683.1	978.4	220	5.53	4 569.0
文教、工美、体育和娱乐用品制造业	656	1 937.3	1 242.5	694.9	1 723.0	415	8.72	5 019.9
石油加工、炼焦和核燃料加工业	1 383	79 586.8	67 925.9	11 660.9	199 199.6	15 38	438.97	508 389.5
化学原料和化学制品制造业	12 219	288 331.4	233 687.9	54 643.5	524 258.1	101 92	2 554.58	997 768.8
医药制造业	3 197	48 586.0	28 558.4	20 027.5	43 057.4	25 52	194.87	146 961.0
化学纤维制造业	490	41 428.1	34 146.1	7 282.0	38 184.3	409	178.90	90 074.4
橡胶和塑料制品业	2 521	12 155.5	6 762.8	5 392.6	10 371.9	909	83.36	23 317.4
非金属矿物制品业	35 208	26 075.4	22 568.3	3 507.1	61 778.1	51 62	548.28	131 412.2
黑色金属冶炼和压延加工业	4 913	121 036.8	108 881.0	12 155.7	2 461 163.1	45 32	9 587.78	1 208 044.6
有色金属冶炼和压延加工业	4 505	33 545.0	30 432.0	3 113.0	164 554.6	32 37	523.85	138 992.9
金属制品业	8 793	29 912.0	20 278.2	9 633.7	50 075.3	57 04	300.63	229 905.6
通用设备制造业	3 201	11 973.4	7 748.8	4 224.6	8 147.9	12 49	40.92	185 318.2
专用设备制造业	1 374	6 454.4	2 710.3	3 744.2	4 148.7	620	25.53	10 383.6
汽车制造业	1 773	15 069.2	6 802.9	8 266.3	13 912.4	12 60	298.73	59 425.6
铁路、船舶、航空航天和其他运输设备制造业	1 023	13 326.5	9 228.7	4 097.8	9 091.9	894	42.15	18 581.1
电气机械和器材制造业	1 983	9 631.4	4 375.4	5 256.0	8 193.2	13 57	47.06	35 425.0
计算机、通信和其他电子设备制造业	2 546	44 961.3	20 615.6	24 345.7	46 152.2	62 63	237.65	205 835.3
仪器仪表制造业	391	2 242.1	1 273.8	968.3	2 491.4	323	20.04	7 495.2
其他制造业	1 130	3 997.2	2 680.7	1 316.5	3 654.6	444	15.80	10 064.8
废弃资源综合利用业	454	2 069.2	1 697.5	371.7	1 456.6	188	11.25	8 380.5
金属制品、机械和设备修理业	188	1 309.5	803.6	505.9	1 295.0	130	6.82	4 479.1
电力、热力生产和供应业	5 668	158 927.7	148 507.9	10 419.8	367 984.0	32 63	6 772.58	248 143.1
燃气生产和供应业	60	989.1	907.3	81.8	1 247.6	30	5.16	4 468.7
水的生产和供应业	58	3 558.9	2 960.5	598.4	12 928.6	43	61.50	12 177.2
其他行业	114	840.5	829.3	11.2	747.3	93	7.61	625.1

按行业分重点调查工业废水排放及处理情况（二）

（2011）　　　　　　　　　　　　　　　　　　　　　　　　　　　　　　单位：吨

行业名称	工业废水中污染物产生量				
	化学需氧量	氨氮	石油类	挥发酚	氰化物
行业总计	34 116 095.8	1 714 649.6	375 971.8	70 560.5	6 220.9
煤炭开采和洗选业	365 725.4	6 827.1	8 369.6	273.0	164.2
石油和天然气开采业	256 607.1	5 402.4	45 940.3	67.2	0
黑色金属矿采选业	168 579.0	4 934.4	400.2	7.0	10.0
有色金属矿采选业	168 850.8	3 340.3	35.4	0.1	35.9
非金属矿采选业	67 417.7	1 052.8	306.4	0.8	0
开采辅助活动	9 658.6	114.8	2 982.6	12.2	0
其他采矿业	619.4	4.8	2.9	0	0
农副食品加工业	2 577 875.9	65 860.0	839.7	4.4	11.7
食品制造业	1 215 119.7	110 224.5	190.7	0.1	3.1
酒、饮料和精制茶制造业	3 508 402.3	45 041.7	119.5	5.3	2.4
烟草制品业	8 379.5	383.6	13.9	…	0.1
纺织业	2 504 513.1	69 732.0	916.3	14.7	0.8
纺织服装、服饰业	104 803.2	3 328.1	19.4	2.3	0
皮革、毛皮、羽毛及其制品和制鞋业	434 286.6	24 688.4	3 092.5	0.9	102.2
木材加工和木、竹、藤、棕、草制品业	67 140.0	493.4	6.6	1.1	0
家具制造业	3 414.0	135.5	45.2	…	4.5
造纸和纸制品业	6 666 336.8	60 705.1	262.8	266.1	0.2
印刷和记录媒介复制业	6 439.5	363.4	35.9	…	0.1
文教、工美、体育和娱乐用品制造业	7 308.6	450.2	86.4	0.7	32.2
石油加工、炼焦和核燃料加工业	1 255 373.0	217 298.2	145 862.2	46 097.0	1 665.6
化学原料和化学制品制造业	10 803 628.9	837 835.6	19 515.1	4 709.5	371.7
医药制造业	801 752.2	29 686.8	2 402.5	35.4	1.7
化学纤维制造业	585 084.4	9 163.7	292.6	0.6	0.6
橡胶和塑料制品业	48 468.7	2 421.4	215.4	0.1	1.4
非金属矿物制品业	146 026.1	5 475.1	2 243.9	36.8	0.7
黑色金属冶炼和压延加工业	1 216 563.2	67 594.8	113 198.5	17 305.3	1 742.8
有色金属冶炼和压延加工业	174 388.2	99 256.9	2 832.4	1 169.4	38.4
金属制品业	116 359.0	6 018.6	6 613.0	7.7	1 656.3
通用设备制造业	37 728.8	1 520.1	3 193.8	34.7	30.2
专用设备制造业	18 449.7	1 186.6	868.3	…	3.3
汽车制造业	97 395.4	2 695.4	7 436.8	3.8	26.4
铁路、船舶、航空航天和其他运输设备制造业	42 335.8	1 844.3	3 175.6	0.7	9.7
电气机械和器材制造业	32 443.3	1 548.9	319.4	0.4	10.3
计算机、通信和其他电子设备制造业	201 467.7	10 958.8	896.6	2.5	146.6
仪器仪表制造业	7 535.9	444.3	47.4	0.1	6.0
其他制造业	32 749.9	697.4	253.2	0.2	113.6
废弃资源综合利用业	10 355.7	2 902.4	59.1	0.1	2.0
金属制品、机械和设备修理业	4 720.3	200.8	939.9	0.2	0.7
电力、热力生产和供应业	317 405.7	11 438.8	1 496.2	36.2	0.2
燃气生产和供应业	17 055.0	1 024.8	272.8	463.6	18.6
水的生产和供应业	7 331.7	353.5	170.9	0.3	6.6
其他行业	30 925.6	670.2	14.4	0	0

按行业分重点调查工业废水排放及处理情况（三）

（2011）

单位：吨

行业名称	工业废水中污染物产生量				
	汞	镉	总铬	铅	砷
行业总计	23.916	2 580.017	8 456.407	3 793.586	9 977.714
煤炭开采和洗选业	⋯	0	0.216	0.150	3.118
石油和天然气开采业	0	0	0.012	0	0
黑色金属矿采选业	0.008	1.008	9.202	8.671	28.388
有色金属矿采选业	2.238	21.123	5.845	251.569	170.332
非金属矿采选业	0.136	0.360	0.141	2.191	2.683
开采辅助活动	0	0.049	0	0.385	0
其他采矿业	0	0	0	0.010	0.002
农副食品加工业	0	0.012	0	0.372	0.001
食品制造业	0	0	0	⋯	0
酒、饮料和精制茶制造业	0	0	0	0	0
烟草制品业	0	0	0	0	0
纺织业	0	0	6.734	⋯	0.007
纺织服装、服饰业	0	0	0.262	⋯	0.020
皮革、毛皮、羽毛及其制品和制鞋业	⋯	0.792	1 133.116	0.066	⋯
木材加工和木、竹、藤、棕、草制品业	0	0	0.005	0	0.001
家具制造业	0	0	3.491	0	⋯
造纸和纸制品业	0	0	0.075	0.099	0.675
印刷和记录媒介复制业	⋯	0	4.178	0.014	0.027
文教、工美、体育和娱乐用品制造业	0	0.003	74.847	0.012	⋯
石油加工、炼焦和核燃料加工业	0.007	0.182	0.769	0.425	23.557
化学原料和化学制品制造业	13.256	119.997	226.556	251.346	1 844.126
医药制造业	⋯	0	2.416	0	16.827
化学纤维制造业	⋯	0	0.003	0	⋯
橡胶和塑料制品业	⋯	0	25.830	0.005	0.003
非金属矿物制品业	0.299	0.214	0.412	1.069	0.565
黑色金属冶炼和压延加工业	0.088	16.609	87.209	30.254	39.995
有色金属冶炼和压延加工业	7.010	2 412.661	156.580	3 129.518	7 829.067
金属制品业	0.008	1.516	5 956.072	26.963	2.685
通用设备制造业	⋯	0.027	80.072	0.942	0.010
专用设备制造业	0.403	0.358	57.176	0.305	0.040
汽车制造业	⋯	0.079	217.775	0.623	⋯
铁路、船舶、航空航天和其他运输设备制造业	0	0.606	71.023	0.013	0
电气机械和器材制造业	0.093	3.868	27.338	76.185	0.927
计算机、通信和其他电子设备制造业	0.261	0.354	225.279	9.076	0.923
仪器仪表制造业	0.001	0.031	38.365	0.583	0.341
其他制造业	0	0.006	27.933	1.191	0
废弃资源综合利用业	0.001	0.101	0.358	1.391	0.078
金属制品、机械和设备修理业	0	0.028	14.824	0.002	0
电力、热力生产和供应业	0.109	0.031	0.138	0.155	13.316
燃气生产和供应业	0	0	0.350	0	0
水的生产和供应业	0	0	1.804	0	0
其他行业	0	0	0	0	0

按行业分重点调查工业废水排放及处理情况（四）

（2011）

单位：吨

行业名称	工业废水中污染物排放量				
	化学需氧量	氨氮	石油类	挥发酚	氰化物
行业总计	3 219 679.3	262 370.2	20 589.1	2 410.5	215.4
煤炭开采和洗选业	111 716.0	3 175.9	3 523.9	0.2	1.9
石油和天然气开采业	11 576.0	875.6	662.8	2.6	0
黑色金属矿采选业	8 204.6	1 571.9	111.8	…	0
有色金属矿采选业	41 191.7	982.1	12.2	0.1	0.5
非金属矿采选业	5 819.6	375.3	12.5	…	0
开采辅助活动	866.8	66.6	27.8	0.6	0
其他采矿业	149.0	0.8	…	0	0
农副食品加工业	553 036.3	21 027.6	243.6	4.4	0.3
食品制造业	112 483.8	9 374.9	55.2	0.1	2.1
酒、饮料和精制茶制造业	250 539.7	10 021.4	34.2	4.9	…
烟草制品业	2 254.3	143.9	5.5	…	…
纺织业	292 227.2	20 169.5	171.7	2.1	0.1
纺织服装、服饰业	18 387.4	1 728.1	5.0	0.2	0
皮革、毛皮、羽毛及其制品和制鞋业	65 114.2	6 712.8	469.4	…	…
木材加工和木、竹、藤、棕、草制品业	16 753.1	313.0	2.3	0.2	0
家具制造业	766.9	49.0	2.8	0	…
造纸和纸制品业	741 672.3	25 052.7	66.5	90.7	0.1
印刷和记录媒介复制业	1 834.0	144.8	15.6	…	…
文教、工美、体育和娱乐用品制造业	1 973.8	134.8	26.5	0.3	0.3
石油加工、炼焦和核燃料加工业	76 995.2	15 706.4	2 473.5	1 901.7	56.5
化学原料和化学制品制造业	327 926.1	92 651.1	2 454.2	143.2	63.6
医药制造业	96 791.6	7 239.8	461.1	4.8	…
化学纤维制造业	150 747.7	4 844.9	47.2	0.5	0.2
橡胶和塑料制品业	13 805.0	1 040.8	102.1	0.1	0.2
非金属矿物制品业	30 330.9	1 434.7	246.1	18.2	0.1
黑色金属冶炼和压延加工业	79 834.7	7 556.6	3 322.6	216.0	38.3
有色金属冶炼和压延加工业	29 676.9	18 702.0	453.9	5.3	2.4
金属制品业	31 576.1	2 237.9	1 653.4	2.9	39.4
通用设备制造业	11 274.1	617.9	745.8	2.2	0.8
专用设备制造业	5 592.1	550.3	308.0	…	0.1
汽车制造业	14 875.5	813.8	886.8	3.8	0.4
铁路、船舶、航空航天和其他运输设备制造业	15 672.4	875.2	1 140.3	0.3	0.3
电气机械和器材制造业	7 018.6	560.1	106.5	0.4	0.5
计算机、通信和其他电子设备制造业	31 841.7	2 582.6	197.3	2.3	4.2
仪器仪表制造业	1 600.4	109.8	19.2	0.1	0.2
其他制造业	5 131.6	263.0	47.8	…	0.4
废弃资源综合利用业	3 378.8	192.0	39.8	0.1	…
金属制品、机械和设备修理业	1 291.6	61.4	135.5	0.2	…
电力、热力生产和供应业	25 373.6	1 299.8	265.0	0.8	0.1
燃气生产和供应业	1 154.4	187.2	13.7	1.1	0.7
水的生产和供应业	1 851.0	280.4	18.9	0.3	1.8
其他行业	19 372.6	641.9	1.2	0	0

按行业分重点调查工业废水排放及处理情况（五）

（2011）   单位：吨

行业名称	工业废水中污染物排放量				
	汞	镉	总铬	铅	砷
行业总计	1.214	35.096	290.264	150.831	145.169
煤炭开采和洗选业	…	0	0.037	0.150	0.024
石油和天然气开采业	0	0	0	0	0
黑色金属矿采选业	…	0.024	2.366	1.166	0.272
有色金属矿采选业	0.332	3.683	1.777	46.387	34.906
非金属矿采选业	0.065	0.245	0.074	0.700	1.714
开采辅助活动	0	0.049	0	0.385	0
其他采矿业	0	0	0	0	0
农副食品加工业	0	0.012	0	0.372	0.001
食品制造业	0	0	0	…	0
酒、饮料和精制茶制造业	0	0	0	0	0
烟草制品业	0	0	0	0	0
纺织业	0	0	0.977	…	0.006
纺织服装、服饰业	0	0	0.052	…	0.002
皮革、毛皮、羽毛及其制品和制鞋业	…	0.785	120.123	0.025	…
木材加工和木、竹、藤、棕、草制品业	0	0	0.003	0	0
家具制造业	0	0	0.420	0	…
造纸和纸制品业	0	0	0.016	0.098	0.675
印刷和记录媒介复制业	…	0	0.057	0.008	0
文教、工美、体育和娱乐用品制造业	0	…	0.248	0.008	…
石油加工、炼焦和核燃料加工业	0.007	0.031	0.473	0.162	0.070
化学原料和化学制品制造业	0.496	1.993	6.605	10.494	48.122
医药制造业	…	0	0.086	0	0.035
化学纤维制造业	…	0	0.003	0	…
橡胶和塑料制品业	…	0	0.291	0.005	0.001
非金属矿物制品业	0.013	0.004	0.179	0.224	0.260
黑色金属冶炼和压延加工业	0.010	0.474	4.673	1.775	5.109
有色金属冶炼和压延加工业	0.249	27.003	15.710	77.736	50.428
金属制品业	0.002	0.600	106.809	2.652	0.269
通用设备制造业	…	0.024	1.963	0.136	0.010
专用设备制造业	0.007	0.016	1.024	0.153	0.040
汽车制造业	0	0.002	12.959	0.081	…
铁路、船舶、航空航天和其他运输设备制造业	0	0.019	1.082	0.010	0
电气机械和器材制造业	0.015	0.049	0.635	3.931	0.013
计算机、通信和其他电子设备制造业	0.011	0.043	9.852	3.794	0.065
仪器仪表制造业	…	0.011	0.572	0.017	0.341
其他制造业	0	0.002	0.520	0.146	0
废弃资源综合利用业	0.001	0.009	0.021	0.123	0.008
金属制品、机械和设备修理业	0	0.002	0.231	0.001	0
电力、热力生产和供应业	0.006	0.015	0.060	0.088	2.797
燃气生产和供应业	0	0	0.031	0	0
水的生产和供应业	0	0	0.331	0	0
其他行业	0	0	0	0	0

按行业分重点调查工业废气排放及处理情况（一）

（2011）

行业名称	工业废气排放总量（标态）/亿米³	废气治理设施数/套	废气治理设施处理能力（标态）/（万米³/时）	废气治理设施运行费用/万元
行业总计	674 509.3	216 457	1 568 591.7	15 794 757.8
煤炭开采和洗选业	2 039.0	5 120	6 077.2	44 291.5
石油和天然气开采业	1 342.1	187	289.9	12 274.1
黑色金属矿采选业	2 864.9	1 218	1 945.0	26 643.3
有色金属矿采选业	242.9	534	637.9	5 393.2
非金属矿采选业	612.9	387	4 057.4	5 718.8
开采辅助活动	75.0	87	275.6	2 574.5
其他采矿业	9.9	5	1.0	118.6
农副食品加工业	5 473.2	6 827	25 697.9	78 069.7
食品制造业	2 351.0	3 332	7 034.2	85 109.8
酒、饮料和精制茶制造业	2 217.7	2 846	35 776.7	39 696.0
烟草制品业	542.3	1 005	2 143.9	17 644.8
纺织业	4 342.4	8 856	12 234.4	93 243.3
纺织服装、服饰业	643.5	1 169	1 062.0	7 526.8
皮革、毛皮、羽毛及其制品和制鞋业	408.9	1 523	1 363.8	6 487.9
木材加工和木、竹、藤、棕、草制品业	3 258.4	2 872	6 093.4	33 262.6
家具制造业	284.7	723	2 438.9	3 240.5
造纸和纸制品业	17 093.9	6 052	20 627.2	246 844.0
印刷和记录媒介复制业	251.4	296	435.7	3 239.9
文教、工美、体育和娱乐用品制造业	217.4	533	650.9	3 044.9
石油加工、炼焦和核燃料加工业	21 762.4	3 291	31 877.8	700 242.2
化学原料和化学制品制造业	31 205.3	20 854	61 510.4	652 507.8
医药制造业	3 604.0	4 163	9 498.3	40 215.8
化学纤维制造业	2 069.3	1 058	4 620.8	33 679.8
橡胶和塑料制品业	4 128.7	4 047	7 013.8	56 312.4
非金属矿物制品业	129 851.0	61 449	205 117.4	1 187 253.8
黑色金属冶炼和压延加工业	173 215.4	16 640	360 018.6	2 872 875.4
有色金属冶炼和压延加工业	31 891.6	8 794	60 608.8	688 684.7
金属制品业	8 871.1	7 318	11 941.6	121 626.1
通用设备制造业	1 630.8	2 665	4 383.9	16 050.7
专用设备制造业	3 071.3	1 744	4 678.2	25 027.9
汽车制造业	3 780.2	2 475	16 674.7	51 712.8
铁路、船舶、航空航天和其他运输设备制造业	2 166.3	4 709	7 718.2	31 568.9
电气机械和器材制造业	1 523.5	3 937	4 702.9	144 203.5
计算机、通信和其他电子设备制造业	6 153.2	5 762	10 641.3	94 187.8
仪器仪表制造业	100.8	305	430.8	1 855.0
其他制造业	667.5	892	1 318.3	6 737.8
废弃资源综合利用业	227.4	400	696.6	158 843.1
金属制品、机械和设备修理业	1 135.0	290	944.5	2 110.5
电力、热力生产和供应业	202 905.6	21 881	634 672.2	8 190 221.9
燃气生产和供应业	261.4	187	665.4	4 330.7
水的生产和供应业	7.0	10	9.3	35.9
其他行业	9.1	14	4.7	49.0

按行业分重点调查工业废气排放及处理情况（二）

（2011）

单位：万吨

行业名称	二氧化硫产生量	二氧化硫排放量	氮氧化物产生量	氮氧化物排放量	烟（粉）尘产生量	烟（粉）尘排放量
行业总计	5 864.0	1 896.5	1 746.3	1 660.1	81 891.0	1 028.0
煤炭开采和洗选业	16.7	12.9	4.5	4.5	147.3	20.9
石油和天然气开采业	4.8	2.5	2.3	2.3	5.1	1.7
黑色金属矿采选业	2.9	2.6	0.8	0.8	41.7	12.2
有色金属矿采选业	4.3	1.8	0.4	0.4	5.7	1.5
非金属矿采选业	5.1	4.6	1.2	1.2	35.4	5.6
开采辅助活动	10.5	0.5	0.2	0.2	1.3	0.3
其他采矿业	0.1	…	…	…	…	…
农副食品加工业	30.5	24.0	9.8	9.8	187.6	20.2
食品制造业	18.3	14.2	4.8	4.7	107.8	7.0
酒、饮料和精制茶制造业	17.0	13.4	4.1	4.0	103.6	7.4
烟草制品业	1.7	1.1	0.5	0.5	66.6	0.6
纺织业	31.6	27.2	7.8	7.7	153.2	10.1
纺织服装、服饰业	2.2	1.9	0.6	0.6	7.0	0.9
皮革、毛皮、羽毛及其制品和制鞋业	2.8	2.6	0.6	0.6	6.0	1.3
木材加工和木、竹、藤、棕、草制品业	4.8	4.7	1.4	1.4	297.6	20.5
家具制造业	0.3	0.3	0.1	0.1	1.5	0.3
造纸和纸制品业	85.1	54.3	22.5	22.1	551.8	20.7
印刷和记录媒介复制业	0.4	0.4	0.1	0.1	1.3	0.2
文教、工美、体育和娱乐用品制造业	0.3	0.2	0.1	0.1	0.7	0.2
石油加工、炼焦和核燃料加工业	225.9	80.8	37.6	36.5	1 111.1	45.9
化学原料和化学制品制造业	251.1	127.5	54.4	51.5	1 720.3	65.7
医药制造业	13.5	10.4	3.1	2.9	94.7	4.8
化学纤维制造业	17.3	12.1	5.4	5.3	157.2	3.3
橡胶和塑料制品业	10.8	8.1	2.8	2.7	37.8	3.0
非金属矿物制品业	229.2	201.7	272.3	269.4	22 968.6	279.1
黑色金属冶炼和压延加工业	319.0	251.4	97.8	95.1	7 541.1	206.2
有色金属冶炼和压延加工业	1 213.3	114.6	20.5	20.2	1 733.3	34.8
金属制品业	9.1	5.8	1.9	1.8	40.2	8.8
通用设备制造业	3.1	2.7	0.9	0.9	42.1	4.9
专用设备制造业	2.1	1.6	0.9	0.9	16.2	2.2
汽车制造业	1.8	1.3	0.7	0.7	35.3	10.0
铁路、船舶、航空航天和其他运输设备制造业	2.4	1.7	0.9	0.9	33.5	5.9
电气机械和器材制造业	1.1	0.9	0.4	0.4	7.8	0.6
计算机、通信和其他电子设备制造业	1.0	0.8	0.9	0.6	21.5	0.5
仪器仪表制造业	0.1	0.1	…	…	0.3	0.1
其他制造业	1.6	1.4	1.0	1.0	8.1	2.2
废弃资源综合利用业	0.6	0.4	0.2	0.2	71.8	0.5
金属制品、机械和设备修理业	0.2	0.2	0.1	0.1	8.3	1.0
电力、热力生产和供应业	3 319.0	901.2	1 181.7	1 106.8	44 499.4	215.6
燃气生产和供应业	1.9	1.6	1.2	1.1	20.7	1.2
水的生产和供应业	0.4	0.4	…	…	0.2	…
其他行业	0.1	0.1	…	…	0.2	0.1

按行业分重点调查一般工业固体废物产生及处置利用情况

(2011)   单位：万吨

行业名称	一般工业固体废物产生量	一般工业固体废物综合利用量	一般工业固体废物处置量	一般工业固体废物贮存量	一般工业固体废物倾倒丢弃量	一般工业固体废弃物综合利用率/%
行业总计	306 133	183 727	67 562	58 747	389	59.4
煤炭开采和洗选业	34 988	26 730	6 857	2 581	130	74.8
石油和天然气开采业	124	85	37	4	…	67.7
黑色金属矿采选业	69 085	13 365	28 999	27 005	56	19.3
有色金属矿采选业	37 419	13 508	12 264	12 081	90	36.0
非金属矿采选业	3 737	1 922	1 109	964	5	48.3
开采辅助活动	267	40	226	2	0	14.8
其他采矿业	9	6	2	0	0	71.6
农副食品加工业	1 986	1 846	133	4	7	92.9
食品制造业	610	585	25	3	1	95.6
酒、饮料和精制茶制造业	1 006	930	74	2	2	92.4
烟草制品业	58	44	13	…	…	76.3
纺织业	673	579	93	1	1	85.9
纺织服装、服饰业	47	42	4	…	…	90.6
皮革、毛皮、羽毛及其制品和制鞋业	64	48	16	…	…	75.3
木材加工和木、竹、藤、棕、草制品业	344	339	5	…	…	98.4
家具制造业	13	12	1	…	…	94.3
造纸和纸制品业	2 482	2 133	321	38	1	85.9
印刷和记录媒介复制业	22	16	7	…	…	69.4
文教、工美、体育和娱乐用品制造业	6	5	1	…	…	79.6
石油加工、炼焦和核燃料加工业	3 951	3 113	691	154	8	78.5
化学原料和化学制品制造业	26 548	16 927	4 028	5 800	20	63.3
医药制造业	309	273	32	3	1	88.5
化学纤维制造业	365	316	36	5	9	86.4
橡胶和塑料制品业	208	197	10	1	…	94.6
非金属矿物制品业	5 949	5 625	279	138	22	92.9
黑色金属冶炼和压延加工业	42 344	38 139	2 492	2 234	19	89.0
有色金属冶炼和压延加工业	10 304	5 149	2 869	2 346	4	49.7
金属制品业	472	335	119	18	…	70.9
通用设备制造业	211	153	42	10	6	72.7
专用设备制造业	151	126	24	…	…	83.5
汽车制造业	329	294	35	1	…	89.1
铁路、船舶、航空航天和其他运输设备制造业	244	206	38	…	…	84.4
电气机械和器材制造业	67	56	11	…	…	83.5
计算机、通信和其他电子设备制造业	97	63	33	…	…	65.4
仪器仪表制造业	5	3	1	…	…	69.4
其他制造业	114	110	4	…	…	96.2
废弃资源综合利用业	354	336	15	3	…	94.7
金属制品、机械和设备修理业	29	28	1	…	…	97.0
电力、热力生产和供应业	61 061	49 964	6 612	5 350	5	80.7
燃气生产和供应业	68	66	2	…	0	97.2
水的生产和供应业	10	10	…	…	0	99.6
其他行业	2	2	…	0	0	97.7

按行业分重点调查危险废物产生及处置利用情况

（2011）

单位：万吨

行业名称	危险废物产生量	危险废物综合利用量	危险废物处置量	危险废物贮存量	危险废物倾倒丢弃量
行业总计	3 431	1 773	916	824	0
煤炭开采和洗选业	…	…	…	…	0
石油和天然气开采业	36	14	22	…	0
黑色金属矿采选业	7	…	…	7	0
有色金属矿采选业	110	88	11	23	0
非金属矿采选业	723	37	…	686	0
开采辅助活动	1	…	1	0	0
其他采矿业	0	0	0	0	0
农副食品加工业	…	…	…	…	0
食品制造业	2	2	…	…	0
酒、饮料和精制茶制造业	…	…	…	…	0
烟草制品业	…	…	…	…	0
纺织业	4	2	2	…	0
纺织服装、服饰业	…	…	…	…	0
皮革、毛皮、羽毛及其制品和制鞋业	3	1	2	…	0
木材加工和木、竹、藤、棕、草制品业	…	…	…	…	0
家具制造业	…	…	…	…	0
造纸和纸制品业	746	487	259	…	0
印刷和记录媒介复制业	1	…	1	…	0
文教、工美、体育和娱乐用品制造业	1	…	1	…	0
石油加工、炼焦和核燃料加工业	192	146	45	1	0
化学原料和化学制品制造业	643	442	225	28	0
医药制造业	56	43	13	…	0
化学纤维制造业	31	1	29	…	0
橡胶和塑料制品业	7	3	4	…	0
非金属矿物制品业	28	14	14	…	0
黑色金属冶炼和压延加工业	154	123	33	5	0
有色金属冶炼和压延加工业	373	228	82	69	0
金属制品业	41	16	24	1	0
通用设备制造业	10	2	7	1	0
专用设备制造业	3	1	2	…	0
汽车制造业	22	4	18	…	0
铁路、船舶、航空航天和其他运输设备制造业	12	7	5	…	0
电气机械和器材制造业	18	7	12	…	0
计算机、通信和其他电子设备制造业	139	82	57	…	0
仪器仪表制造业	10	8	2	…	0
其他制造业	3	…	2	…	0
废弃资源综合利用业	6	3	3	…	0
金属制品、机械和设备修理业	1	…	1	…	0
电力、热力生产和供应业	47	12	37	1	0
燃气生产和供应业	2	…	1	…	0
水的生产和供应业	…	…	…	…	0
其他行业	0	0	0	0	0

按行业分重点调查工业汇总情况（一）

（2011）

行业名称	汇总工业企业数/个	工业总产值（现价）/万元	工业煤炭消耗量/万吨	燃料煤	工业用水总量/万吨	取水量	重复用水量
行业总计	153 027	450 660.2	361 734.6	272 993.1	37 816 661.2	6 279 697.5	31 536 963.7
煤炭开采和洗选业	6 587	13 207.8	15 808.7	1 230.8	228 014.6	87 380.3	140 634.3
石油和天然气开采业	247	5 608.8	71.6	71.5	128 467.1	29 263.0	99 204.1
黑色金属矿采选业	4 662	2 481.0	178.5	166.7	438 076.3	81 190.7	356 885.6
有色金属矿采选业	3 609	1 974.8	81.9	77.5	223 453.4	67 299.7	156 153.7
非金属矿采选业	1 082	379.2	296.7	294.6	33 099.5	11 017.4	22 082.1
开采辅助活动	61	522.2	171.0	32.3	2 093.7	489.8	1 603.9
其他采矿业	33	2.4	1.8	1.8	346.9	305.8	41.0
农副食品加工业	11 979	40 688.2	2 156.6	2 139.3	386 682.2	171 548.5	215 133.7
食品制造业	3 813	5 290.5	1 257.5	1 201.7	158 488.3	69 192.4	89 296.0
酒、饮料和精制茶制造业	4 390	6 279.2	1 133.4	1 113.9	224 320.5	100 912.1	123 408.4
烟草制品业	154	6 868.8	85.6	85.3	13 358.2	3 647.3	9 710.9
纺织业	8 929	12 683.1	2 294.6	2 254.8	355 263.9	291 821.3	6 3442.6
纺织服装、服饰业	1 537	1 066.4	150.3	148.6	29 121.3	23 517.2	5 604.1
皮革、毛皮、羽毛及其制品和制鞋业	2 342	1 702.2	183.9	181.3	34 150.0	31 458.5	2 691.5
木材加工和木、竹、藤、棕、草制品业	2 707	2 100.5	245.9	234.4	7 685.4	5 388.7	2 296.6
家具制造业	449	502.6	15.9	15.7	1 268.6	1 159.9	108.7
造纸和纸制品业	5 871	6 764.5	5 285.8	5 198.6	1287 691.7	455 926.1	831 765.6
印刷和记录媒介复制业	727	775.3	27.1	27.0	3 864.1	1 952.1	1 912.0
文教、工美、体育和娱乐用品制造业	656	1 072.4	14.0	13.8	3 159.6	2 540.0	619.6
石油加工、炼焦和核燃料加工业	1 383	34 354.1	38 108.9	7 585.9	2 377 099.3	162 267.6	2 214 831.7
化学原料和化学制品制造业	12 219	79 752.4	18 441.1	10 493.3	5 512 068.1	507 803.7	5 004 264.4
医药制造业	3197	6 486.1	855.5	826.0	286 250.7	70 470.9	215 779.9
化学纤维制造业	490	3 916.9	1 159.2	1 120.4	404 963.1	72 853.7	332 109.4
橡胶和塑料制品业	2 521	5 757.7	1 919.1	718.0	83 337.6	17 686.0	65 651.5
非金属矿物制品业	35 208	15 663.0	30 469.5	19 605.7	425 478.7	81 866.1	343 612.6
黑色金属冶炼和压延加工业	4 913	50 998.6	29 459.2	10 262.6	7 799 396.1	347 517.5	7 451 878.6
有色金属冶炼和压延加工业	4 505	18 637.7	5 380.9	4 478.2	730 310.0	82 790.3	647 519.7
金属制品业	8 793	8 726.1	393.4	350.5	145 170.5	40 654.4	104 516.1
通用设备制造业	3 201	11 489.4	187.5	163.0	43 584.7	20 464.6	23 120.0
专用设备制造业	1 374	5 770.3	146.1	136.5	34 712.8	9 528.1	25 184.7
汽车制造业	1 773	27 630.9	129.4	124.8	119 231.2	21 182.9	98 048.2
铁路、船舶、航空航天和其他运输设备制造业	1 023	8 551.4	208.6	203.6	40 092.0	18 478.0	21 614.0
电气机械和器材制造业	1 983	11 041.3	76.4	74.0	47 888.5	13 372.7	34 515.8
计算机、通信和其他电子设备制造业	2 546	23 133.5	57.2	54.2	248 492.2	58 306.3	190 186.0
仪器仪表制造业	391	1 255.9	6.3	6.3	12 194.1	3 599.2	8 594.9
其他制造业	1 130	1 547.1	262.9	102.5	27 841.9	6 239.5	21 602.4
废弃资源综合利用业	454	436.2	25.1	22.3	5 041.9	3 375.9	1 666.0
金属制品、机械和设备修理业	188	607.8	12.9	12.9	2 513.8	1 853.4	660.4
电力、热力生产和供应业	5 668	24 427.6	202 823.4	201 974.2	15 691 011.3	3 281 797.7	12 409 213.6
燃气生产和供应业	60	216.5	2 140.0	177.8	60 204.7	2 589.4	57 615.3
水的生产和供应业	58	164.5	4.1	4.1	160 255.7	18 184.1	142 071.6
其他行业	114	125.7	7.1	6.9	916.8	804.4	112.4

按行业分重点调查工业汇总情况（二）

（2011）

行业名称	工业炉窑数/台	工业锅炉数/台	35 蒸吨及以上的	20（含）～35 蒸吨的	10（含）～20 蒸吨的	10 蒸吨以下的
行业总计	101 705	104 009	13 013	7 619	13 984	69 393
煤炭开采和洗选业	240	7 098	104	180	976	5 838
石油和天然气开采业	3 349	2 970	83	527	196	2 164
黑色金属矿采选业	192	507	14	25	66	402
有色金属矿采选业	110	494	2	16	39	437
非金属矿采选业	375	132	23	12	3	94
开采辅助活动	65	165	6	21	21	117
其他采矿业	3	4	0		2	2
农副食品加工业	502	8 818	596	543	626	7 053
食品制造业	130	4 242	166	238	542	3 296
酒、饮料和精制茶制造业	239	4 258	135	307	734	3 082
烟草制品业	16	348	37	62	146	103
纺织业	255	9 083	130	299	1 477	7 177
纺织服装、服饰业	12	1 494	8	11	123	1 352
皮革、毛皮、羽毛及其制品和制鞋业	10	1 857	3	11	80	1 763
木材加工和木、竹、藤、棕、草制品业	190	2 766	15	60	241	2 450
家具制造业	66	278	0	5	16	257
造纸和纸制品业	471	6 474	486	264	785	4 939
印刷和记录媒介复制业	10	356	0		20	336
文教、工美、体育和娱乐用品制造业	46	249	0		7	242
石油加工、炼焦和核燃料加工业	3 785	1 865	538	276	419	632
化学原料和化学制品制造业	6 278	10 667	1 191	609	1 421	7 446
医药制造业	66	3 297	96	135	388	2 678
化学纤维制造业	76	848	129	96	223	400
橡胶和塑料制品业	77	2 525	68	148	310	1 999
非金属矿物制品业	40 248	3 574	97	170	397	2 910
黑色金属冶炼和压延加工业	12 336	1 818	518	213	213	874
有色金属冶炼和压延加工业	16 326	1 565	178	109	201	1 077
金属制品业	3 471	2 101	18	35	91	1 957
通用设备制造业	2 491	974	12	42	101	819
专用设备制造业	2 311	617	21	51	128	417
汽车制造业	1 360	1 025	23	70	116	816
铁路、船舶、航空航天和其他运输设备制造业	4 502	835	61	136	159	479
电气机械和器材制造业	291	685	4	18	54	609
计算机、通信和其他电子设备制造业	519	1 013	10	22	107	874
仪器仪表制造业	44	126	0	3	19	104
其他制造业	760	476	7	14	51	404
废弃资源综合利用业	150	111	3	4	5	99
金属制品、机械和设备修理业	88	133	0	1	12	120
电力、热力生产和供应业	117	18 006	8 201	2 866	3 449	3 490
燃气生产和供应业	46	113	28	20	17	48
水的生产和供应业	4	12	2		0	10
其他行业	78	30	0		3	27

各地区火电行业污染排放及处理情况（一）

（2011）

地 区 名 称	汇总工业 企业数/个	机组数/台	废气治理 设施数/套	脱硫 设施数	脱硝 设施数	除尘 设施数
全 国	1 828	52 29	8 975	3 379	274	5 285
北 京	11	40	61	14	25	22
天 津	17	53	110	50	4	56
河 北	102	264	499	211	10	276
山 西	110	309	617	188	11	418
内蒙古	94	276	464	169	8	286
辽 宁	74	239	396	120	8	250
吉 林	40	118	192	46	8	138
黑龙江	77	231	374	69	2	297
上 海	24	64	93	38	8	47
江 苏	187	488	1 054	467	31	553
浙 江	147	431	925	410	18	495
安 徽	51	125	210	76	4	127
福 建	25	68	127	47	19	61
江 西	22	51	96	28	7	61
山 东	288	782	1 560	644	12	897
河 南	90	195	373	151	14	208
湖 北	38	97	171	65	5	99
湖 南	29	67	109	41	7	61
广 东	86	225	346	139	49	177
广 西	17	37	61	26	1	33
海 南	7	20	19	8	2	9
重 庆	36	75	139	60	0	79
四 川	47	91	135	47	4	75
贵 州	24	77	127	54	0	73
云 南	16	39	83	37	1	45
西 藏	2	22	0	0	0	0
陕 西	53	149	206	62	5	139
甘 肃	20	60	98	29	6	63
青 海	6	23	28	4	0	24
宁 夏	23	57	113	43	5	65
新 疆	65	456	189	36	0	151

各地区火电行业污染排放及处理情况（二）

（2011）

地 区名 称	燃料煤消耗量/万吨	燃料煤平均含硫量/%	燃料油消耗量/万吨	燃料油平均含硫量/%
全 国	193 920.1	0.67	75.1	0.56
北 京	946.6	0.56	11.2	0.08
天 津	2 952.4	0.76	0.1	0.09
河 北	10 738.3	1.00	1.5	0.39
山 西	11 918.6	1.08	2.5	0.21
内蒙古	18 368.0	0.77	2.1	0.74
辽 宁	8 646.7	0.65	5.1	0.33
吉 林	4 603.7	0.42	0.9	1.60
黑龙江	5 475.4	0.34	0.5	9.38
上 海	3 437.0	0.61	14.5	0.83
江 苏	16 350.9	0.77	0.4	0.23
浙 江	10 352.4	0.73	9.8	0.92
安 徽	7 474.4	0.51	0.2	0.59
福 建	4 981.6	0.66	1.3	0.59
江 西	3 056.4	1.11	0.6	1.06
山 东	16 225.4	1.10	1.9	0.35
河 南	12 195.2	0.88	1.5	0.46
湖 北	4 368.1	1.37	0.5	0.33
湖 南	4 203.9	1.08	0.8	0.24
广 东	12 037.9	0.71	2.0	0.54
广 西	2 565.6	1.85	0.2	0.31
海 南	510.0	0.79	0.1	0.02
重 庆	2 389.8	2.79	0.3	0.15
四 川	3 030.1	0.07	1.0	0.89
贵 州	5 670.1	2.44	0.7	1.36
云 南	3 364.0	1.45	1.5	0.45
西 藏	0.0	0.00	11.2	0.00
陕 西	5 850.4	1.22	0.9	0.39
甘 肃	3 345.7	0.74	0.4	0.08
青 海	588.4	0.75	0.5	0.00
宁 夏	5 165.7	1.15	0.3	0.26
新 疆	3 107.3	1.27	0.4	5.98

各地区火电行业污染排放及处理情况（三）

（2011）　　　　　　　　　　　　　　　　　　　　　　　单位：吨

地 区 名 称	工业废气 排放量/ 亿米³	工业二氧化硫 产生量	工业二氧化硫 排放量	工业氮氧化物 产生量	工业氮氧化物 排放量	工业烟粉尘 产生量	工业烟粉尘 排放量
全 国	192 044	32 129 141	8 190 310	11 472 174	10 726 068	438 033 632	1 728 108
北 京	2 865	104 767	13 152	59 074	48 856	1 820 735	5 432
天 津	2 425	378 923	69 619	209 649	200 989	5 699 590	10 439
河 北	9 898	2 315 821	352 865	717 796	678 904	25 984 434	68 053
山 西	13 070	2 165 884	728 243	777 578	647 178	31 571 540	210 444
内蒙古	15 767	2 391 557	630 153	904 690	874 372	55 084 519	162 046
辽 宁	7 571	948 544	372 297	462 349	443 141	16 497 056	78 405
吉 林	4 155	326 451	156 494	272 797	265 934	11 030 794	148 338
黑龙江	4 114	338 774	202 015	335 847	332 077	13 205 600	185 562
上 海	3 317	352 276	64 503	256 781	227 210	4 773 180	13 717
江 苏	14 459	2 122 652	454 959	881 178	796 793	23 432 452	78 976
浙 江	10 059	1 347 086	285 772	530 198	474 844	11 003 876	43 070
安 徽	6 219	619 738	143 573	416 289	411 733	18 995 995	39 803
福 建	4 448	558 659	94 982	256 389	206 573	3 503 200	9 709
江 西	2 934	585 080	162 873	235 176	216 468	7 983 290	34 352
山 东	15 554	3 211 798	755 921	774 822	764 251	34 728 353	116 057
河 南	17 238	1 873 936	460 645	756 693	752 944	35 235 086	103 171
湖 北	5 165	987 020	232 290	288 975	273 432	11 939 356	27 117
湖 南	3 441	760 696	168 634	343 737	321 173	11 999 916	43 480
广 东	11 740	1 437 180	300 612	603 715	495 038	14 151 480	30 256
广 西	4 456	820 596	188 929	152 294	152 294	4 784 740	11 033
海 南	614	74 435	14 894	55 862	31 220	646 984	2 267
重 庆	2 063	1 105 749	237 104	183 063	183 063	8 605 006	34 664
四 川	4 365	927 394	293 574	185 985	175 240	16 698 896	40 049
贵 州	3 622	2 143 124	627 018	342 748	342 748	16 172 958	29 377
云 南	3 603	1 088 091	216 960	181 026	179 819	9 025 196	22 699
西 藏	13	473	473	737	737	28	28
陕 西	6 356	1 254 757	419 341	466 931	436 914	11 894 106	44 815
甘 肃	3 677	463 323	155 818	231 354	226 821	12 865 777	25 456
青 海	662	82 800	36 873	41 750	41 750	980 770	9 323
宁 夏	4 883	973 308	163 804	309 592	286 453	13 410 169	41 092
新 疆	3 291	368 248	185 918	237 099	237 099	4 308 551	58 876

5 流域及入海陆源废水排放统计

LIUYU JI RUHAI LUYUAN FEISHUI PAIFANG TONGJI

ANNUAL STATISTIC REPORT ON ENVIRONMENT IN CHINA

2011

流域接纳工业废水及处理情况（一）

（2011）

流域	地区名称	工业废水排放量/万吨	直接排入环境的	排入污水处理厂的	工业废水处理量/万吨	废水治理设施数/套	废水治理设施治理能力/（万吨/日）	废水治理设施运行费用/万元
	总　计	2 308 743	1 837 708	469 108	5 805 511	91 506	31 406	7 321 459.5
辽河	内蒙古	5 757	3 668	2 088	7 950	187	42	9 427.0
	辽宁	90 457	75 930	14 527	351 993	2 287	6 391	302 283.4
	吉林	11 988	11 507	481	32 622	234	98	16 621.2
	合计	108 202	91 105	17 097	392 564	2 708	6 532	328 331.6
海河	北京	8 633	2 827	5 806	11 294	508	60	37 124.0
	天津	19 795	8 413	11 382	35 198	957	136	134 315.1
	河北	117 305	84 505	32 799	933 200	4 760	3 521	468 218.3
	山西	17 586	15 890	1 696	62 425	1 169	218	55 388.3
	内蒙古	1 319	1 319	0	987	20	3	898.0
	山东	28 319	21 215	7 104	38 877	580	171	78 795.9
	河南	34 955	33 418	1 536	65 320	827	391	73 277.6
	合计	227 912	167 587	60 325	1 147 302	8 821	4 500	848 017.2
淮河	江苏	77 998	66 245	11 753	122 558	2 236	885	201 573.5
	安徽	35 668	30 800	4 868	40 487	850	288	53 945.4
	山东	149 576	98 189	51 388	320 086	4 513	1 391	678 550.1
	河南	67 364	60 575	6 789	71 792	1 271	448	83 810.0
	合计	330 607	255 809	74 798	554 922	8 870	3 013	1 017 879.0
松花江	内蒙古	6 853	4 646	2 206	8 611	146	50	9 516.4
	吉林	29 896	23 197	6 699	32 948	451	252	46 445.7
	黑龙江	44 072	33 055	11 017	109 965	1 171	465	241 107.4
	合计	80 821	60 899	19 922	151 525	1 768	766	297 069.5
珠江	江西	428	428	0	179	13	1	598.7
	湖南	2 920	2 920	0	3 085	82	13	66 970.0
	广东	178 626	151 662	26 963	258 541	10 143	1 408	468 752.7
	广西	100 540	97 256	3 283	349 710	2 244	1 059	162 630.4
	贵州	6 888	6 835	54	13 888	475	101	114 991.1
	云南	10 199	10 155	44	68 011	1 016	411	49 448.6
	海南	6 820	6 231	589	8 817	299	40	35 788.1
	福建	4 771	4 578	193	5 517	229	45	11 246.5
	合计	311 192	280 066	31 126	707 747	14 501	3 079	910 426.1

流域接纳工业废水及处理情况（一）（续表）

（2011）

流域	地名区称	工业废水排放量/万吨	直接排入环境的	排入污水处理厂的	工业废水处理量/万吨	废水治理设施数/套	废水治理设施治理能力/（万吨/日）	废水治理设施运行费用/万元
	上 海	44 626	20 366	24 260	71 099	5 872	305	398 206.8
	江 苏	168 300	101 102	67 198	281 513	5 019	2 157	513 557.7
	浙 江	39 882	11 128	28 755	86 291	1 957	433	116 250.7
	安 徽	33 622	30 052	3 570	171 478	1 377	610	127 804.4
	江 西	70 768	69 307	1 461	138 651	2 935	745	145 686.6
	河 南	8 197	8 197	0	8 800	232	76	15 789.2
	湖 北	104 434	93 061	11 373	325 234	5 296	1 322	344 256.0
	湖 南	94 277	90 639	3 639	236 396	3 029	657	200 445.2
长 江	广 西	694	694	0	979	85	6	557.7
	重 庆	33 954	32 777	1 176	37 328	1 507	338	67 100.1
	四 川	80 390	73 844	6 545	286 330	4 067	1 116	622 460.1
	贵 州	13 737	13 082	655	86 371	1 173	428	53 353.4
	云 南	9 509	8 937	572	35 350	718	167	40 130.5
	西 藏	0	0	0	0	0	0	…
	陕 西	4 651	4 337	314	17047	443	72	12 462.1
	甘 肃	1 058	1 058	0	1939	69	65	1 307.1
	青 海	0	0	0	0	0	0	…
	合 计	708 100	558 581	149 518	1 784 804	33 779	8 497	2 659 367.6
	山 西	22 079	20 173	1 906	180 053	2 232	782	303 341.3
	内蒙古	18 121	8 418	9 704	72 727	560	350	66 168.8
	山 东	9 349	7 496	1 853	64 637	351	297	54 507.5
	河 南	28 138	25 435	2 703	61 466	969	309	49 938.2
黄 河	四 川	30	30	0	25	6	0	32.9
	陕 西	36 156	30 011	6 144	56 044	1 831	316	107 646.3
	甘 肃	10 677	9 362	1 315	11 010	299	95	16 402.3
	青 海	3 765	3 765	0	15 006	95	56	6 099.5
	宁 夏	19 285	17 996	1 289	26 135	397	165	35 041.1
	合 计	147 600	122 686	24 913	487 101	6 740	2 368	639 177.9

流域接纳工业废水及处理情况（一）（续表）

（2011）

流域	地名	区称	工业废水排放量/万吨	直接排入环境的	排入污水处理厂的	工业废水处理量/万吨	废水治理设施数/套	废水治理设施治理能力/（万吨/日）	废水治理设施运行费用/万元
东南诸河	浙	江	142 357	71 074	71 283	217 967	6 505	988	350 313.3
	安	徽	1 429	1 351	78	1 370	94	5	2 135.3
	福	建	172 414	158 757	13 655	245 800	3 200	917	148 113.6
	合	计	316 201	231 183	85 016	465 137	9 799	1 910	500 562.2
西北诸河	河	北	1 201	658	542	319	9	1	305.8
	内蒙古		7 359	6 363	978	5 008	107	30	4 294.8
	西	藏	0	0	0	0	0	0	…
	甘	肃	7 985	6 046	35	9 598	185	66	12 559.7
	青	海	4 912	4 912	0	1 771	81	8	3 209.9
	新	疆	28 769	24 673	4 096	55 440	985	279	67 954.8
	合	计	50 226	42 652	5 650	72 137	1 367	384	88 325.0
西南诸河	云	南	27 520	26 792	728	41 774	3 123	355	32 071.6
	西	藏	363	347	15	498	30	2	231.7
	青	海	0	0	0	0	0	0	…
	合	计	27 882	27 140	743	42 272	3 153	357	32 303.3

流域接纳工业废水及处理情况（二）

（2011）
单位：吨

流域	地区名称	工业废水中污染物产生量				
		化学需氧量	氨氮	石油类	挥发酚	氰化物
总计		34 147 021.5	1 715 319.8	375 986.2	70 560.5	6 220.9
辽河	内蒙古	89 157.8	3 406.5	34.8	10.2	0
	辽宁	543 311.4	60 573.2	18 210.6	1 958.2	563.1
	吉林	95 274.6	2 544.7	417.4	664.9	6.7
	合计	727 743.8	66 524.4	18 662.8	2 633.4	569.8
海河	北京	80 590.7	3 153.1	2 730.4	877	6.9
	天津	154 687.5	6 591.3	1 251.3	3.2	1.4
	河北	1 727 740.5	75 759.7	22 083.6	11 626.9	696.6
	山西	196 096.9	23 238	2 543.2	3 169.1	98.2
	内蒙古	1 687.7	35.2	1	0	0
	山东	526 096.3	34 750.9	2 327.3	0.5	2.8
	河南	694 376.7	16 467.6	2 278.4	1 009.8	58.9
	合计	3 381 276.4	159 995.9	33 215.2	16 686.4	864.8
淮河	江苏	1 054 553.5	33 307.2	4 055.9	823.4	94.4
	安徽	334 188.4	65 461.9	1 230.1	44.4	14.7
	山东	2 180 334.1	93 889.6	24 712.4	4 015.8	122.6
	河南	704 981.1	36 316.8	3 443.9	931.3	57.8
	合计	4 274 057.1	228 975.6	33 442.3	5 814.8	289.5
松花江	内蒙古	64 990.7	1 822.4	139.9	10.1	7.9
	吉林	521 074	20 136.4	1 553.1	195.4	13
	黑龙江	1 579 834.2	68 745.1	48 926.8	5 602.1	102.1
	合计	2 165 898.9	90 703.9	50 619.7	5 807.6	123.1
珠江	江西	2 250.7	11 135.3	1	0	0
	湖南	15 289.4	1 124.2	128.3	0	1.7
	广东	1 639 681.6	58 217.4	5 008.8	157.1	643.6
	广西	1 092 554.3	54 212.9	3 417.1	1 568	26.8
	贵州	30 044.2	3 642	447.6	196.8	5.3
	云南	203 212.1	37 244.4	3 057.9	2 579.7	114
	海南	111 959.1	1 868.7	232.1	0.1	0.2
	福建	14 710	3 488.9	123.9	8.2	6.2
	合计	3 109 701.4	170 933.7	12 416.7	4 509.9	797.8

流域接纳工业废水及处理情况（二）（续表）

（2011）　　　　　　　　　　　　　　　　　　　　　　单位：吨

| 流域 | 地区名称 | 工业废水中污染物产生量 | | | | |
		化学需氧量	氨氮	石油类	挥发酚	氰化物
长江	上 海	318 556.6	12 447.1	9 509.3	1 119.2	243.9
	江 苏	1 168 456.1	54 842.7	12 472.2	1 196.4	345.8
	浙 江	445 538.6	9 581.8	870.1	300.9	3.6
	安 徽	427 492.4	31 255.8	6 912.8	1 942.2	285.4
	福 建	0	0	0	0	0
	江 西	550 370.3	39 499.9	6 152.1	1 821.3	60.7
	河 南	355 478.9	3 235.8	909.2	0.7	1.3
	湖 北	8 783 004.2	48 799.9	4 459.6	928.8	69
	湖 南	558 386.8	64 431.9	4 152.5	1 255	363.1
	广 东	0	0	0	0	0
	广 西	5 027.4	25.4	0	0	0
	重 庆	253 405.9	28 558.7	3 536	1 219.5	88.4
	四 川	856 910.8	66 965.1	6 216.9	1 861.6	106.6
	贵 州	160 814.4	75 539.2	1 488.3	447.2	134.1
	云 南	101 856.5	32 296.6	888.8	706.6	25.7
	西 藏	1.3	0.1	0	0	0
	陕 西	37 416.3	5 188.8	238.9	0.1	9.5
	甘 肃	5 859.1	30.9	5.4	0	0
	青 海	0	0	0	0	0
	合 计	14 028 575.7	472 699.6	57 812	12 799.6	1 737
黄河	山 西	294 214.1	123 560.6	7 288.3	8 633.6	498.4
	内蒙古	221 931.2	29 042.4	3 027.9	1 069.6	20.1
	山 东	128 638	2 417.4	3 996.7	1 360.9	148.7
	河 南	246 140.8	25 704.3	51 586.3	423.3	19.3
	四 川	227.5	13.1	0	0	0
	陕 西	510 712.1	34 741.2	10 930.6	7 534.6	226.4
	甘 肃	122 788.1	13 418.3	2 140.7	197.5	1.4
	青 海	42 699.5	2 613.1	535.2	4.3	0
	宁 夏	502 992.3	90 173.4	4 250	564.6	24.4
	合 计	2 070 343.7	321 683.9	83 755.8	19 788.2	938.6

流域接纳工业废水及处理情况（二）（续表）

（2011）

流域	地区名称	工业废水中污染物产生量				
		化学需氧量	氨氮	石油类	挥发酚	氰化物
东南诸河	浙 江	2 192 971.8	66 455.4	21 970.4	80.9	701
	安 徽	7 490.3	920.9	44.9	2.6	0.5
	福 建	1 025 936.1	40 684	12 137.1	253.7	135.4
	合 计	3 226 398.3	108 060.3	34 152.4	337.3	836.8
西北诸河	河 北	9 435.7	371.7	5.1	0	0
	内蒙古	34 003.3	7 565.2	260.5	0.1	3.7
	西 藏	0	0	0	0	0
	甘 肃	96 360.8	12 016.5	345.2	91.7	0.4
	青 海	28 097.1	1 317.3	556.6	1	4.9
	新 疆	548 769	71 268.7	50 319.8	2 084.6	44.3
	合 计	716 666	92 539.4	51 487.1	2 177.5	53.4
西南诸河	云 南	440 279.4	3 048.2	417.2	5.9	10
	西 藏	6 080.8	154.9	4.9	0	0
	青 海	0	0	0	0	0
	合 计	446 360.2	3 203.1	422.1	5.9	10

流域接纳工业废水及处理情况（三）

（2011）

单位：吨

流域	地 区名 称	工业废水中污染物产生量				
		汞	镉	总铬	铅	砷
总 计		23.916	2 580.017	8 456.407	9 977.714	9 977.714
辽河	内蒙古	0.092	65.006	0.061	59.071	183.128
	辽 宁	0.168	56.314	20.190	56.704	48.902
	吉 林	0.339	0.004	0.103	0.253	0.119
	合 计	0.598	121.324	20.355	116.028	232.149
海河	北 京	…	0.001	14.134	0.337	3.821
	天 津	0.004	0.011	21.094	20.290	0.054
	河 北	0.051	16.732	220.003	34.432	79.796
	山 西	0.164	0.792	1.724	0.861	1.369
	内蒙古	…	…	…	…	…
	山 东	0.119	217.212	49.966	242.357	640.122
	河 南	0.358	81.348	15.252	26.253	21.949
	合 计	0.696	316.096	322.173	324.530	747.111
淮河	江 苏	0.550	0.159	119.396	17.476	1.261
	安 徽	0.001	0.007	10.694	1.686	2.120
	山 东	0.803	23.983	97.793	22.580	225.652
	河 南	0.003	3.038	141.978	7.355	8.388
	合 计	1.357	27.187	369.861	49.097	237.421
松花江	内蒙古	0.018	0.064	0.002	0.957	0.292
	吉 林	0.005	0.019	11.909	0.168	2.836
	黑龙江	0.063	0.014	12.683	1.602	17.345
	合 计	0.085	0.097	24.595	2.727	20.474
珠江	江 西	0.005	0.088	0.170	0.928	0.520
	湖 南	0.004	0.536	8.166	6.539	3.306
	广 东	2.146	15.383	3 213.163	42.083	86.908
	广 西	0.513	104.375	127.599	155.043	203.764
	贵 州	0.103	0.043	0.013	0.399	0.397
	云 南	0.069	61.678	4.385	88.153	245.870
	海 南	0.001	0.005	16.896	0.040	0.027
	福 建	0.004	3.145	0.534	18.689	0.372
	合 计	2.846	185.254	3 370.926	311.875	541.164

流域接纳工业废水及处理情况（三）（续表）

（2011）

单位：吨

流域	地 区名 称	工业废水中污染物产生量				
		汞	镉	总铬	铅	砷
长江	上 海	0.003	0.069	216.112	1.042	0.828
	江 苏	0.021	2.017	282.235	11.764	4.657
	浙 江	...	0.016	209.381	2.805	0.018
	安 徽	2.315	56.178	19.427	56.319	568.700
	江 西	0.365	376.148	57.513	85.289	4 022.313
	河 南	0.001	0.005	9.922	0.132	0.128
	湖 北	0.459	3.425	98.825	23.669	108.610
	湖 南	4.479	203.557	84.999	491.157	516.718
	广 西	0.014	0.008	0.008	0.098	3.675
	重 庆	...	0.004	220.530	6.095	3.529
	四 川	1.601	52.321	73.041	78.529	326.686
	贵 州	0.037	4.307	8.021	7.445	22.074
	云 南	0.376	64.294	0.911	119.388	192.629
	西 藏	...	...	...	...	...
	陕 西	0.356	112.409	3.761	400.165	47.564
	甘 肃	0.035	8.497	...	4.956	3.441
	青 海	...	...	...	...	...
	合 计	10.063	883.256	1 284.686	1 288.852	5 821.571
黄河	山 西	0.003	14.805	14.937	34.728	87.127
	内蒙古	0.800	16.591	7.180	25.958	63.411
	山 东	0.082	0.114	1.361	0.753	0.081
	河 南	0.989	176.069	56.910	270.472	107.567
	四 川	...	...	...	...	...
	陕 西	0.043	0.416	6.386	2.912	45.860
	甘 肃	0.919	164.018	42.921	393.910	404.985
	青 海	0.342	7.582	7.702	6.094	8.533
	宁 夏	0.723	5.254	1.512	6.331	152.012
	合 计	3.901	384.848	138.909	741.159	869.575

流域接纳工业废水及处理情况（三）（续表）

（2011）

单位：吨

流域	地区名称	工业废水中污染物产生量				
		汞	镉	总铬	铅	砷
东南诸河	浙 江	0.010	2.802	2 014.261	6.410	32.669
	安 徽	0.008	0.033	0.457	0.017	0.012
	福 建	0.383	23.441	888.291	76.307	32.282
	合 计	0.401	26.276	2 903.010	82.734	64.963
西北诸河	河 北	…	…	…	…	…
	内蒙古	0.002	11.198	0.010	8.643	10.283
	西 藏	…	…	…	…	…
	甘 肃	0.306	403.636	6.316	515.005	1 085.690
	青 海	0.093	22.746	0.003	17.002	10.392
	新 疆	3.315	12.565	12.383	11.400	40.329
	合 计	3.717	450.145	18.711	552.050	1 146.693
西南诸河	云 南	0.251	185.157	0.040	322.216	281.005
	西 藏	…	0.376	3.140	2.316	15.589
	青 海	…	…	…	…	…
	合 计	0.251	185.534	3.180	324.532	296.594

流域接纳工业废水及处理情况（四）

（2011）

单位：吨

流域	地区名称	工业废水中污染物排放量				
		化学需氧量	氨氮	石油类	挥发酚	氰化物
总 计		3 547 997.6	281 201.7	20 589.1	2 410.5	215.4
辽河	内蒙古	27 985.8	3 034.9	14.5	0	0
	辽 宁	110 369.4	9 142.4	880.4	69.1	4.3
	吉 林	20 292.6	918.5	117	2.2	2.1
	合 计	158 647.9	13 095.9	1 012	71.2	6.4
海河	北 京	7 117.7	433.8	77.5	0.2	0.2
	天 津	24 294.2	3 253.1	199.4	1.3	0
	河 北	191 214.6	17 632.3	1 289.5	307.3	16.6
	山 西	34 280.6	3 817.9	679.5	497.2	17
	内蒙古	787.6	10.9	1	0	0
	山 东	23 639.3	1 968	34.3	0.1	0.1
	河 南	52 401.5	3 317.1	140.2	1	3.9
	合 计	333 735.5	30 433.1	2 421.3	807	37.8
淮河	江 苏	119 444	9 046.2	743.2	28.4	6.8
	安 徽	41 039.1	5 789.3	369.9	3.2	2.1
	山 东	109 195	9 499	589.7	46.3	5.2
	河 南	91 484.4	6 477.3	772.1	145.9	11.2
	合 计	361 162.5	30 811.8	2 474.9	223.9	25.3
松花江	内蒙古	18 655.8	1 089.6	27.8	3.7	0
	吉 林	58 408.6	3 581.3	157.2	0.3	0.1
	黑龙江	102 372	6 251.2	1 251.4	25.6	2.1
	合 计	179 436.4	10 922	1 436.4	29.5	2.2
珠江	江 西	829.8	1 914.2	0.5	0	0
	湖 南	12 705.2	954.5	114.5	0	0.7
	广 东	243 959.2	15 559.7	830.5	13.2	14.8
	广 西	202 357.5	8 452.2	311.3	64.4	5.1
	贵 州	14 685.5	503.7	121.1	0	0.1
	云 南	28 624.4	2 225.1	148.5	1	1.1
	海 南	12 499.6	811.9	4.3	0.1	0
	福 建	3 450.4	735	12.9	7.8	0
	合 计	519 111.6	31 156.3	1 543.6	86.4	21.9

流域接纳工业废水及处理情况（四）（续表）

（2011）

单位：吨

流域	地 区名 称	工业废水中污染物排放量				
		化学需氧量	氨氮	石油类	挥发酚	氰化物
长江	上 海	27 357	2 611.4	773.8	5.7	3.5
	江 苏	119 874.7	7 664.1	829.8	34.5	9.6
	浙 江	35 221.7	3 455.5	83	2.6	0.4
	安 徽	50 595.3	2 788.5	389.9	3.5	4.6
	江 西	116 310.1	10 220.4	696.1	75.7	10.2
	河 南	13 093.5	1 294	64.9	0.1	0
	湖 北	144 718.5	17 091.3	1 136.4	20.8	9.1
	湖 南	158 241.4	26 914	753.1	102.5	12.1
	广 西	2 556.5	20.7	0	0	0
	重 庆	58 028.2	3 205.3	521.1	35	2.7
	四 川	129 045.4	5 673.5	547.4	55.4	1.8
	贵 州	50 821.3	2 951.4	360.4	1	8
	云 南	14 132.3	625.1	132.5	0.5	0.4
	西 藏	1.3	0.1	0	0	0
	陕 西	13 662.1	2 556.7	11.7	0	0.2
	甘 肃	2 939	20.5	5.4	0	0
	青 海	0	0	0	0	0
	合 计	936 598.4	87 092.6	6 305.5	337.4	62.7
黄河	山 西	56 827.7	5 050.8	805.4	550.5	24.2
	内蒙古	26 887.8	5 065.3	573	1.5	1.3
	山 东	7 951.9	312.8	93.9	1.6	0.1
	河 南	36 622.3	2 839.4	282.5	11.6	2.6
	四 川	80.3	6.8	0	0	0
	陕 西	92 827.5	6 137.8	836.3	199.1	7.7
	甘 肃	45 261.8	8 022.8	245.3	8.7	0
	青 海	24 231	644.5	65.5	0.8	0
	宁 夏	111 549.6	9 254.4	145.8	9.5	1.5
	合 计	402 240	37 334.5	3 047.6	783.3	37.5

流域接纳工业废水及处理情况（四）（续表）

（2011）

单位：吨

流域	地区名称	工业废水中污染物排放量				
		化学需氧量	氨氮	石油类	挥发酚	氰化物
东南诸河	浙 江	155 926.9	9 141.7	893	20.4	10.6
	安 徽	2 454.4	393.2	17	0.1	0
	福 建	92 464.9	6 965.2	428.3	8.7	7.3
	合 计	250 846.2	16 500	1 338.4	29.1	18
西北诸河	河 北	2 787.7	104.8	5.1	0	0
	内蒙古	17 957.3	2 097.6	186.9	0.1	0
	西 藏	0	0	0	0	0
	甘 肃	47 871.2	5 996.9	81.4	1.4	0.3
	青 海	16 354.6	1 262.3	254.9	0.7	0
	新 疆	187 629.6	12 133.2	275.4	39.9	3.3
	合 计	272 600.3	21 594.8	803.7	42.2	3.6
西南诸河	云 南	132 599	2 195	205.4	0.4	0
	西 藏	1 019.7	65.6	0.4	0	0
	青 海	0	0	0	0	0
	合 计	133 618.7	2 260.6	205.8	0.4	0

流域接纳工业废水及处理情况（五）

（2011）

单位：吨

流域	地区名称	工业废水中污染物排放量				
		汞	镉	总铬	铅	砷
	总　计	1.214	35.096	290.264	150.831	145.169
辽河	内蒙古	0.035	0.122	0.060	1.284	1.903
	辽　宁	0.002	0.072	0.624	0.889	0.459
	吉　林	0.002	0.001	0.026	0.084	0.034
	合　计	0.039	0.195	0.710	2.257	2.396
海河	北　京	…	0.001	0.446	0.068	0.013
	天　津	0.001	0.010	0.285	1.454	0.023
	河　北	…	0.007	8.429	0.440	0.045
	山　西	0.002	0.784	0.374	0.396	0.511
	内蒙古	…	…	…	…	…
	山　东	0.005	0.978	11.321	0.446	1.836
	河　南	0.001	0.229	3.591	0.749	0.061
	合　计	0.009	2.009	24.447	3.553	2.488
淮河	江　苏	0.091	0.109	5.664	2.333	0.238
	安　徽	0.001	0.004	0.445	0.184	2.114
	山　东	0.024	0.008	2.343	0.226	0.178
	河　南	0.003	0.617	9.181	2.100	0.666
	合　计	0.118	0.739	17.633	4.843	3.196
松花江	内蒙古	0.003	0.014	0.002	0.284	0.137
	吉　林	0.001	0.002	0.110	0.039	0.967
	黑龙江	…	0.001	0.867	0.023	0.072
	合　计	0.004	0.018	0.980	0.346	1.176
珠江	江　西	…	0.053	0.022	0.239	0.040
	湖　南	0.001	0.034	8.166	0.827	1.393
	广　东	0.045	1.104	74.665	11.164	2.126
	广　西	0.077	2.485	4.623	15.595	9.053
	贵　州	0.024	0.014	0.001	0.216	0.370
	云　南	0.005	0.353	0.030	5.702	1.487
	海　南	…	0.005	0.131	0.025	0.027
	福　建	0.003	0.052	0.039	0.686	0.133
	合　计	0.155	4.100	87.677	34.455	14.628

流域接纳工业废水及处理情况（五）（续表）

（2011）

单位：吨

流域	地区名称	工业废水中污染物排放量				
		汞	镉	总铬	铅	砷
长江	上　海	…	0.003	2.505	0.104	0.010
	江　苏	0.002	0.017	6.557	1.160	0.531
	浙　江	…	0.016	5.398	0.170	0.018
	安　徽	0.002	0.734	6.391	2.686	4.938
	江　西	0.087	2.705	22.519	8.984	10.755
	河　南	…	0.001	0.359	0.033	0.027
	湖　北	0.212	0.810	17.365	4.024	11.921
	湖　南	0.257	14.375	25.821	40.180	54.000
	广　西	0.001	0.002	0.008	0.001	0.005
	重　庆	…	…	0.592	0.075	1.377
	四　川	0.065	0.160	1.680	1.575	3.500
	贵　州	0.014	0.100	0.236	0.277	0.238
	云　南	0.035	2.058	0.024	11.974	7.412
	西　藏	…	…	…	…	…
	陕　西	0.030	0.922	0.024	2.322	0.723
	甘　肃	0.009	0.128	…	1.691	1.161
	青　海	…	…	…	…	…
	合　计	0.714	22.031	89.479	75.257	96.616
黄河	山　西	0.002	0.041	0.133	0.241	0.239
	内蒙古	0.003	0.270	0.025	1.132	2.244
	山　东	…	0.001	0.105	0.014	0.066
	河　南	0.017	1.908	24.195	4.138	1.021
	四　川	…	…	…	…	…
	陕　西	0.001	0.032	1.600	1.879	0.231
	甘　肃	0.046	0.354	0.703	1.844	1.451
	青　海	0.003	0.076	1.025	0.145	0.651
	宁　夏	0.004	0.028	0.448	0.102	0.211
	合　计	0.077	2.710	28.236	9.496	6.114

流域接纳工业废水及处理情况（五）（续表）

（2011）

单位：吨

流域	地区名称	工业废水中污染物排放量				
		汞	镉	总铬	铅	砷
东南诸河	浙 江	0.002	0.245	16.188	0.284	0.169
	安 徽	0.001	0.007	0.016	0.003	0.012
	福 建	0.042	0.280	15.237	4.348	1.351
	合 计	0.045	0.532	31.441	4.635	1.532
西北诸河	河 北	…	…	…	…	…
	内蒙古	…	0.130	0.001	0.333	0.628
	西 藏	…	…	…	…	…
	甘 肃	0.016	0.937	3.957	3.323	3.117
	青 海	0.007	0.068	…	0.955	1.412
	新 疆	0.019	0.717	5.697	1.213	2.615
	合 计	0.043	1.851	9.655	5.824	7.771
西南诸河	云 南	0.011	0.910	0.004	10.167	6.532
	西 藏	…	…	0.001	…	2.721
	青 海	…	…	…	…	…
	合 计	0.011	0.910	0.005	10.167	9.253

流域工业废水污染防治投资情况

（2011）

流域	地区名称	工业废水治理施工项目数/个	工业废水治理竣工项目数/个	工业废水治理项目完成投资/万元	废水治理竣工项目新增处理能力/（万吨/日）
	总　计	3 738	2 633	1 577 471.1	1 493.8
辽河	内蒙古	13	12	12 529.8	0.8
	辽　宁	75	41	63 953.5	19.1
	吉　林	6	4	2 523.6	1.1
	合　计	94	57	79 006.9	20.9
海河	北　京	25	21	5 892.6	2.4
	天　津	40	33	31 317.9	9.7
	河　北	90	71	70 720.7	14.9
	山　西	33	28	22 628.4	8.2
	内蒙古	0	0	…	…
	山　东	59	18	42 850.7	193.6
	河　南	34	23	13 049.7	14.9
	合　计	281	194	186 459.9	243.7
淮河	江　苏	78	60	22 930.8	67.8
	安　徽	40	33	20 196.7	6.6
	山　东	283	223	249 810.0	106.1
	河　南	35	28	25 550.9	13.0
	合　计	436	344	318 488.4	193.5
松花江	内蒙古	5	2	2 416.0	0.4
	吉　林	30	26	22 535.6	4.9
	黑龙江	25	13	18 565.3	5.1
	合　计	60	41	43 516.9	10.3
珠江	江　西	1	1	440.0	0.1
	湖　南	14	14	930.0	0.7
	广　东	365	243	55 886.1	92.2
	广　西	139	98	45 998.3	31.4
	贵　州	31	24	3 383.6	2.5
	云　南	93	79	16 558.6	14.6
	海　南	28	19	18 403.8	7.3
	福　建	14	11	868.6	1.4
	合　计	685	489	142 468.9	150.2

流域工业废水污染防治投资情况（续表）

（2011）

流域	地名 区称	工业废水治理施工项目数/个	工业废水治理竣工项目数/个	工业废水治理项目完成投资/万元	废水治理竣工项目新增处理能力/（万吨/日）
长江	上 海	33	22	4 929.7	1.2
	江 苏	304	256	102 549.7	45.3
	浙 江	64	47	17 226.3	5.0
	安 徽	51	46	6 668.3	2.4
	江 西	102	66	24 000.6	47.7
	河 南	5	3	3 975.0	4.5
	湖 北	74	59	35 041.4	28.1
	湖 南	99	85	49 285.4	32.4
	广 西	0	0	…	…
	重 庆	115	73	24 915.1	5.4
	四 川	167	126	65 497.0	45.3
	贵 州	64	49	4 489.3	3.3
	云 南	38	25	6 775.5	11.3
	西 藏	0	0	…	…
	陕 西	15	13	2 311.5	0.9
	甘 肃	12	9	5 978.5	2.1
	青 海	0	0	…	…
	合 计	1 143	879	353 643.4	234.7
黄河	山 西	77	51	26 229.2	16.2
	内蒙古	36	12	19 114.0	66.6
	山 东	11	9	2 879.5	6.5
	河 南	43	39	8 283.7	4.7
	四 川	3	3	639.2	…
	陕 西	85	48	155 400.5	21.0
	甘 肃	8	3	6 788.0	0.6
	青 海	6	6	8 338.0	2.1
	宁 夏	17	11	17 416.9	17.6
	合 计	286	182	245 089.0	135.3

流域工业废水污染防治投资情况（续表）

（2011）

流域	地区名称	工业废水治理施工项目数/个	工业废水治理竣工项目数/个	工业废水治理项目完成投资/万元	废水治理竣工项目新增处理能力/（万吨/日）
东南诸河	浙 江	399	289	70 871.5	54.9
	安 徽	0	0	…	…
	福 建	179	100	53 399.2	40.2
	合 计	578	389	124 270.7	95.1
西北诸河	河 北	0	0	…	…
	内蒙古	6	3	6 747.7	0.7
	西 藏	0	0	…	…
	甘 肃	23	18	16 318.5	34.5
	青 海	5	1	172.0	…
	新 疆	78	10	43 093.3	151.5
	合 计	112	32	66 331	187
西南诸河	云 南	59	25	17 535.5	193.3
	西 藏	4	1	660.0	30.0
	青 海	0	0	…	…
	合 计	63	26	18 195.5	223.3

流域农业污染排放情况

（2011）　　　　　　　　　　　　　　　　　　　　　　　　单位：万吨

| 流域 | 地区名称 | 农业污染物排放（流失）总量 | | | |
		化学需氧量	总氮	总磷	氨氮
	总　计	1 186.1	454.7	56.3	82.7
辽河	内蒙古	16.9	5.0	0.5	0.4
	辽　宁	88.3	20.0	2.8	3.6
	吉　林	12.8	3.2	0.4	0.5
	合　计	118.1	28.1	3.7	4.5
海河	北　京	8.2	3.3	0.4	0.5
	天　津	11.5	3.7	0.5	0.6
	河　北	92.9	44.6	6.8	4.5
	山　西	6.8	3.1	0.4	0.4
	内蒙古	2.8	0.6	…	…
	山　东	36.7	22.5	2.1	1.8
	河　南	16.4	7.2	1.0	1.3
	合　计	175.2	85.0	11.2	9.2
淮河	江　苏	30.1	12.5	1.3	2.8
	安　徽	28.5	12.8	1.9	2.7
	山　东	91.5	30.1	3.9	5.3
	河　南	48.5	24.9	2.9	4.0
	合　计	198.6	80.3	10.0	14.8
松花江	内蒙古	17.0	3.7	0.5	0.4
	吉　林	40.0	7.8	1.0	1.3
	黑龙江	110.7	26.9	2.9	3.6
	合　计	167.7	38.5	4.4	5.2
珠江	江　西	0.6	0.2	…	0.1
	湖　南	0.4	0.2	…	0.1
	广　东	62.0	20.1	3.1	5.9
	广　西	21.5	11.2	1.4	2.7
	贵　州	1.0	1.1	0.1	0.2
	云　南	2.5	2.2	0.2	0.4
	海　南	10.7	3.9	0.6	0.9
	福　建	3.6	1.2	0.2	0.5
	合　计	102.4	40.2	5.7	10.7

流域农业污染排放情况（续表）

（2011）

单位：万吨

流域	地区名称	农业污染物排放（流失）总量			
		化学需氧量	总氮	总磷	氨氮
长江	上 海	3.4	1.5	0.2	0.3
	江 苏	9.8	5.1	0.6	1.2
	浙 江	6.4	2.5	0.3	0.8
	安 徽	10.9	5.8	1.1	1.1
	江 西	24.3	9.2	1.3	3.0
	河 南	6.1	4.6	0.5	0.6
	湖 北	48.1	19.8	2.5	4.7
	湖 南	58.1	22.9	2.8	6.4
	广 西	0.7	0.3	…	0.1
	重 庆	12.7	5.5	0.7	1.4
	四 川	54.9	23.0	2.7	5.9
	贵 州	4.6	3.3	0.3	0.6
	云 南	2.0	2.7	0.3	0.4
	西 藏	…	…	…	…
	陕 西	4.5	2.3	0.3	0.6
	甘 肃	2.3	0.8	0.1	0.1
	青 海	…	…	…	…
	合 计	249.0	109.4	13.6	27.2
黄河	山 西	11.6	5.7	0.6	0.9
	内蒙古	21.5	5.0	0.5	0.4
	山 东	9.8	3.4	0.4	0.4
	河 南	11.2	4.8	0.6	0.8
	四 川	…	…	…	…
	陕 西	15.6	7.9	0.8	0.9
	甘 肃	6.6	2.6	0.3	0.3
	青 海	2.0	0.7	0.1	0.1
	宁 夏	10.3	3.1	0.3	0.2
	合 计	88.7	33.1	3.5	4.0

流域农业污染排放情况（续表）

（2011）

单位：万吨

| 流域 | 地 区名 称 | 农业污染物排放（流失）总量 | | | |
		化学需氧量	总氮	总磷	氨氮
东南诸河	浙 江	14.9	6.7	0.9	2.0
	安 徽	0.5	0.3	…	0.1
	福 建	19.0	8.3	1.1	2.9
	合 计	34.3	15.3	2.0	5.0
西北诸河	河 北	1.2	0.4	…	…
	内蒙古	6.5	1.7	0.2	0.1
	西 藏	…	…	…	…
	甘 肃	5.8	1.8	0.1	0.2
	青 海	0.2	0.1	…	…
	新 疆	35.1	17.1	1.4	1.2
	合 计	48.8	21.0	1.8	1.5
西南诸河	云 南	2.8	3.2	0.4	0.4
	西 藏	0.5	0.5	0.1	…
	青 海	…	…	…	…
	合 计	3.2	3.7	0.5	0.5

流域城镇生活污染排放及处理情况（一）

（2011）

流域	地区名称	城镇人口/万人	生活用水总量/万吨	污水处理厂数/座	污水处理厂设计处理能力/（万吨/日）	生活污水实际处理量/万吨
	总　计	69 258	4 983 750	3 974	13 991	3 497 278
辽河	内蒙古	303	15 637	22	70	12 070
	辽宁	2 807	176 838	121	568	127 162
	吉林	303	17 843	8	40	8 794
	合计	3 413	210 318	151	678	148 025
海河	北京	1 741	151 935	116	410	116 527
	天津	1 091	52 580	52	235	49 342
	河北	3 237	186 652	232	803	155 886
	山西	583	31 900	57	95	20 914
	内蒙古	27	1 119	1	1	42
	山东	610	34 942	38	145	26 670
	河南	679	51 046	28	148	37 858
	合计	7 968	510 175	524	1 837	407 239
淮河	江苏	2 148	149 661	174	368	77 561
	安徽	1 379	93 249	47	194	48 576
	山东	3 931	244 512	218	865	393 173
	河南	2 147	167 640	84	403	110 225
	合计	9 604	655 062	523	1 829	629 535
松花江	内蒙古	253	14 043	12	34	4 765
	吉林	1 165	71 217	39	211	44 004
	黑龙江	2 176	126 892	52	260	50 549
	合计	3 594	212 152	103	505	99 318
珠江	江西	21	1 458	2	2	589
	湖南	31	2 138	2	3	796
	广东	6 986	716 001	342	1 742	483 085
	广西	1 919	135 979	106	363	82 849
	贵州	216	13 454	26	18	3 139
	云南	458	29 654	17	49	10 194
	海南	443	37 087	44	129	35 024
	福建	98	7 410	4	21	5 703
	合计	10 172	943 181	543	2 328	621 378

流域城镇生活污染排放及处理情况（一）（续表）

（2011）

流域	地区名称	城镇人口/万人	生活用水总量/万吨	污水处理厂数/座	污水处理厂设计处理能力/（万吨/日）	生活污水实际处理量/万吨
长江	上 海	2 096	188 100	55	670	155 435
	江 苏	2 744	199 348	386	953	192 802
	浙 江	682	58 530	59	174	40 684
	安 徽	1 240	93 655	53	239	68 183
	江 西	2 030	140 909	110	283	83 256
	河 南	361	24 714	12	38	9 829
	湖 北	2 914	214 147	135	542	139 023
	湖 南	2 944	217 215	126	486	129 035
	广 西	23	1 828	4	4	374
	重 庆	1 606	114 536	117	261	74 780
	四 川	3 401	235 344	227	455	128 195
	贵 州	997	70 779	77	147	40 112
	云 南	682	56 924	26	130	44 116
	西 藏	1	53	0	0	0
	陕 西	308	16 853	6	23	4 921
	甘 肃	58	3 654	2	3	330
	青 海	9	511	0	0	0
	合 计	22 096	1 637 099	1 395	4 406	1 111 077
黄河	山 西	1 176	76 742	103	179	41 441
	内蒙古	671	38 965	47	140	24 127
	山 东	369	13 203	14	56	13 319
	河 南	623	44 336	37	119	31 303
	四 川	3	160	0	0	0
	陕 西	1 463	87 704	101	249	51 969
	甘 肃	700	35 934	14	68	13 906
	青 海	206	11 107	14	31	6 026
	宁 夏	319	22 687	18	74	16 157
	合 计	5 529	330 838	348	916	198 248

流域城镇生活污染排放及处理情况（一）（续表）

流域	地区名称	城镇人口/万人	生活用水总量/万吨	污水处理厂数/座	污水处理厂设计处理能力/（万吨/日）	生活污水实际处理量/万吨
东南诸河	浙 江	2 721	204 052	165	803	138 819
	安 徽	55	3 868	5	13	2 991
	福 建	2 062	152 544	117	361	89 947
	合 计	4 838	360 464	287	1 176	231 757
西北诸河	河 北	23	1 221	3	6	703
	内蒙古	151	6 179	16	22	2 097
	西 藏	2	91	0	0	0
	甘 肃	196	10 587	12	54	6 343
	青 海	45	3 896	4	6	1 696
	新 疆	963	66 529	50	200	34 076
	合 计	1 378	88 503	85	287	44 914
西南诸河	云 南	566	30 870	13	28	5 551
	西 藏	96	5 024	2	1	236
	青 海	4	66	0	0	0
	合 计	666	35 960	15	29	5 787

流域城镇生活污染排放及处理情况（二）

（2011）

流域	地区名称	城镇生活污水排放量/万吨	城镇生活化学需氧量产生量/吨	城镇生活氨氮产生量/吨	城镇生活化学需氧量排放量/吨	城镇生活氨氮排放量/吨
	总 计	4 279 159	16 984 202	2 166 719	9 388 423	1 476 626
辽河	内蒙古	14 234	65 651	8 248	27 716	7 531
	辽宁	141 699	641 690	88 513	341 578	65 354
	吉林	15 363	67 899	8 737	48 215	7 864
	合计	171 296	775 240	105 499	417 510	80 749
海河	北京	136 741	501 931	61 629	99 097	15 513
	天津	47 322	274 010	37 356	96 422	17 128
	河北	158 930	719 634	95 678	243 589	49 723
	山西	24 844	140 333	18 596	70 403	12 275
	内蒙古	947	4 692	673	4 601	656
	山东	31 191	145 559	18 721	68 684	11 187
	河南	42 776	152 339	19 538	62 580	12 207
	合计	442 752	1 938 498	252 191	645 377	118 690
淮河	江苏	142 118	545 021	70 545	377 013	54 210
	安徽	85 142	326 288	38 738	243 230	31 172
	山东	206 437	925 073	123 889	344 085	66 159
	河南	138 012	503 821	64 464	249 785	42 917
	合计	571 709	2 300 202	297 636	1 214 113	194 458
松花江	内蒙古	12 526	52 435	7 203	35 331	5 577
	吉林	58 850	267 640	34 366	156 606	26 222
	黑龙江	106 574	486 354	64 693	365 134	54 365
	合计	177 950	806 429	106 262	557 071	86 164
珠江	江西	1 239	5 136	604	4 363	489
	湖南	1 391	7 522	936	6 181	829
	广东	606 589	1 784 936	229 524	1 006 295	154 633
	广西	119 554	462 446	58 706	358 231	46 701
	贵州	10 450	50 621	6 312	42 229	5 394
	云南	22 894	110 929	13 827	95 431	12 607
	海南	28 858	122 194	14 542	78 858	12 365
	福建	6 301	23 985	3 115	15 495	2 499
	合计	797 277	2 567 770	327 565	1 607 082	235 516

流域城镇生活污染排放及处理情况（二）（续表）

（2011）

流域	地 区名 称	城镇生活污水排放量/万吨	城镇生活化学需氧量产生量/吨	城镇生活氨氮产生量/吨	城镇生活化学需氧量排放量/吨	城镇生活氨氮排放量/吨
长江	上 海	169 290	581 430	74 209	181 866	43 963
	江 苏	204 134	701 208	90 518	225 239	45 643
	浙 江	48 979	172 532	21 691	58 152	13 866
	安 徽	84 012	295 075	34 859	197 069	28 144
	江 西	121 758	640 100	73 709	385 338	48 593
	河 南	20 843	70 359	10 817	46 614	7 194
	湖 北	188 384	729 005	85 902	461 323	64 645
	湖 南	179 852	708 704	86 032	525 670	70 383
	广 西	1 554	5 656	720	3 441	604
	重 庆	97 356	398 609	50 999	230 769	38 069
	四 川	199 104	870 355	106 638	609 658	78 486
	贵 州	46 811	225 131	28 644	172 657	22 777
	云 南	49 134	178 407	20 005	61 560	11 389
	西 藏	38	217	28	186	25
	陕 西	14 972	69 439	9 200	56 282	7 870
	甘 肃	2 499	11 643	1 605	11 342	1 566
	青 海	340	1 975	253	1 975	253
	合 计	1 429 060	5 659 843	695 828	3 229 141	483 471
黄河	山 西	51 595	277 820	36 872	141 001	25 028
	内蒙古	28 483	159 995	18 773	85 912	12 735
	山 东	18 242	84 491	11 278	44 501	7 523
	河 南	38 367	136 373	17 888	51 261	10 174
	四 川	141	702	87	702	87
	陕 西	65 961	336 160	44 615	186 989	30 271
	甘 肃	28 677	150 923	19 591	119 739	17 527
	青 海	9 203	45 329	5 850	29 276	5 169
	宁 夏	20 138	71 650	9 591	18 568	6 398
	合 计	260 808	1 263 444	164 544	677 951	114 910

流域城镇生活污染排放及处理情况（二）（续表）

（2011）

流域	地名	区称	城镇生活污水排放量/万吨	城镇生活化学需氧量产生量/吨	城镇生活氨氮产生量/吨	城镇生活化学需氧量排放量/吨	城镇生活氨氮排放量/吨
东南诸河	浙	江	188 515	687 831	91 852	348 788	60 670
	安	徽	3 230	12 983	1 538	10 210	1 260
	福	建	132 428	504 447	65 516	338 698	50 761
	合	计	324 172	1 205 261	158 906	697 697	112 690
西北诸河	河	北	1 038	4 862	664	2 531	396
	内蒙古		4 782	27 518	3 707	23 563	3 261
	西	藏	77	418	54	418	54
	甘	肃	8 315	41 438	5 204	21 222	3 419
	青	海	2 910	7 772	1 265	7 063	1 180
	新	疆	54 522	217 675	26 998	129 744	22 084
	合	计	71 645	299 683	37 891	184 542	30 395
西南诸河	云	南	28 180	145 874	17 579	136 482	16 803
	西	藏	4 157	21 070	2 704	20 568	2 666
	青	海	153	889	114	889	114
	合	计	32 489	167 832	20 397	157 939	19 583

湖泊水库接纳工业废水及处理情况（一）

（2011）

流域	地区名称	工业废水排放量/万吨	直接排入环境的	排入污水处理厂的	工业废水处理量/万吨	废水治理设施数/套	废水治理设施治理能力/（万吨/日）	废水治理设施运行费用/万元
总　计		341 842	254 253	87 589	653 937	16 506	4 088	967 043.5
滇　池	云南	902	603	300	2 041	143	30	5 772.8
巢　湖	安徽	6 663	6 240	424	27 065	302	86	15 147.9
洞庭湖	江西	1 929	1 929	0	28 940	125	86	8 722.9
	湖北	827	810	17	552	12	6	676.2
	湖南	93 614	90 130	3 484	237 285	3 037	658	261 918.1
	广西	1 296	1 295	0	1 760	114	9	828.2
	合计	97 666	94 164	3 502	268 537	3 288	760	272 145.4
鄱阳湖	安徽	28	28	0	9	7	0	23.7
	江西	68 060	66 599	1 461	109 069	2 797	656	135 786.1
	合计	68 088	66 626	1 461	109 078	2 804	657	135 809.8
太　湖	上海	1 822	330	1 492	1 097	3 921	7	67 826.6
	江苏	115 033	64 217	50 816	132 773	3 278	1 693	324 818.4
	浙江	40 371	11 123	29 248	86 732	1 946	435	117 489.2
	合计	157 227	75 671	81 556	220 603	9 145	2 134	510 134.2
丹江口	河南	4 323	4 323	0	7 061	216	68	11 145.1
	湖北	2 192	2 192	0	2 021	150	257	4 631.6
	陕西	4 781	4 435	346	17 531	458	98	12 256.7
	合计	11 296	10 949	346	26 613	824	422	28 033.4

湖泊水库接纳工业废水及处理情况（二）

（2011）

单位：吨

流域	地区名称	工业废水中污染物排放量				
		汞	镉	总铬	铅	砷
总　计		0.378	18.588	77.699	52.588	69.097
滇　池	云　南	…	0.047	…	0.178	0.756
巢　湖	安　徽	…	0.004	0.028	0.061	0.092
洞庭湖	江　西	…	…	0.061	0.012	…
	湖　北	…	…	…	…	0.098
	湖　南	0.256	14.396	33.984	41.007	55.304
	广　西	0.001	0.005	0.010	0.003	0.006
	合　计	0.258	14.400	34.055	41.022	55.408
鄱阳湖	安　徽	…	…	…	0.023	…
	江　西	0.087	2.705	22.432	8.972	10.747
	合　计	0.087	2.705	22.432	8.995	10.747
太　湖	上　海	…	…	0.171	0.023	…
	江　苏	…	0.005	3.598	0.323	0.019
	浙　江	…	0.016	5.385	0.170	0.018
	合　计	…	0.021	9.153	0.516	0.037
丹江口	河　南	0.002	0.767	2.401	0.556	0.330
	湖　北	0.001	0.010	9.605	0.030	1.252
	陕　西	0.030	0.634	0.023	1.230	0.475
	合　计	0.033	1.411	12.029	1.817	2.057

湖泊水库接纳工业废水及处理情况（三）

（2011）

単位：吨

流域	地区名称	工业废水中污染物排放量				
		氰化物	化学需氧量	石油类	氨氮	挥发酚
总　计		31.5	431 132.3	2 270.3	50 247.4	196.2
滇　池	云　南	0.1	1 224.3	48.7	61.1	0.1
巢　湖	安　徽	…	12 291.2	52.6	698.2	…
洞庭湖	江　西	…	5 387.8	33.5	152.5	1.1
	湖　北	…	2 797.2	5.4	56.3	2.2
	湖　南	12.8	161 983.6	852.8	27 598.1	99.9
	广　西	…	3 508.3	15.7	208.0	…
	合　计	12.8	173 676.9	907.4	28 014.9	103.2
鄱阳湖	安　徽	…	44.1	5.5	1.2	…
	江　西	10.2	109 033.4	658.4	9 923.7	74.6
	合　计	10.2	109 077.4	663.9	9 924.9	74.6
太　湖	上　海	0.3	918.8	13.5	124.3	…
	江　苏	6.9	70 854.3	316.9	4 459.2	13.8
	浙　江	0.4	35 368.1	82.2	3 485.0	2.6
	合　计	7.7	107 141.2	412.6	8 068.6	16.4
丹江口	河　南	…	4 738.1	22.1	253.0	0.1
	湖　北	0.5	9 407.4	151.8	749.3	1.7
	陕　西	0.1	13 575.9	11.2	2 477.4	…
	合　计	0.7	27 721.3	185.1	3 479.8	1.8

湖泊水库接纳工业废水及处理情况（四）

（2011）

单位：吨

流域	地区名称	工业废水中污染物产生量				
		汞	镉	总铬	铅	砷
总　计		5.220	672.505	578.767	4 589.288	4 589.288
滇　池	云　南	…	0.047	0.008	0.178	0.769
巢　湖	安　徽	…	0.059	1.656	0.648	0.475
洞庭湖	江　西	…	0.002	0.061	0.033	0.033
	湖　北	…	…	…	…	0.098
	湖　南	4.482	204.078	93.092	497.694	517.998
	广　西	0.014	0.024	0.100	0.104	3.683
	合　计	4.496	204.104	93.253	497.832	521.811
鄱阳湖	安　徽	…	…	…	0.209	…
	江　西	0.365	376.146	57.031	85.256	4 022.250
	合　计	0.365	376.146	57.031	85.465	4 022.250
太　湖	上　海	…	…	18.755	0.118	0.001
	江　苏	…	1.822	175.961	3.750	3.886
	浙　江	…	0.016	209.211	2.805	0.018
	合　计	…	1.838	403.927	6.673	3.904
丹江口	河　南	0.002	0.880	6.745	9.892	1.499
	湖　北	0.001	0.010	12.530	0.030	1.252
	陕　西	0.354	89.422	3.616	239.823	37.328
	合　计	0.358	90.312	22.891	249.745	40.079

湖泊水库接纳工业废水及处理情况（五）

（2011）

单位：吨

流域	地区名称	工业废水中污染物产生量				
		氰化物	化学需氧量	石油类	氨氮	挥发酚
总　计		544.2	2 484 485.4	18 185.7	171 362.2	3 586.1
滇　池	云　南	1.2	20 461.8	218.3	1 021.5	5.0
巢　湖	安　徽	3.2	73 747.2	1 044.2	23 070.2	0.1
洞庭湖	江　西	9.4	25 069.2	1 172.6	1 126.7	136.8
	湖　北	…	16 680.8	6.1	63.1	35.2
	湖　南	364.7	552 786.8	4 262.1	65 252.8	1 248.0
	广　西	…	7 214.7	15.7	187.0	…
	合　计	374.1	601 751.5	5 456.5	66 629.6	1 420.0
鄱阳湖	安　徽	…	871.4	132.7	1.1	…
	江　西	50.9	511 528.6	4 957.5	38 031.2	1 684.6
	合　计	50.9	512 400.0	5 090.2	38 032.3	1 684.6
太　湖	上　海	15.1	10 519.7	90.7	330.6	…
	江　苏	91.9	722 691.5	4 573.6	25 024.3	169.9
	浙　江	3.6	445 406.7	867.6	9 736.4	300.9
	合　计	110.6	1 178 618.0	5 531.8	35 091.3	470.8
丹江口	河　南	1.3	46 517.7	93.4	693.2	0.6
	湖　北	1.7	15 376.4	514.9	1 811.4	5.0
	陕　西	1.1	35 612.9	236.4	5 012.7	0.1
	合　计	4.1	97 506.9	844.7	7 517.3	5.6

湖泊水库工业污染防治投资情况

（2011）

流域	地区名称		工业废水治理施工项目数/个	工业废水治理竣工项目数/个	工业废水治理项目完成投资/万元	废水治理竣工项目新增处理能力/（万吨/日）
总　计			478	380	163 562.1	118.2
滇　池	云　南		8	5	1 069.5	0.2
巢　湖	安　徽		10	6	1 789.3	0.5
洞庭湖	江　西		3	3	126.0	3.0
	湖　北		6	5	950.0	0.1
	湖　南		100	94	48 977.4	29.5
	广　西		0	0	…	…
	合　计		117	102	50 053.4	32.6
鄱阳湖	安　徽		0	0	…	…
	江　西		99	63	23 874.6	44.6
	合　计		99	63	23 874.6	44.6
太　湖	上　海		1	1	30.0	…
	江　苏		151	131	64 483.6	31.1
	浙　江		65	48	17 475.0	5.1
	合　计		217	180	81 988.7	36.2
丹江口	河　南		1	1	850.0	3.0
	湖　北		9	8	1 860.7	0.4
	陕　西		17	15	2 076.0	0.8
	合　计		27	24	4 786.7	4.1

湖泊水库农业污染排放情况

（2011）

单位：万吨

流域	地 区名 称	农业污染物排放/流失总量			
		化学需氧量	总氮	总磷	氨氮
总 计		108.0	43.7	5.7	12.3
滇 池	云 南	0.4	0.2	0.1	0.1
巢 湖	安 徽	4.7	1.9	0.4	0.4
洞庭湖	江 西	0.6	0.2	…	0.1
	湖 北	0.5	0.3	…	0.1
	湖 南	57.4	22.6	2.8	6.3
	广 西	1.0	0.4	0.1	0.1
	合 计	59.4	23.6	2.9	6.6
鄱阳湖	安 徽	…	…	…	…
	江 西	23.6	9.0	1.3	2.9
	合 计	23.6	9.0	1.3	2.9
太 湖	上 海	0.2	0.1	…	…
	江 苏	4.9	2.2	0.3	0.5
	浙 江	6.4	2.5	0.3	0.8
	合 计	11.5	4.9	0.6	1.3
丹江口	河 南	2.3	1.1	0.1	0.2
	湖 北	1.1	0.6	0.1	0.1
	陕 西	5.0	2.5	0.3	0.7
	合 计	8.4	4.2	0.5	1.0

湖泊水库城镇生活污染排放及处理情况（一）

（2011）

流域	地区名称	城镇人口/万人	生活用水总量/万吨	污水处理厂数/座	污水处理厂设计处理能力/（万吨/日）	生活污水实际处理量/万吨
总计		8 795	644 376	603	1 823	439 172
滇池	云南	313	33 623	9	74	24 993
巢湖	安徽	484	38 952	15	95	32 579
洞庭湖	江西	91	4 534	3	11	2 656
	湖北	33	2 264	1	2	595
	湖南	2 916	215 060	125	476	129 353
	广西	34	2 679	5	7	929
	合计	3 074	224 538	134	495	133 532
鄱阳湖	安徽	7	400	1	1	256
	江西	1 918	134 864	105	270	80 197
	合计	1 925	135 264	106	271	80 453
太湖	上海	77	6 906	12	25	6 440
	江苏	1 497	95 439	247	619	106 183
	浙江	833	72 258	57	166	39 389
	合计	2 406	174 603	316	810	152 012
丹江口	河南	123	10 406	6	15	3 511
	湖北	154	9 885	13	41	7 317
	陕西	317	17 104	4	22	4 775
	合计	593	37 396	23	78	15 602

湖泊水库城镇生活污染排放及处理情况（二）

（2011）

流域	地区名称	城镇生活污水排放量/万吨	城镇生活化学需氧量产生量/吨	城镇生活氨氮产生量/吨	城镇生活化学需氧量排放量/吨	城镇生活氨氮排放量/吨
	总　计	571 953	2 292 409	27 7781	1 227 912	181 151
滇　池	云南	28 580	99 074	9 539	4 730	2 610
巢　湖	安徽	36 717	113 738	13 408	68 336	9 695
洞庭湖	江西	3 627	22 512	2 648	18 164	2 258
	湖北	1 927	7 950	901	6 198	840
	湖南	177 932	702 102	85 252	519 762	69 653
	广西	2 278	8 288	1 056	5 044	867
	合计	185 763	740 852	89 857	549 166	73 619
鄱阳湖	安徽	448	1 732	205	585	74
	江西	116 851	600 246	69 137	363 458	45 862
	合计	117 298	601 977	69 342	364 043	45 936
太　湖	上海	6 215	21 347	2 725	6 677	1 614
	江苏	105 104	378 321	48 940	75 315	17 853
	浙江	60 181	203 719	26 345	61 624	15 082
	合计	171 501	603 388	78 010	143 617	34 549
丹江口	河南	7 722	25 886	3 696	16 796	2 749
	湖北	9 212	36 084	4 507	22 447	3 852
	陕西	15 160	71 411	9 422	58 776	8 141
	合计	32 094	133 380	17 625	98 020	14 742

三峡地区接纳工业废水及处理情况（一）

（2011）

流域	地区名称	工业废水排放量/万吨	直接排入环境的	排入污水处理厂的	工业废水处理量/万吨	废水治理设施数/套	废水治理设施治理能力/（万吨/日）	废水治理设施运行费用/万元
	总 计	144 477	135 564	8 912	402 163	7 297	1 882	768 893.3
库 区	湖北	1 460	1 365	95	1 392	52	10	1 471.0
	重庆	16 034	15 003	1 030	21 266	823	242	39 259.7
	合计	17 494	16 368	1 126	22 658	875	253	40 730.7
影响区	湖北	8 330	7 398	931	7 375	134	30	5 750.4
	重庆	17 211	17 111	101	15 383	570	90	25 573.6
	四川	14 598	14 345	253	14 897	710	119	29 840.7
	贵州	1 503	1 503	0	1 039	77	8	3 638.3
	合计	41 642	40 357	1 285	38 693	1 491	247	64 803.0
上游区	重庆	413	413	0	371	64	3	724.0
	四川	65 109	59 529	5 580	270 662	3 304	994	590 776.9
	贵州	10 686	10 221	465	28 075	743	215	29 250.1
	云南	9 133	8 676	457	41 703	820	170	42 608.6
	合计	85 341	78 839	6 501	340 811	4 931	1 383	663 359.6

三峡地区接纳工业废水及处理情况（二）

（2011）

单位：吨

流域	地区名称	工业废水中污染物排放量				
		汞	镉	总铬	铅	砷
	总 计	0.106	2.391	2.395	14.667	13.029
库 区	湖 北	…	…	…	…	…
	重 庆	…	…	0.470	0.074	1.323
	合 计	…	…	0.470	0.074	1.323
影响区	湖 北	0.002	…	0.052	…	…
	重 庆	…	…	0.058	0.001	0.052
	四 川	0.005	…	0.308	0.060	0.130
	贵 州	…	…	…	…	…
	合 计	0.007	…	0.417	0.061	0.182
上游区	重 庆	…	…	0.059	…	0.002
	四 川	0.060	0.160	1.372	1.515	3.369
	贵 州	0.003	0.092	0.075	0.181	0.082
	云 南	0.035	2.139	0.001	12.836	8.070
	合 计	0.099	2.390	1.508	14.531	11.524

三峡地区接纳工业废水及处理情况（三）

（2011）

单位：吨

流域	地区名称	工业废水中污染物排放量				
		氰化物	化学需氧量	石油类	氨氮	挥发酚
总　计		8.5	250 471.0	1 619.8	12 186.3	91.3
库　区	湖　北	…	4 525.3	2.8	191.2	…
	重　庆	0.7	29 072.2	257.5	1 824.2	1.7
	合　计	0.7	33 597.5	260.3	2 015.4	1.7
影响区	湖　北	…	8 873.1	3.9	1 239.9	…
	重　庆	1.5	27 895.8	259.9	875.0	33.3
	四　川	0.7	35 713.5	113.8	1 694.0	0.1
	贵　州	…	14 568.7	…	303.8	…
	合　计	2.2	87 051.1	377.6	4 112.7	33.4
上游区	重　庆	0.6	841.6	3.1	490.7	…
	四　川	1.1	92 575.1	431.8	3 941.3	55.4
	贵　州	3.6	22 914.8	277.2	1 078.8	0.1
	云　南	0.3	13 490.8	269.8	547.4	0.7
	合　计	5.6	129 822.3	981.9	6 058.2	56.2

三峡地区接纳工业废水及处理情况（四）

（2011）

单位：吨

流域	地区名称	工业废水中污染物产生量				
		汞	镉	总铬	铅	砷
总　计		2.141	153.470	302.420	565.024	565.024
库　区	湖　北	…	…	…	…	…
	重　庆	…	0.001	212.103	5.685	1.439
	合　计	…	0.001	212.103	5.685	1.439
影响区	湖　北	0.022	…	5.127	0.001	…
	重　庆	…	0.003	8.079	0.409	2.088
	四　川	0.747	0.077	11.565	1.022	64.867
	贵　州	…	…	…	…	…
	合　计	0.768	0.080	24.771	1.432	66.956
上游区	重　庆	…	…	0.314	…	0.002
	四　川	0.855	52.245	61.301	77.507	261.819
	贵　州	0.004	0.460	3.048	0.529	1.031
	云　南	0.514	100.685	0.882	188.059	233.778
	合　计	1.372	153.389	65.546	266.095	496.630

三峡地区接纳工业废水及处理情况（五）

（2011）

单位：吨

流域	地 区名 称	工业废水中污染物产生量				
		氰化物	化学需氧量	石油类	氨氮	挥发酚
总 计		278.5	1 360 054.8	12 098.9	155 332.8	4 090.3
库 区	湖 北	…	5 781.1	2.9	267.1	…
	重 庆	65.9	148 302.5	2 704.6	25 662.7	931.2
	合 计	65.9	154 083.6	2 707.4	25 929.8	931.2
影响区	湖 北	4.2	41 668.8	94.0	2 001.8	67.5
	重 庆	21.6	101 308.8	783.7	2 308.7	288.2
	四 川	4.5	311 002.3	454.8	36 486.3	81.3
	贵 州	…	38 984.0	…	12 272.3	…
	合 计	30.3	492 963.9	1 332.6	53 069.2	437.0
上游区	重 庆	0.8	1 847.4	3.4	498.4	…
	四 川	57.9	545 279.3	5 760.2	30 448.1	1 780.3
	贵 州	98.1	65 640.4	1 157.9	13 227.7	239.2
	云 南	25.3	100 240.1	1 137.4	32 159.5	702.4
	合 计	182.2	713 007.2	8 058.9	76 333.8	2 722.0

三峡地区工业污染防治投资情况

（2011）

流域	地 区名 称	工业废水治理施工项目数/个	工业废水治理竣工项目数/个	工业废水治理项目完成投资/万元	废水治理竣工项目新增处理能力/（万吨/日）
总 计		401	290	109 429.1	67.7
库区	湖 北	3	3	5 461.0	1.5
	重 庆	73	52	16 875.0	3.6
	合 计	76	55	22 336.0	5.0
影响区	湖 北	4	4	1 880.0	…
	重 庆	37	18	7 674.1	1.6
	四 川	39	32	20 392.4	8.8
	贵 州	2	0	…	…
	合 计	82	54	29 946.5	10.4
上游区	重 庆	1	0	50.0	0.1
	四 川	131	97	45 743.8	36.6
	贵 州	58	46	2 933.8	3.1
	云 南	53	38	8 419.0	12.5
	合 计	243	181	57 146.6	52.2

三峡地区农业污染排放情况

（2011）

流域	地 区名 称	农业污染物排放/流失总量			
		化学需氧量	总氮	总磷	氨氮
总 计		75.0	35.1	4.0	8.3
库区	湖 北	0.5	0.3	...	0.1
	重 庆	6.4	2.8	0.4	0.7
	合 计	6.9	3.1	0.4	0.7
影响区	湖 北	1.2	0.7	0.1	0.1
	重 庆	5.7	2.4	0.3	0.7
	四 川	10.0	4.0	0.5	1.1
	贵 州	0.1	0.1	...	...
	合 计	17.1	7.3	0.9	2.0
上游区	重 庆	0.6	0.3	...	...
	四 川	44.9	19.0	2.2	4.7
	贵 州	3.2	2.5	0.3	0.4
	云 南	2.3	2.9	0.3	0.4
	合 计	50.9	24.7	2.8	5.6

三峡地区城镇生活污染排放及处理情况（一）

（2011）

流域	地 区名 称	城镇人口/万人	生活用水总量/万吨	污水处理厂数/座	污水处理厂设计处理能力/（万吨/日）	生活污水实际处理量/万吨
总 计		6 563	464 137	442	908	255 287
库区	湖 北	37	2 316	14	9	2 070
	重 庆	1 096	79 019	80	214	61 424
	合 计	1 133	81 335	94	223	63 493
影响区	湖 北	114	7 240	11	34	8 212
	重 庆	439	30 607	36	45	12 923
	四 川	615	37 301	23	48	14 211
	贵 州	57	2 934	4	6	1 122
	合 计	1 225	78 082	74	132	36 468
上游区	重 庆	35	2 355	1	1	433
	四 川	2 734	192 948	200	352	93 891
	贵 州	770	57 069	50	101	28 715
	云 南	665	52 348	23	97	32 287
	合 计	4 205	304 720	274	552	155 326

三峡地区城镇生活污染排放及处理情况（二）

（2011）

流域	地 区名 称	城镇生活污水排放量/万吨	城镇生活化学需氧量产生量/吨	城镇生活氨氮产生量/吨	城镇生活化学需氧量排放量/吨	城镇生活氨氮排放量/吨
总 计		383 528	1 614 384	199 028	1 063 240	148 656
库区	湖北	2 149	8 582	1 073	4 257	616
	重庆	67 166	272 104	34 813	138 041	24 681
	合计	69 315	280 686	35 886	142 299	25 296
影响区	湖北	6 947	26 725	3 340	12 696	2 627
	重庆	26 016	109 064	13 954	83 645	11 612
	四川	31 320	156 276	19 162	124 608	15 896
	贵州	2 347	12 679	1 618	10 782	1 413
	合计	66 631	304 744	38 074	231 731	31 548
上游区	重庆	2 002	8 799	1 126	7 700	1 000
	四川	163 433	672 175	82 496	480 902	62 033
	贵州	36 270	172 470	21 955	130 283	17 036
	云南	45 877	175 511	19 492	70 325	11 743
	合计	247 582	1 028 954	125 068	689 211	91 811

入海陆源接纳工业废水及处理情况（一）

（2011）

流域	地区名称	工业废水排放量/万吨	直接排入环境的	排入污水处理厂的	工业废水处理量/万吨	废水治理设施数/套	废水治理设施治理能力/（万吨/日）	废水治理设施运行费用/万元
总　计		458 398	344 004	114 392	769 777	14 315	8 381	1 353 341.7
渤海	天津	9 881	2 990	6 890	10 368	265	40	94 494.1
	河北	10 624	8 264	2 361	111 264	267	450	22 997.7
	辽宁	31 983	30 531	1 452	40 223	381	5 281	35 363.6
	山东	28 088	20 367	7 721	52 843	601	396	88 744.7
	合计	80 576	62 152	18 424	214 697	1 514	6 167	241 600.1
黄海	辽宁	5 049	4 543	507	4 532	180	22	5 516.8
	江苏	26 949	21 919	5 030	34 299	894	143	80 284.7
	山东	15 768	5 311	10 457	34 722	756	185	211 373.0
	合计	47 766	31 772	15 994	73 553	1 830	350	297 174.5
东海	上海	30 628	18 493	12 135	60 266	1 238	245	258 718.5
	浙江	75 580	29 798	45 782	72 902	3 519	435	231 047.7
	福建	138 242	125 231	13 010	152 018	1 466	570	101 867.9
	广东	2 031	1 986	44	1 505	103	17	3 924.9
	合计	246 482	175 509	70 971	286 691	6 326	1 267	595 559.0
南海	广东	74 840	66 937	7 903	85 452	4 281	500	198 895.0
	广西	5 051	4 501	550	103 789	139	73	13 885.2
	海南	3 683	3 133	550	5 595	225	25	6 227.9
	合计	83 574	74 571	9 003	194 836	4 645	597	219 008.1

入海陆源接纳工业废水及处理情况（二）

（2011）

单位：吨

流域	地区名称	工业废水中污染物排放量				
		汞	镉	总铬	铅	砷
总 计		0.035	0.346	42.185	4.350	0.832
渤海	天 津	…	…	0.215	1.444	0.019
	河 北	…	…	0.167	…	…
	辽 宁	…	…	0.052	0.161	0.041
	山 东	…	0.001	5.439	0.004	0.002
	合 计	0.001	0.001	5.872	1.609	0.061
黄海	辽 宁	…	0.017	0.066	0.158	0.069
	江 苏	0.001	0.006	2.403	0.530	0.354
	山 东	0.012	0.007	0.439	0.033	0.036
	合 计	0.013	0.030	2.909	0.721	0.459
东海	上 海	…	0.002	1.194	0.050	0.002
	浙 江	…	0.220	11.091	0.134	0.029
	福 建	…	0.010	14.860	0.653	0.183
	广 东	…	…	…	0.023	…
	合 计	0.001	0.231	27.145	0.859	0.214
南海	广 东	0.020	0.079	5.967	1.135	0.069
	广 西	0.001	…	0.161	…	0.001
	海 南	…	0.005	0.131	0.025	0.027
	合 计	0.021	0.084	6.259	1.161	0.097

入海陆源接纳工业废水及处理情况（三）

（2011）

单位：吨

流域	地区名称	工业废水中污染物排放量				
		氰化物	化学需氧量	石油类	氨氮	挥发酚
总 计		17.0	424 821.1	2 481.6	32 195.7	66.2
渤海	天 津	…	13 318.6	59.7	1 743.9	…
	河 北	…	18 779.0	125.0	1 414.9	0.2
	辽 宁	…	28 129.3	260.7	3 045.5	12.5
	山 东	0.7	23 200.7	83.0	2 411.2	3.9
	合 计	0.7	83 427.6	528.4	8 615.5	16.5
黄海	辽 宁	…	15 126.3	31.4	1 603.9	…
	江 苏	1.3	41 214.8	230.8	3 583.0	21.7
	山 东	0.2	11 214.6	61.5	996.3	0.1
	合 计	1.5	67 555.7	323.7	6 183.3	21.8
东海	上 海	2.8	21 487.3	722.1	1 487.5	4.7
	浙 江	4.1	87 642.1	367.6	6 016.1	20.2
	福 建	2.2	48 421.2	266.6	3 212.5	…
	广 东	…	2 632.8	1.3	98.2	…
	合 计	9.1	160 183.4	1 357.6	10 814.2	25.0
南海	广 东	5.2	90 425.6	206.2	5 841.5	2.2
	广 西	0.4	17 264.2	61.9	451.3	0.6
	海 南	…	5 964.6	3.8	289.8	…
	合 计	5.6	113 654.4	271.9	6 582.7	2.8

入海陆源接纳工业废水及处理情况（四）

（2011）

单位：吨

流域	地区名称	工业废水中污染物产生量				
		汞	镉	总铬	铅	砷
总　计		1.604	71.117	3 209.777	98.340	278.261
渤海	天　津	0.003	…	17.723	20.013	0.049
	河　北	…	…	11.596	…	3.565
	辽　宁	0.091	46.427	1.133	41.748	32.090
	山　东	0.003	0.019	23.033	1.274	110.106
	合　计	0.097	46.445	53.484	63.035	145.811
黄海	辽　宁	…	0.017	1.831	0.158	0.069
	江　苏	0.020	0.032	73.461	6.415	0.430
	山　东	0.727	23.772	28.657	17.839	104.443
	合　计	0.747	23.821	103.948	24.412	104.942
东海	上　海	0.003	0.040	51.540	0.265	0.696
	浙　江	0.001	0.271	1 064.015	3.475	16.840
	福　建	0.001	0.029	868.458	2.422	0.625
	广　东	…	…	0.003	0.308	…
	合　计	0.005	0.339	1 984.016	6.471	18.160
南海	广　东	0.750	0.505	1 002.159	4.336	9.313
	广　西	0.004	0.001	49.274	0.046	0.008
	海　南	0.001	0.005	16.896	0.040	0.027
	合　计	0.755	0.511	1 068.328	4.422	9.348

入海陆源接纳工业废水及处理情况（五）

（2011）

单位：吨

流域	地 区 名 称	工业废水中污染物产生量				
		氰化物	化学需氧量	石油类	氨氮	挥发酚
总 计		1 213.6	3 909 568.4	62 747.7	153 597.2	3 038.5
渤 海	天 津	0.1	38 109.0	551.1	2 636.2	0.1
	河 北	207.5	207 075.5	3 899.0	3 166.7	854.9
	辽 宁	1.8	101 699.9	4 587.4	25 697.8	251.8
	山 东	4.6	348 758.1	10 131.7	9 815.9	73.5
	合 计	213.9	695 642.5	19 169.2	41 316.5	1 180.2
黄 海	辽 宁	0.4	36 634.1	64.9	2 040.7	…
	江 苏	67.1	447 420.4	1 863.2	8 381.2	247.3
	山 东	13.6	145 637.4	1 392.7	5 879.1	5.8
	合 计	81.1	629 691.9	3 320.8	16 301.1	253.1
东 海	上 海	217.7	245 139.6	8 395.2	9 046.6	886.6
	浙 江	291.5	877 322.6	20 291.6	45 893.2	96.4
	福 建	113.0	624 839.8	9 402.9	20 781.2	182.6
	广 东	…	10 758.4	1.3	209.7	…
	合 计	622.2	1 758 060.3	38 091.0	75 930.7	1 165.6
南 海	广 东	293.3	678 526.3	1 429.8	18 011.5	17.4
	广 西	2.8	96 936.2	718.5	1 080.0	422.1
	海 南	0.2	50 711.2	18.4	957.4	…
	合 计	296.4	826 173.7	2 166.6	20 048.8	439.4

入海陆源工业废水污染防治投资情况

（2011）

流域	地区名称	工业废水治理施工项目数/个	工业废水治理竣工项目数/个	工业废水治理项目完成投资/万元	废水治理竣工项目新增处理能力/（万吨/日）
	总　计	666	425	279 982.5	278.8
渤海	天　津	15	9	15 047.0	1.1
	河　北	11	9	11 212.4	4.1
	辽　宁	16	14	12 296.7	2.3
	山　东	78	46	99 470.7	198.0
	合　计	120	78	138 026.8	205.5
黄海	辽　宁	24	4	13 465.0	2.8
	江　苏	71	62	15 846.6	4.1
	山　东	15	10	6 683.3	11.4
	合　计	110	76	35 994.9	18.3
东海	上　海	16	10	2 864.1	0.6
	浙　江	166	108	36 878.2	16.8
	福　建	99	49	42 740.5	20.1
	广　东	5	2	40.0	0.5
	合　计	286	169	82 522.8	38.0
南海	广　东	126	87	19 660.2	10.3
	广　西	2	1	450.0	…
	海　南	22	14	3 327.9	6.8
	合　计	150	102	23 438.1	17.0

入海陆源农业污染排放情况

（2011）

单位：万吨

流域	地区名称	农业污染物排放/流失总量			
		化学需氧量	总氮	总磷	氨氮
总 计		110.6	38.6	5.4	9.6
渤海	天 津	0.6	0.2	…	…
	河 北	7.8	2.7	0.4	0.5
	辽 宁	11.5	2.7	0.5	0.5
	山 东	14.1	4.6	0.7	0.7
	合 计	34.0	10.1	1.6	1.8
黄海	辽 宁	6.7	1.3	0.2	0.3
	江 苏	11.9	5.2	0.5	1.2
	山 东	14.2	3.8	0.6	0.7
	合 计	32.8	10.3	1.3	2.3
东海	上 海	2.9	1.2	0.2	0.3
	浙 江	7.0	3.3	0.4	1.0
	福 建	10.0	4.2	0.6	1.8
	广 东	1.6	0.6	0.1	0.2
	合 计	21.4	9.3	1.2	3.3
南海	广 东	13.3	5.5	0.8	1.5
	广 西	1.9	0.8	0.1	0.1
	海 南	7.3	2.5	0.3	0.6
	合 计	22.4	8.9	1.2	2.2

入海陆源城镇生活污染排放及处理情况（一）

（2011）

流域	地区名称	城镇人口/万人	生活用水总量/万吨	污水处理厂数/座	污水处理厂设计处理能力/（万吨/日）	生活污水实际处理量/万吨
	总 计	8 820	675 417	585	2 755	658 294
渤海	天 津	244	11 750	21	54	5 752
	河 北	221	13 043	16	72	12 051
	辽 宁	426	29 013	16	80	15 484
	山 东	361	24 913	34	88	17 926
	合 计	1 252	78 719	87	294	51 213
黄海	辽 宁	276	18 502	12	60	18 786
	江 苏	554	43 189	88	84	16 630
	山 东	898	55 686	45	203	48 024
	合 计	1 729	117 376	145	348	83 439
东海	上 海	834	74 856	24	536	126 152
	浙 江	1 347	101 369	73	399	72 596
	福 建	1 172	84 573	68	240	55 951
	广 东	105	7 776	4	9	2 901
	合 计	3 458	268 574	169	1 184	257 601
南海	广 东	2 038	185 049	142	777	225 046
	广 西	151	11 809	5	30	6 887
	海 南	192	13 889	37	123	34 109
	合 计	2 382	210 747	184	930	266 041

入海陆源城镇生活污染排放及处理情况（二）

（2011）

流域	地 区 名 称	城镇生活污水 排放量/ 万吨	城镇生活 化学需氧量 产生量/ 吨	城镇生活 氨氮产生量/吨	城镇生活化学 需氧量排放量/ 吨	城镇生活 氨氮排放量/ 吨
	总 计	597 533	2 220 871	289 734	1 096 592	184 811
渤 海	天 津	10 575	61 328	8 358	35 727	6 437
	河 北	10 835	54 583	7 103	9 899	3 359
	辽 宁	22 681	99 585	13 537	56 156	9 380
	山 东	21 482	85 846	11 222	33 283	6 067
	合 计	65 573	301 342	40 220	135 065	25 242
黄 海	辽 宁	15 715	64 540	8 773	31 109	4 539
	江 苏	38 041	142 174	18 226	87 340	14 303
	山 东	47 166	212 956	28 504	51 407	12 062
	合 计	100 922	419 671	55 504	169 856	30 904
东 海	上 海	67 370	231 384	29 532	72 375	17 495
	浙 江	93 924	341 224	46 164	175 001	30 459
	福 建	73 847	286 690	37 240	181 068	28 955
	广 东	6 612	23 422	3 099	17 647	2 644
	合 计	241 753	882 720	116 035	446 091	79 554
南 海	广 东	167 625	522 815	66 641	265 625	39 733
	广 西	9 778	36 237	4 605	26 688	3 138
	海 南	11 883	58 086	6 728	53 267	6 240
	合 计	189 286	617 138	77 975	345 580	49 111

环境管理统计

HUANJING GUANLI TONGJI

ANNUAL STATISTIC REPORT ON ENVIRONMENT IN CHINA

2011

各地区环保机构情况（一）

（2011）

单位：个

地区名称	环保系统机构总数	国家级、省级机构数	环保行政主管部门机构数	环境监测站机构数	环境监察机构数	科研机构数	宣教机构数	信息机构数	其他机构数
总　计	13 482	442	32	44	47	33	32	29	225
国家级	44	44	1	1	12	3	1	1	25
北　京	103	13	1	1	1	1	1	1	7
天　津	86	13	1	1	1	1	1	1	7
河　北	847	16	1	1	2	1	1	1	9
山　西	594	19	1	2	1	1	1	1	12
内蒙古	414	12	1	2	3	1	1	0	4
辽　宁	511	21	1	1	1	1	1	1	15
吉　林	265	11	1	1	1	1	1	1	5
黑龙江	436	16	1	1	1	1	1	1	10
上　海	72	8	1	1	1	1	1	1	2
江　苏	735	17	1	1	1	1	1	1	11
浙　江	545	9	1	2	1	1	1	1	2
安　徽	428	14	1	1	1	1	1	1	8
福　建	409	14	1	4	1	1	1	1	5
江　西	428	10	1	1	1	1	1	1	4
山　东	887	14	1	1	1	3	1	1	8
河　南	891	8	1	1	1	1	1	1	2
湖　北	519	10	1	1	1	1	1	1	4
湖　南	503	12	1	3	1	1	1	1	4
广　东	861	12	1	1	1	1	1	1	6
广　西	371	12	1	3	1	1	1	0	5
海　南	251	8	1	1	1	1	1	1	2
重　庆	174	11	1	1	1	0	1	1	6
四　川	745	14	1	2	1	1	1	1	7
贵　州	407	14	1	1	1	1	1	1	8
云　南	471	442	1	1	1	33	1	1	6
西　藏	120	8	1	1	1	0	1	0	4
陕　西	375	15	1	1	1	1	1	1	9
甘　肃	379	14	1	1	1	1	1	1	8
青　海	137	10	1	1	1	1	1	1	4
宁　夏	78	15	1	1	1	1	1	1	9
新　疆	396	16	1	2	2	2	1	1	7

各地区环保机构情况（二）

（2011）

单位：个

地 区 名 称	地市级环 保机构数	环保行政 主管部门 机构数	环境监测 站机构数	环境监察 机构数	科研 机构数	宣教 机构数	信息 机构数	其他 机构数
总 计	2 109	332	346	356	211	126	141	597
国家级	—	—	—	—	—	—	—	—
北 京	0	0	0	0	0	0	0	0
天 津	0	0	0	0	0	0	0	0
河 北	90	11	11	14	11	6	8	29
山 西	108	11	11	11	10	8	9	48
内蒙古	78	12	13	12	8	6	4	23
辽 宁	126	14	14	14	14	13	8	49
吉 林	67	9	9	10	6	9	4	20
黑龙江	87	13	14	14	9	5	4	28
上 海	0	0	0	0	0	0	0	0
江 苏	90	13	13	14	10	8	6	26
浙 江	64	11	11	11	6	4	4	17
安 徽	85	16	17	16	12	4	13	7
福 建	60	9	10	9	9	8	7	8
江 西	78	11	11	11	10	5	4	26
山 东	118	17	17	18	9	9	7	41
河 南	113	17	17	20	14	6	4	35
湖 北	87	13	13	20	12	4	4	21
湖 南	85	14	14	15	10	2	3	27
广 东	160	21	21	19	19	11	15	54
广 西	82	14	17	15	12	2	4	18
海 南	19	2	2	2	2	0	2	9
重 庆	0	0	0	0	0	0	0	0
四 川	94	21	27	21	6	1	11	7
贵 州	59	9	10	11	1	3	3	22
云 南	70	16	16	16	6	3	3	10
西 藏	28	7	7	7	0	1	0	6
陕 西	59	10	10	11	5	4	0	19
甘 肃	87	14	15	15	4	0	8	31
青 海	27	8	6	9	1	0	1	2
宁 夏	25	5	6	6	1	0	1	6
新 疆	63	14	14	15	4	4	4	8

各地区环保机构情况（三）

（2011）

单位：个

地 区 名 称	县级环保机构数	环保行政主管部门机构数	环境监测站机构数	环境监察机构数	其他机构数	乡镇环保机构数
总　计	8 973	2 809	2 313	2 718	1 133	1 958
国家级	—	—	—	—	—	—
北　京	90	16	18	22	34	0
天　津	73	16	20	21	16	0
河　北	542	172	141	150	79	199
山　西	441	119	112	137	73	26
内蒙古	309	101	96	97	15	15
辽　宁	334	100	81	88	65	30
吉　林	187	57	54	59	17	0
黑龙江	332	113	95	105	19	1
上　海	64	17	18	17	12	0
江　苏	386	101	94	132	59	242
浙　江	304	90	73	77	64	168
安　徽	293	104	66	95	28	36
福　建	280	84	79	91	26	55
江　西	338	100	97	98	43	2
山　东	571	140	136	155	140	184
河　南	557	158	126	188	85	213
湖　北	343	99	85	93	66	79
湖　南	364	121	103	113	27	42
广　东	451	113	106	95	137	238
广　西	264	105	71	86	2	13
海　南	53	20	15	16	2	171
重　庆	141	40	40	40	21	22
四　川	542	181	143	182	36	95
贵　州	297	88	84	89	36	37
云　南	365	129	118	113	5	24
西　藏	84	73	3	7	1	0
陕　西	301	107	82	108	4	0
甘　肃	231	86	55	86	4	47
青　海	100	43	13	41	3	0
宁　夏	38	17	11	10	0	0
新　疆	298	99	78	107	14	19

各地区环保机构情况（四）

（2011）

地 区 名 称	环保系统 总人数	国家级、 省级环保 系统人数	环保行政主 管部门人数	环境监测 站人数	环境监 察人数	科研机 构人数	宣教机 构人数	信息机 构人数	其他机 构人数
总　　计	201 161	16 110	2 965	3 308	1 601	2 970	515	361	4 390
国家级	3 020	3 020	318	176	445	550	35	34	1 462
北　京	2 609	837	129	149	71	237	50	10	191
天　津	1 958	648	101	181	60	161	21	22	102
河　北	17 454	436	83	79	56	50	16	11	141
山　西	11 546	422	92	84	40	42	9	10	145
内蒙古	5 442	304	70	91	42	24	14	0	63
辽　宁	8 734	471	64	78	30	102	14	9	174
吉　林	5 242	350	77	87	17	71	16	12	70
黑龙江	5 165	415	76	83	30	70	13	7	136
上　海	2 294	666	104	177	60	213	27	16	69
江　苏	10 324	508	128	73	17	69	13	23	185
浙　江	6 630	408	73	204	25	71	7	6	22
安　徽	5 740	352	70	87	39	72	12	12	60
福　建	4 048	392	85	120	39	43	9	7	89
江　西	5 265	403	61	89	44	108	10	17	74
山　东	13 807	476	98	63	33	66	24	23	169
河　南	22 608	348	106	66	46	40	22	31	37
湖　北	7 580	300	81	60	30	63	19	5	42
湖　南	9 117	453	73	173	23	126	24	5	29
广　东	11 604	499	100	97	49	40	16	9	188
广　西	3 614	305	51	133	23	43	17	0	38
海　南	1 580	224	93	13	25	74	2	13	4
重　庆	2 699	530	137	198	64	0	18	10	103
四　川	8 271	490	101	155	35	131	13	10	45
贵　州	3 219	380	75	72	47	81	8	9	88
云　南	4 889	476	92	94	25	154	15	11	85
西　藏	658	131	53	34	9	0	3	0	32
陕　西	6 002	464	104	82	54	44	13	16	151
甘　肃	4 314	359	90	77	15	72	11	8	86
青　海	918	221	52	64	22	19	7	7	50
宁　夏	1 143	346	58	61	34	27	23	0	143
新　疆	3 667	476	70	108	52	107	14	8	117

各地区环保机构情况（五）

（2011）

单位：人

地 区 名 称	地市级 环保系统 人数	环保行政主 管部门人数	环境监测站 人数	环境监察 人数	科研机构 人数	宣教机构 人数	信息机构 人数	其他机构 人数
总 计	45 019	9 032	15 881	9 343	3 539	828	805	5 591
国家级	—	—	—	—	—	—	—	—
北 京	0	0	0	0	0	0	0	0
天 津	0	0	0	0	0	0	0	0
河 北	2 679	505	825	469	273	82	54	471
山 西	1 962	357	563	383	151	39	79	390
内蒙古	1 497	270	574	264	151	31	31	176
辽 宁	3 449	501	995	612	505	118	66	652
吉 林	998	206	282	266	62	34	15	133
黑龙江	1 541	351	517	339	114	42	17	161
上 海	0	0	0	0	0	0	0	0
江 苏	2 567	512	1 074	544	176	45	26	190
浙 江	1 498	293	605	326	130	31	20	93
安 徽	1 639	326	664	424	108	13	56	48
福 建	1 063	172	378	275	143	35	30	30
江 西	1 271	255	405	305	110	30	16	150
山 东	2 770	538	995	538	154	77	77	391
河 南	3 346	660	1 171	878	210	64	26	337
湖 北	1 716	363	574	398	217	25	16	123
湖 南	2 104	443	811	431	208	18	15	178
广 东	4 465	836	1 056	754	418	60	72	1 269
广 西	1 348	314	621	275	70	10	24	34
海 南	253	39	84	56	17	0	32	25
重 庆	0	0	0	0	0	0	0	0
四 川	2 189	545	999	475	63	14	38	55
贵 州	782	175	324	173	12	15	15	68
云 南	1 202	347	485	210	64	12	20	64
西 藏	227	116	41	31	0	1	0	38
陕 西	1 427	290	636	188	106	23	0	184
甘 肃	1 201	285	482	254	14	0	37	129
青 海	204	64	66	54	6	0	8	6
宁 夏	496	71	155	106	12	0	4	148
新 疆	1 125	198	499	315	45	9	11	48

各地区环保机构情况（六）

（2011）

单位：人

地 区名 称	县级环保系统人数	环保行政主管部门人数	环境监测站人数	环境监察人数	其他机构人数	乡镇环保机构人数
总 计	132 596	34 131	37 037	53 482	7 946	7 436
国家级	—	—	—	—	—	—
北 京	1 772	405	359	289	719	0
天 津	1 310	315	644	179	172	0
河 北	12 581	3 719	3 007	4 771	1 084	1 758
山 西	8 677	1 274	2 920	3 797	686	485
内蒙古	3 626	1 230	871	1 412	113	15
辽 宁	4 814	904	1 290	2 058	562	0
吉 林	3 894	502	1 135	2 091	166	0
黑龙江	3 205	793	936	1 395	81	4
上 海	1 628	343	772	427	86	0
江 苏	6 689	1 658	2 333	2 415	283	560
浙 江	4 165	1 073	1 374	1 398	320	559
安 徽	3 691	794	983	1 750	164	58
福 建	2 590	563	838	1 061	128	3
江 西	3 578	1 226	908	1 303	141	13
山 东	9 598	2 389	3 040	3 329	840	963
河 南	17 931	2 873	4 187	10 182	689	983
湖 北	5 186	1 521	1 268	1 993	404	378
湖 南	6 428	2 031	1 715	2 534	148	132
广 东	5 950	2 172	1 729	1 257	792	690
广 西	1 945	797	564	580	4	16
海 南	741	351	142	241	7	362
重 庆	2 113	540	771	756	46	56
四 川	5 519	1 612	1 684	2 112	111	73
贵 州	1 983	747	475	689	72	74
云 南	3 130	1 394	903	789	44	81
西 藏	300	274	3	22	1	0
陕 西	4 111	888	1 007	2 199	17	0
甘 肃	2 675	942	501	1 221	11	79
青 海	493	145	131	207	10	0
宁 夏	301	79	105	117	0	0
新 疆	1 972	577	442	908	45	94

各地区环境信访与环境法制情况（一）

（2011）

地 区 名 称	当年受理行政 复议案件数/ 件	本级行政处 罚案件数/ 件	承办的人大 建议数/ 件	承办的政协 提案数/ 件	当年备案的地 方环境标准数/ 件	累积备案的地方 环境标准数/ 件
总 计	838	119 333	5 522	7 452	15	76
国家级	189	16	257	195	—	—
北 京	2	5 096	65	46	4	27
天 津	0	164	11	39	1	2
河 北	6	5 904	194	266	0	0
山 西	0	2 086	128	114	0	0
内蒙古	0	1 618	125	190	0	0
辽 宁	5	15 150	217	278	0	1
吉 林	0	828	52	43	0	0
黑龙江	1	30 844	112	153	0	2
上 海	24	1 158	60	73	0	8
江 苏	29	6 172	349	602	0	0
浙 江	17	9 936	464	591	0	1
安 徽	11	690	138	259	0	0
福 建	37	2 714	245	333	0	2
江 西	1	1 930	200	304	0	0
山 东	14	7 022	280	470	0	10
河 南	12	2 114	203	376	1	2
湖 北	1	1 609	283	368	0	0
湖 南	3	2 676	360	412	0	0
广 东	68	11 405	278	414	6	13
广 西	4	883	141	180	0	0
海 南	2	235	16	23	0	0
重 庆	28	1 548	229	285	1	6
四 川	14	1 769	306	435	0	0
贵 州	3	456	94	111	0	0
云 南	3	1 423	238	322	0	0
西 藏	0	3	17	18	0	0
陕 西	5	1 442	93	161	1	1
甘 肃	1	550	109	150	1	1
青 海	0	68	44	17	0	0
宁 夏	355	102	79	67	0	0
新 疆	3	1 722	135	157	0	0

各地区环境信访与环境法制情况（二）

（2011）

地 区 名 称	电话/网络投 诉数/件	电话/网络投 诉办结数/件	来信总件数/ 件	来访批次/ 批	来访总数人数/ 人	来信、来访已办 结数量/件
总　计	852 700	834 588	201 631	53 505	107 597	251 607
国家级	1 281	1 281	2 412	490	1 500	2 284
北　京	21 119	21 115	664	176	576	807
天　津	13 026	13 001	13 349	227	410	13 498
河　北	18 529	17 592	2 184	2 350	3 971	5 183
山　西	11 604	11 136	3 272	491	1 775	3 586
内蒙古	6 978	6 884	9 696	1 837	3 178	11 203
辽　宁	29 242	28 923	4 635	3 016	6 920	8 770
吉　林	15 880	15 872	2 758	1 331	3 079	4 050
黑龙江	16 821	15 808	2 022	6 852	16 373	6 268
上　海	30 152	28 935	2 279	552	1 467	2 700
江　苏	103 705	103 328	11 068	3 605	6 430	15 562
浙　江	46 733	46 297	8 511	2 261	5 189	10 115
安　徽	22 449	22 474	8 131	1 512	2 640	9 357
福　建	32 419	32 030	5 134	1 202	1 912	5 598
江　西	9 221	9 135	4 412	2 354	5 076	6 799
山　东	74 064	76 270	22 338	1 434	3 430	25 265
河　南	21 792	21 281	4 073	1 902	3 499	5 910
湖　北	32 219	31 761	10 991	2 907	4 907	15 725
湖　南	20 599	20 452	6 976	4 376	9 702	10 439
广　东	100 315	99 038	25 272	2 427	4 388	26 234
广　西	27 032	26 585	2 337	1 485	2 636	3 505
海　南	5 848	5 812	125	58	123	169
重　庆	74 560	71 927	5 083	1 596	2 819	5 980
四　川	23 671	22 911	9 743	3 047	5 188	13 198
贵　州	2 802	2 776	2 711	882	1 739	3 423
云　南	52 532	43 942	4 674	2 271	3 905	6 764
西　藏	214	214	39	21	28	49
陕　西	22 018	22 018	18 629	1 084	1 735	19 424
甘　肃	4 402	4 401	1 778	546	967	2 853
青　海	1 190	1 190	1 394	75	114	1 451
宁　夏	2 596	2 596	3 435	409	618	3 836
新　疆	7 687	7 603	1 506	729	1 303	1 602

各地区环境污染源控制与管理情况

（2011）

地 区 名 称	清洁生产审核当年完成企业数/家	强制性审核当年完成数/家	应开展监测的重金属污染防控重点企业数/家	重金属排放达标的重点企业数/家	已发放危险废物经营许可证数/个	具有医疗废物经营范围的许可证数/个
总　计	8 138	4 526	4 677	3 932	1 222	208
北　京	29	29	6	6	12	2
天　津	62	29	31	31	15	1
河　北	695	573	163	158	51	12
山　西	119	87	61	39	13	7
内蒙古	156	118	78	77	17	9
辽　宁	193	151	96	59	51	5
吉　林	64	49	37	37	66	4
黑龙江	98	75	64	64	17	5
上　海	304	114	168	154	52	2
江　苏	1 522	886	596	394	211	18
浙　江	1 285	244	234	226	88	11
安　徽	109	47	97	87	30	8
福　建	163	129	294	293	17	5
江　西	93	69	225	191	27	7
山　东	835	637	164	161	75	17
河　南	189	171	430	430	56	13
湖　北	93	76	81	65	42	3
湖　南	245	188	251	225	83	9
广　东	974	292	726	700	112	15
广　西	122	102	79	76	20	4
海　南	20	11	9	9	6	2
重　庆	89	82	71	71	18	4
四　川	150	91	159	132	62	16
贵　州	30	30	56	36	4	1
云　南	220	83	159	35	2	2
西　藏	1	1	13	0	0	0
陕　西	69	53	101	37	19	4
甘　肃	86	84	192	106	15	9
青　海	9	4	13	13	5	1
宁　夏	19	2	10	8	7	3
新　疆	95	19	13	12	29	9

各地区环境监测情况（一）

（2011）

地区名称	监测用房面积/米²	监测业务经费/万元	环境空气监测点位数/个	国控监测点位数/个	酸雨监测点位/个	沙尘天气影响环境质量监测点位数/个	开展污染源监督性监测的重点企业数/个
总　计	2 395 469	1 712 063	2 941	1 126	1 217	99	56 684
国家级	5 381	18 988	—	—	—	—	—
北　京	20 087	3 576	34	13	3	1	476
天　津	22 146	7 894	112	14	9	1	588
河　北	115 224	566 233	80	37	22	3	3 098
山　西	83 375	22 766	139	52	22	2	1 849
内蒙古	92 183	8 855	120	14	31	22	699
辽　宁	92 191	335 686	95	61	61	18	2 191
吉　林	51 835	6 840	81	24	23	5	545
黑龙江	51 742	4 435	65	55	3I	0	820
上　海	94 812	5 774	158	20	40	0	1 721
江　苏	149 087	71 206	178	60	103	0	6 460
浙　江	114 875	64 243	138	43	75	0	3 575
安　徽	60 940	37 829	62	42	40	0	960
福　建	59 096	7 171	72	29	36	0	1 781
江　西	87 917	78 299	57	27	22	0	619
山　东	235 705	32 808	185	63	71	1	4 908
河　南	83 737	8 340	93	79	50	0	967
湖　北	75 076	14 332	91	37	58	5	2 692
湖　南	90 834	13 511	101	66	38	0	2 271
广　东	135 755	76 335	206	72	65	0	9 254
广　西	52 463	32 408	71	42	39	0	1 740
海　南	10 906	3 333	43	8	24	2	232
重　庆	108 766	9 179	220	27	82	0	1 965
四　川	191 786	19 803	138	94	113	0	1 886
贵　州	23 551	13 368	35	11	16	0	631
云　南	72 665	39 069	74	14	36	0	1 121
西　藏	11 650	235	16	9	2	0	10
陕　西	77 807	21 168	87	29	29	11	1 147
甘　肃	41 192	198 474	65	24	27	9	501
青　海	10 023	1 563	19	7	8	3	537
宁　夏	14 726	1 567	15	11	9	3	263
新　疆	63 317	5 763	91	42	32	13	1 177

各地区环境监测情况（二）

（2011）

单位：个

地 区 名 称	地表水水质 监测断面 （点位）数	国控断面 （点位）数	饮用水水 源地监测 点位数	地表水监测 点位数	地下水监测 点位数	近岸海域监 测点位数	近岸海域 环境功能区 点位数	近岸海域 环境质量 点位数
总　　计	8 960	1 040	3 856	2 995	881	724	482	387
北　京	199	8	8	2	6	0	0	0
天　津	216	16	5	1	4	17	12	10
河　北	248	49	49	15	29	20	9	11
山　西	105	10	43	2	38	0	0	0
内蒙古	203	33	140	12	128	0	0	0
辽　宁	228	34	65	37	28	129	51	29
吉　林	127	28	50	24	26	0	0	0
黑龙江	193	61	57	31	26	0	0	0
上　海	931	16	123	138	8	6	6	6
江　苏	1 066	102	98	92	12	24	20	24
浙　江	814	47	198	194	1	81	54	40
安　徽	266	61	72	53	20	0	0	0
福　建	206	18	57	49	8	79	56	18
江　西	228	41	38	38	0	0	0	0
山　东	537	53	207	143	64	186	136	140
河　南	146	46	49	15	34	0	0	0
湖　北	431	60	160	158	0	0	0	0
湖　南	164	40	91	73	10	0	0	0
广　东	439	20	128	119	8	106	78	46
广　西	119	21	45	38	28	8	8	4
海　南	149	4	45	44	1	68	52	59
重　庆	453	37	1 400	1 270	130	0	0	0
四　川	321	24	197	163	46	0	0	0
贵　州	111	13	84	67	17	0	0	0
云　南	428	60	105	101	4	0	0	0
西　藏	25	9	49	10	9	0	0	0
陕　西	174	20	44	32	17	0	0	0
甘　肃	71	12	73	27	46	0	0	0
青　海	37	3	17	10	7	0	0	0
宁　夏	46	20	23	7	16	0	0	0
新　疆	279	74	136	30	110	0	0	0

各地区污染源自动监控情况

（2011）

单位：个

地 区 名 称	已实施自动 监控国家重 点监控企业数	已实施自动监控国家重点 监控企业中排放口数		已实施自动监控国家重点监控企业中 监控设备与环保部门稳定联网数			
		水排放口数	气排放口数	COD 监控设备 与环保部门稳 定联网数	NH₃-N 监控设 备与环保部门 稳定联网数	SO₂ 监控设备 与环保部门稳 定联网数	NOₓ 监控设备 与环保部门稳 定联网数
总　计	7 990	6 224	6 015	3 947	1 488	2 160	1 961
北　京	33	24	11	0	0	7	0
天　津	59	42	94	27	10	24	24
河　北	516	470	645	300	26	141	122
山　西	481	393	685	202	73	178	148
内蒙古	240	136	341	136	39	137	137
辽　宁	188	110	225	106	13	156	149
吉　林	143	92	135	76	45	54	52
黑龙江	174	119	140	104	18	57	56
上　海	76	63	43	53	40	2	2
江　苏	728	589	350	527	249	159	155
浙　江	349	249	159	249	45	90	90
安　徽	247	207	211	18	8	19	19
福　建	169	136	100	0	0	62	54
江　西	227	167	165	125	52	79	62
山　东	710	493	416	28	22	42	41
河　南	598	612	459	327	259	7	7
湖　北	250	203	117	203	106	102	94
湖　南	414	343	205	304	27	132	129
广　东	445	382	244	0	0	5	3
广　西	441	344	218	306	83	117	118
海　南	36	30	20	24	6	9	9
重　庆	121	88	73	60		43	4
四　川	391	299	175	268	145	136	134
贵　州	115	78	100	78	78	96	87
云　南	105	90	79	81	12	25	24
西　藏	0	0	0	0	0	0	0
陕　西	353	219	231	139	56	64	56
甘　肃	101	95	139	73	37	46	41
青　海	46	20	52	20	13	26	17
宁　夏	81	41	70	41	14	47	36
新　疆	153	90	113	72	12	98	91

各地区排污费征收情况

（2011）

地 区 名 称	排污费解缴入库户数/ 户	排污费解缴入库户金额/ 万元
总 计	370 700	1 898 958.0
北 京	1 588	2 848.6
天 津	3 733	19 643.8
河 北	23 396	148 737.3
山 西	10 111	191 457.4
内 蒙 古	5 131	91 842.3
辽 宁	26 369	133 549.5
吉 林	17 618	34 791.4
黑 龙 江	11 513	43 711.7
上 海	5 104	23 497.0
江 苏	32 157	200 731.9
浙 江	26 685	95 439.3
安 徽	10 753	47 146.6
福 建	19 429	37 984.7
江 西	8 364	63 766.3
山 东	15 356	139 894.8
河 南	16 359	92 886.6
湖 北	7 452	38 672.7
湖 南	12 805	60 153.2
广 东	56 761	92 263.9
广 西	7 720	26 394.4
海 南	1 143	3 468.9
重 庆	6 986	37 170.8
四 川	7 618	60 549.4
贵 州	7 427	44 454.0
云 南	7 403	31 564.8
西 藏	0	0
陕 西	5 213	51 941.4
甘 肃	5 082	26 296.1
青 海	1 013	6 369.9
宁 夏	2 668	15 879.7
新 疆	7 743	35 849.9

各地区自然生态保护与建设情况（一）

（2011）

地 区 名 称	自然保护 区个数/ 个	国家级	省级	市级	县级	自然保护 区面积/ 公顷	国家级	省级	市级	县级
总 计	2 640	335	870	421	1 014	149 711 464	93 152 684	41 525 888	4 723 921	10 308 971
北 京	20	2	12	6	0	133 966	26 403	71 413	36 150	0
天 津	8	3	5	0	0	91 115	37 862	53 253	0	0
河 北	35	12	19	2	2	587 268	238 489	329 797	8 806	10 176
山 西	46	5	41	0	0	1 157 425	82 936	1 074 489	0	0
内蒙古	184	24	60	25	75	13 804 738	3 950 007	6 983 691	315 783	2 555 257
辽 宁	101	13	30	35	23	2 714 937	948 322	874 056	794 093	98 466
吉 林	38	14	15	3	6	2 303 900	946 103	1 314 252	20 564	22 981
黑龙江	221	24	87	33	77	6 604 508	2 399 689	2 984 280	364 642	855 897
上 海	4	2	2	0	0	93 821	66 175	27 646	0	0
江 苏	30	3	10	10	7	564 983	336 211	85 448	122 827	20 497
浙 江	32	10	8	0	14	197 204	146 542	12 754	0	37 908
安 徽	102	7	26	5	64	524 836	139 221	281 985	22 860	80 770
福 建	92	12	26	10	44	445 478	205 521	130 628	37 652	71 677
江 西	195	9	34	2	150	1 191 425	156 027	400 003	2 759	632 636
山 东	86	7	33	24	22	1 097 916	219 828	498 690	244 476	134 922
河 南	34	11	21	0	2	734 658	426 316	306 942	0	1 400
湖 北	64	11	18	24	11	959 447	261 102	374 656	185 025	138 664
湖 南	123	18	32	1	72	1 249 614	517 856	446 619	464	284 675
广 东	368	11	66	114	177	3 552 658	225 534	598 189	379 811	2 349 124
广 西	78	16	50	3	9	1 452 941	308 255	903 794	118 947	121 945
海 南	50	9	24	12	5	2 735 320	106 526	2 614 554	13 395	845
重 庆	57	4	18	1	34	850 386	218 964	275 638	3 686	352 098
四 川	167	24	68	28	47	8 993 646	2 771 195	3 167 173	1 319 730	1 735 548
贵 州	129	8	4	16	101	951 766	243 539	56 965	215 109	436 153
云 南	163	17	42	60	44	2 977 614	1 443 531	866 976	450 738	216 369
西 藏	47	9	14	3	21	41 368 882	37 153 065	4 209 486	4 870	1 461
陕 西	55	14	34	4	3	1 172 115	466 550	609 429	61 534	34 602
甘 肃	59	16	39	0	4	7 346 761	4 825 358	2 406 503	0	114 900
青 海	11	5	6	0	0	21 822 201	20 252 490	1 569 711	0	0
宁 夏	14	6	8	0	0	535 570	426 916	108 654	0	0
新 疆	27	9	18	0	0	21 494 365	13 606 151	7 888 214	0	0

各地区自然生态保护与建设情况（二）

（2011）

单位：个

地 区 名 称	生态市、县 建设个数	国家级生态市 个数	国家级生态县 （区）个数	省级生态市 个数	省级生态县 （区）个数
总　计	296	1	37	28	230
北　京	12	0	1	9	2
天　津	0	0	0	0	0
河　北	0	0	0	0	0
山　西	2	0	0	0	2
内蒙古	2	0	2	0	0
辽　宁	12	0	0	0	12
吉　林	7	0	0	1	6
黑龙江	15	0	1	1	13
上　海	25	0	17	0	8
江　苏	28	0	6	5	17
浙　江	39	0	1	7	31
安　徽	3	0	0	0	3
福　建	0	0	0	0	0
江　西	10	0	1	0	9
山　东	9	0	0	4	5
河　南	2	0	0	0	2
湖　北	0	0	0	0	0
湖　南	12	1	3	0	8
广　东	74	0	0	0	74
广　西	0	0	0	0	0
海　南	2	0	0	1	1
重　庆	6	0	2	0	4
四　川	17	0	0	0	17
贵　州	0	0	0	0	0
云　南	0	0	0	0	0
西　藏	1	0	1	0	0
陕　西	5	0	0	0	5
甘　肃	1	0	0	0	1
青　海	0	0	0	0	0
宁　夏	0	0	0	0	0
新　疆	0	0	0	0	10

各地区自然生态保护与建设情况（三）

（2011）

单位：个

地 区 名 称	农村生态示范建设个数	国家级生态乡镇个数	国家级生态村个数	国家有机食品生产基地数量
总　计	1 797	1 559	238	68
北　京	72	70	2	0
天　津	15	15	0	0
河　北	37	26	11	0
山　西	7	4	3	0
内蒙古	36	32	4	2
辽　宁	85	77	8	14
吉　林	26	23	3	2
黑龙江	30	27	3	1
上　海	53	51	2	2
江　苏	282	238	44	6
浙　江	283	274	9	19
安　徽	91	70	21	2
福　建	21	18	3	0
江　西	50	41	9	1
山　东	223	217	6	10
河　南	35	28	7	0
湖　北	50	28	22	0
湖　南	92	65	27	0
广　东	50	44	6	0
广　西	3	2	1	1
海　南	3	2	1	0
重　庆	5	5	0	0
四　川	105	98	7	0
贵　州	34	26	8	0
云　南	19	16	3	1
西　藏	0	0	0	0
陕　西	36	25	11	0
甘　肃	11	5	6	0
青　海	0	0	0	2
宁　夏	12	8	4	5
新　疆	31	24	7	0

各地区环境影响评价（一）

（2011）

地区名称	当年开工建设的建设项目数/个	执行环境影响评价制度的建设项目数量	当年审批的建设项目投资总额/万元	当年审批的建设项目环保投资总额/万元	当年审查的规划环境影响评价文件数量/个
总 计	378 235	377 823	2 763 686 826	76 925 086	1 350
国家级	291	291	145 590 100	6 695 582	46
北 京	11 198	11 198	85 340 868	3 609 832	10
天 津	3 007	3 004	64 142 969	1 249 272	11
河 北	21 195	21 195	186 253 374	2 809 824	117
山 西	6 898	6 898	52 400 769	1 497 892	7
内蒙古	7 412	7 390	77 552 994	2 742 979	51
辽 宁	14 783	14 783	76 962 643	1 581 758	152
吉 林	4 458	4 458	49 452 385	1 635 375	36
黑龙江	2 117	2 117	50 733 715	974 859	20
上 海	2 299	2 299	96 586 640	4 333 580	1
江 苏	30 520	30 520	362 780 051	11 488 262	75
浙 江	28 404	28 404	215 358 100	2 734 200	60
安 徽	14 100	13 964	109 383 383	2 432 169	19
福 建	17 086	17 086	93 701 848	2 383 233	49
江 西	4 966	4 964	39 212 162	960 842	9
山 东	34 405	34 405	212 658 428	7 036 215	35
河 南	8 858	8 858	100 589 896	1 791 390	79
湖 北	4 657	4 648	73 125 423	1 903 645	51
湖 南	10 222	10 222	52 231 531	1 658 235	35
广 东	64 322	64 322	161 215 141	4 698 591	72
广 西	14 410	14 410	29 596 822	1 014 974	38
海 南	1 307	1 307	30 593 325	749 042	10
重 庆	15 344	15 344	71 956 552	1 637 995	53
四 川	6 018	6 018	62 605 380	1 619 570	110
贵 州	9 052	9 052	30 641 488	648 310	29
云 南	15 270	15 270	59 615 150	2 018 440	71
西 藏	1 334	1 249	3 056 962	51 482	4
陕 西	5 360	5 343	52 355 891	1 133 113	18
甘 肃	2 537	2 537	24 257 384	627 457	21
青 海	1 860	1 860	20 926 109	269 028	4
宁 夏	1 612	1 612	19 381 085	613 406	4
新 疆	12 933	12 795	53 428 258	2 324 535	53

各地区环境影响评价（二）

（2011）

单位：个

地区名称	当年审批的建设项目环境影响评价文件数量	编制报告书的项目数量	填报报告表的项目数量	编制登记表的项目数量
总　计	436 188	28 964	184 068	223 156
国家级	291	264	23	4
北　京	11 198	440	3 509	7 249
天　津	4 152	599	2 457	1 096
河　北	25 317	1 097	13 073	11 147
山　西	6 898	792	3 347	2 759
内蒙古	9 047	732	3 366	4 949
辽　宁	17 392	1 167	7 038	9 187
吉　林	5 298	615	2 590	2 093
黑龙江	6 371	699	3 257	2 415
上　海	13 304	573	6 459	6 272
江　苏	40 740	2 808	19 445	18 487
浙　江	28 926	1 890	14 945	12 091
安　徽	13 988	1 392	6 327	6 269
福　建	17 086	1 096	8 149	7 841
江　西	5 075	739	2 701	1 635
山　东	34 405	2 019	16 669	15 717
河　南	15 543	924	5 875	8 744
湖　北	8 529	1 006	3 908	3 615
湖　南	10 279	1 186	3 080	6 013
广　东	64 322	2 040	30 967	31 315
广　西	14 410	775	4 090	9 545
海　南	2 831	243	1 130	1 458
重　庆	15 344	1 775	5 138	8 431
四　川	6 473	642	2 093	3 738
贵　州	9 052	662	2 817	5 573
云　南	15 270	709	3 065	11 496
西　藏	4 768	78	446	4 244
陕　西	6 028	595	1 857	3 576
甘　肃	4 771	403	1 256	3 112
青　海	1 860	123	485	1 252
宁　夏	2 070	206	837	1 027
新　疆	15 150	675	3 669	10 806

各地区建设项目竣工环境保护验收情况

（2011）

地 区 名 称	当年完成环保验收项目数/个	环保验收一次合格项目数	经限期改正验收合格项目数	当年完成环保验收项目总投资/万元	当年完成环保验收项目环保投资/万元
总　计	127 819	125 139	2 694	674 461 585	21 124 386
国家级	248	240	8	136 691 600	4 334 600
北　京	3 562	3 557	5	9 758 386	197 533
天　津	1 220	1 219	1	10 087 092	543 577
河　北	6 679	6 547	132	24 433 732	1 652 665
山　西	1 756	1 755	1	5 512 410	578 004
内蒙古	1 950	1 931	22	13 436 084	773 243
辽　宁	7 318	7 266	52	21 649 360	514 082
吉　林	2 244	2 235	9	5 973 202	202 202
黑龙江	2 269	2 255	14	7 434 550	554 738
上　海	4 533	4 530	3	20 823 090	694 900
江　苏	9 630	9 486	154	72 230 412	1942 294
浙　江	8 904	8 822	82	26 369 554	876 021
安　徽	3 680	3 596	84	14 221 447	471 838
福　建	5 955	5 397	559	14 527 792	531 305
江　西	1 199	1 180	19	5 918 078	453 825
山　东	10 743	10 668	75	172 149 518	1 287 121
河　南	4 412	4 347	65	8 292 500	427 640
湖　北	3 078	2 977	101	10 992 614	530 645
湖　南	2 834	2 762	72	4 865 533	275 071
广　东	16 555	15 694	861	32 500 319	1 401 991
广　西	5 190	5 114	76	3 887 857	250 373
海　南	742	741	1	1 886 671	97 679
重　庆	2 287	2 226	61	10 492 216	396 702
四　川	5 610	5 571	39	11 555 448	407 386
贵　州	1 522	1 522	0	1 475 377	137 440
云　南	3 900	3 900	0	5 596 890	474 400
西　藏	526	508	18	5 577 674	21 904
陕　西	1 224	1 204	20	3 619 914	280 624
甘　肃	1 704	1 702	2	1 084 526	226 116
青　海	396	396	0	2 119 310	105 374
宁　夏	794	749	45	8 522 513	122 328
新　疆	5 155	5 042	113	775 917	360 765

各地区突发环境事件

（2011）

单位：次

地 区 名 称	突发环境事件次数	特别重大环境事件次数	重大环境事件次数	较大环境事件次数	一般环境事件次数
总 计	542	0	12	12	518
北 京	36	0	0	0	36
天 津	1	0	0	0	1
河 北	16	0	0	2	14
山 西	11	0	0	0	11
内 蒙 古	14	0	0	0	14
辽 宁	2	0	0	0	2
吉 林	2	0	0	0	2
黑 龙 江	5	0	0	0	5
上 海	197	0	0	0	197
江 苏	27	0	0	1	26
浙 江	31	0	2	3	26
安 徽	12	0	0	1	11
福 建	8	0	0	0	8
江 西	8	0	0	1	7
山 东	8	0	0	0	8
河 南	25	0	1	0	24
湖 北	7	0	0	0	7
湖 南	9	0	3	0	6
广 东	26	0	2	1	23
广 西	31	0	1	0	30
海 南	1	0	0	0	1
重 庆	18	0	0	0	18
四 川	25	0	1	2	22
贵 州	7	0	1	1	5
云 南	1	0	1	0	0
西 藏	0	0	0	0	0
陕 西	2	0	0	0	2
甘 肃	4	0	0	0	4
青 海	1	0	0	0	1
宁 夏	1	0	0	0	1
新 疆	6	0	0	0	6

各地区环境宣教情况

（2011）

地 区 名 称	当年开展的社会环境 宣传教育活动数/次	当年开展的社会环境 宣传教育活动人数/人	环境教育基地数/ 个
总 计	13 913	32 146 974	1 620
国家级	18	56 000	0
北 京	226	180 740	16
天 津	618	461 348	4
河 北	481	2 364 418	46
山 西	113	275 235	21
内蒙古	119	184 023	109
辽 宁	235	1 297 700	45
吉 林	161	138 892	1
黑龙江	170	223 610	24
上 海	789	488 050	110
江 苏	648	348 538	349
浙 江	689	789 578	130
安 徽	179	245 017	18
福 建	881	358 035	78
江 西	75	157 960	25
山 东	678	8 847 559	197
河 南	255	179 323	64
湖 北	501	574 111	154
湖 南	463	259 949	7
广 东	821	6 746 823	50
广 西	175	133 148	6
海 南	93	3 238	4
重 庆	2 363	4 199 291	23
四 川	872	913 518	16
贵 州	60	406 850	2
云 南	825	950 147	69
西 藏	160	881	4
陕 西	657	510 564	13
甘 肃	246	494 241	9
青 海	95	92 171	20
宁 夏	87	4 672	6
新 疆	178	261 344	

7

附表

FUBIAO

ANNUAL STATISTIC REPORT ON ENVIRONMENT IN CHINA

2011

全国行政区划

　　　　　　　　　　　　　　　　　　　　　　单位：个

省级区划名称	地级区划数	地级市	县级区划数	县级市	市辖区	县	自治县
全　国	322	284	2 853	369	857	1 456	117
北京市			16		14	2	
天津市			16		13	3	
河北省	11	11	172	22	36	108	6
山西省	11	11	119	11	23	85	
内蒙古自治区	12	9	101	11	21	17	
辽宁省	14	14	100	17	56	19	8
吉林省	9	8	60	20	20	17	3
黑龙江省	13	12	128	18	64	45	1
上海市			17		16	1	
江苏省	13	13	104	25	55	24	
浙江省	11	11	90	22	32	35	1
安徽省	16	16	105	6	43	56	
福建省	9	9	85	14	26	45	
江西省	11	11	100	11	19	70	
山东省	17	17	140	31	49	60	
河南省	17	17	159	21	50	88	
湖北省	13	12	103	24	38	38	2
湖南省	14	13	122	16	35	64	7
广东省	21	21	121	23	54	41	3
广西壮族自治区	14	14	109	7	34	56	12
海南省	2	2	20	6	4	4	6
重庆市			38		19	15	4
四川省	21	18	181	14	44	119	4
贵州省	9	6	88	7	13	56	11
云南省	16	8	129	11	13	76	29
西藏自治区	7	1	73	1	1	71	
陕西省	10	10	107	3	24	80	
甘肃省	14	12	86	4	17	58	7
青海省	8	1	43	2	4	30	7
宁夏回族自治区	5	5	22	2	9	11	
新疆维吾尔自治区	14	2	98	20	11	62	6
香港特别行政区							
澳门特别行政区							
台湾省							

注：数据摘自《中国统计摘要》（中国统计出版社），以下同。

国民经济与社会发展总量指标摘要

指标	单位	2005 年	2006 年	2007 年	2008 年	2009 年	2010 年	2011 年
人口								
年底总人口	万人	130 756	131 448	132 129	132 802	133 450	134 091	134 735
其中：城镇人口	万人	56 212	58 288	60 633	62 403	64 512	66 978	69 079
乡村人口	万人	74 544	73 160	71 496	70 399	68 938	67 113	65 656
国民经济核算								
国内生产总值	亿元	184 937.4	216 314.4	265 810.3	314 045.4	340 902.8	401 512.8	471 563.7
其中：第一产业	亿元	22 420.0	24 040.0	28 627.0	33 702.0	35 226.0	40 533.6	47 712.0
第二产业	亿元	87 598.1	103 719.5	125 831.4	149 003.4	157 638.8	187 383.2	220 591.6
第三产业	亿元	77 230.8	88 554.9	111 351.9	131 340.0	148 038.0	173 596.0	203 260.1
人均国内生产总值	元/人	14 185	16 500	20 169	23 708	25 608	30 015	35 083
支出法国内生产总值	亿元	187 437.4	222 711.9	266 556.7	315 977.2	348 771.3	402 818.7	465 998.7
其中：最终消费	亿元	99 357.5	113 103.8	132 232.9	153 422.5	169 274.8	194 115.0	224 740.8
资本形成总额	亿元	77 856.8	92 954.1	110 943.2	138 325.3	164 463.2	193 603.9	229 102.4
固定资产投资								
全社会固定资产投资总额	亿元	88 773.6	109 998.2	137 323.9	172 828.4	224 599	278 121.9	311 021.9
城镇	亿元	75 095.1	93 368.7	117 464.5	148 738.3	193 920	241 430.9	301 932.8
其中：房地产开发	亿元	15 909.2	19 422.9	25 288.8	31 203.2	36 242	48 259.4	61 739.8
农村	亿元	13 678.5	16 629.5	19 859.5	24 123.0	30 678	36 724.9	
主要农业、工业产品产量								
粮食	万吨	48 402.2	49 804	50 160.3	52 870.9	53 082	54 647.7	57 121
棉花	万吨	571.4	753.3	762.4	749.2	638	596.1	658.9
油料	万吨	3 077.1	2 640.3	2 568.7	2 952.8	3 154	3 230.1	3 306.8
肉类	万吨	6 938.9	7 089.0	6 865.7	7 278.3	7 650	7 925.8	7 957.8
原煤	亿吨	23.50	25.29	26.92	28.02	29.73	32.35	35.20
原油	万吨	18 135.29	18 476.57	18 632	19 043.1	18 949	20 241.4	20 287.6
发电量	亿千瓦·时	25 002.60	28 657.26	32 816	34 957.6	37 146.5	42 071.6	47 000.7
粗钢	万吨	35 324.0	41 914.9	48 929	50 305.8	57 218	63 723.0	68 388.3
水泥	万吨	106 884.8	123 676.5	136 117	142 355.7	164 398	188 191.2	208 500.0

国民经济与社会发展总量指标摘要（续表）

指标	单位	2005 年	2006 年	2007 年	2008 年	2009 年	2010 年	2011 年
国内商业和对外贸易								
社会消费品零售总额	亿元	68 352.6	79 145.2	93 571.6	114 830.1	132 678.4	156 998	183 918.6
进出口总额	亿美元	14 219	17 604	21 765.7	25 632.6	22 075.4	29 740.0	36 420.6
其中：出口额	亿美元	7 620	9 689	12 204.6	14 306.9	12 016.1	15 777.5	18 986.0
进口额	亿美元	6 600	7 915	9 561.1	11 325.6	10 059.2	13 962.4	17 434.6
利用外资								
实际利（使用）用外资额	亿美元	638.1	735.2	783.4	952.6	918.0	1 088.2	1 177.0
其中：外商直接投资	亿美元	603.3	694.7	747.7	924.0	900.3	1 057.3	1 160.1
外商其他投资	亿美元	34.8	40.6	35.7	28.6	17.7	30.9	16.9
旅游								
入境过夜旅游者人数	万人次	4 681	4 991	5 472.0	5 304.9	5 087.5	5 566.5	5 758.1
国内旅游人数	亿人次	12.1	13.9	16.1	17.1	19.0	21.0	26.41
国际旅游外汇收入	亿美元	293.0	339.5	419.2	408.4	396.8	458.1	484.6
国内旅游总收入（国内旅游总花费	亿元	5 285.9	6 229.7	7 770.6	8 749.3	10 183.7	12 579.8	19 305.4
教育、科技、文化、卫生								
普通高等学校学生数	万人	1 561.8	1 738.8	1 884.9	2 021.0	2 144.7	2 231.8	2 308.5
普通中等学校学生数	万人	8 581	8 451.9	8 243.3	8 050.4	7 876.9	7 703.2	7 519.0
小学学生数	万人	10 864	10 711.5	10 564.0	10 331.5	10 071.5	9 940.7	9 926.4
研究与试验发展经费支出	亿元	2 450	3 003.1	3 710.2	4 570.0	5 791.9	7 063	8 610
技术市场成交额	亿元	1 551	1 818.2	2 227	2 665	3 039	3 907	4 764
图书总印数	亿册（张）	64.7	64.0	62.9	70.6	70.4	71.7	76.8
杂志总印数（期刊）	亿册	27.6	28.5	30.4	31.0	31.5	32.2	32.7
报纸总印数	亿份	412.6	424.5	438.0	442.9	439.1	452.1	466.8
医院、卫生院数（医疗卫生机构数）	万个	88.2	91.8	91.2	89.1	91.7	93.7	95.4
医生数	万人	204.2	209.9	212.3	220.2	232.9	241.3	246.6
医院、卫生院床位数（医疗卫生机构床位数	万张	336.8	351.2	370.1	403.9	441.7	478.7	516.0
城市公用事业								
城市房屋集中供热面积	亿米^2	25.2	26.6					
年末供水综合生产能力	万米^3/日	24 720	26 962					
年末污水处理能力	万米^3/日	7 990	9 734					
年末城市道路长度	千米	247 015	241 351					

注：① 由于计算误差的影响，支出法国内生产总值不等于按生产法计算的国内生产总值。

② 本表价值量指标均按当年价格计算。

国民经济与社会发展结构指标摘要

单位：%

指标	2005 年	2006 年	2007 年	2008 年	2009 年	2010 年	2011 年
总人口	100	100	100	100	100	100	100
其中：城镇	43.0	44.3	45.9	47.0	48.3	50.0	51.3
乡村	57.0	5 537	54.1	53.0	51.7	50.0	48.7
国内生产总值	100	100	100	100	100	100	100
其中：第一产业	12.1	11.1	10.8	10.7	10.3	10.1	10.1
第二产业	47.4	47.9	47.3	47.4	46.2	46.7	46.8
第三产业	40.5	40.9	41.9	41.8	43.4	43.2	43.1
固定资产投资总额	100	100	100	100	100	100	100
城镇	84.6	84.9	85.6	86.0	86.3	86.8	
农村	15.4	15.1	14.4	14.0	13.7	13.2	
财政收入	100	100	100	100	100	100	100
其中：中央	52.3	52.8	54.1	53.3	52.4	51.1	49.5
地方	47.7	47.2	45.9	46.7	47.6	48.9	50.5
财政支出	100	100	100	100	100	100	100
其中：中央	25.9	24.7	23.1	21.4	20.0	17.8	15.2
地方	74.1	75.3	76.9	78.6	80.0	82.2	84.8
农、林、牧、渔业产值	100	100	100	100	100	100	100
其中：农业	49.7	51.1	50.4	55.7	50.7	53.3	51.6
林业	3.6	3.9	3.8	3.8	3.9	3.7	3.8
牧业	33.7	29.4	33.0	29.7	32.3	30.0	31.7
渔业	10.2	9.6	9.1	10.9	9.3	9.3	9.3
农、林、牧、渔、服务业	2.8	6.0	3.7		3.8	3.7	
工业总产值	100	100	100	100	100	100	100
其中：轻工业	31.4	30.5					
重工业	68.6	69.5					

注：2002 年起轻重工业结构为国有及规模以上非国有工业企业口径。

自然资源状况

项目	单位	2005 年	2006 年	2007 年	2008 年	2009 年	2010 年	2011 年
国土面积	万千米²	960	960	960	960	960	960	960
海域面积	万千米²	473	473	473	473	473	473	473
大陆岸线长度	万千米	1.80	1.80	1.80	1.80	1.80	1.80	1.8
岛屿面积	万千米²	3.87	3.87	3.87	3.87	3.87	3.87	3.87
降水量								
台湾中部山区	毫米	≥4 000	≥4 000	≥4 000	≥4 000	≥4 000	≥4 000	≥4 000
华南沿海	毫米	1 600～2 000	1 600～2 000	1 600～2 000	1 600～2 000	1 600～2 000	1 600～2 000	1 600～2 000
长江流域	毫米	1 000～1 500	1 000～1 500	1 000～1 500	1 000～1 500	1 000～1 500	1 000～1 500	1 000～1 500
华北、东北	毫米	400～800	400～800	400～800	400～800	400～800	400～800	400～800
西北内陆	毫米	100～200	100～200	100～200	100～200	100～200	100～200	100～200
塔里木盆地、吐鲁番盆地和柴达木盆地	毫米	≤25	≤25	≤25	≤25	≤25	≤25	≤25
耕地面积	万公顷	13 004	13 004	13 004	12 178	12 178	12 172	12 172
荒地面积	万公顷	10 800	10 800	10 800	10 800	10 800		
林业用地面积	万公顷	26 329	28 493	28 493	28 493	30 378	30 590	30 590
草地面积	万公顷	40 000	40 000	40 000	40 000	40 000		
活立木总蓄积量	亿米³	136.2	136.2	136.2	136.2	136.2	149.1	149.1
森林面积	亿公顷	1.75	1.75	1.75	1.75	1.95	1.95	1.95
森林覆盖率	%	18.21	18.21	18.21	18.21	20.36	20.36	20.36
水资源总量	亿米³	28 124	27 430	25 567	24 696	23 763	29 658	24 022
水力资源蕴藏量	亿千瓦	6.76	6.76	6.76	6.76	6.76	6.76	6.76
其中：可开发量	亿千瓦	3.79	3.79	3.79	3.79	3.79	3.79	3.79
内陆水域总面积	万公顷	1 747	1 747	1 747	1 747	1 747	1 747	1 747
其中：可养殖面积	万公顷	675	675	675	675	675	675	675
海水可养殖面积	万公顷	260	260	260	260	260	260	260
其中：已养殖面积	万公顷	109	109	109	109	109	109	109

注：① 森林资源为 1994—1998 年调查数；草地资源为 1991 年调查数；水面资源为 1988 年数；水力资源为 1999 年公报数；耕地面积为 1996 年农业普查数。

② 本表除国土面积、海域面积和降水量外，其余指标均未包括香港特别行政区、澳门特别行政区和台湾省数据。

主要统计指标解释

ZHUYAO TONGJI ZHIBIAO JIESHI

ANNUAL STATISTIC REPORT ON ENVIRONMENT IN CHINA

2011

1. 工业企业污染排放及处理利用情况

【工业用水量】指报告期内企业厂区内用于工业生产活动的水量，它等于取水量与重复用水量之和。

【取水量】指报告期内企业厂区内用于工业生产活动的水量中从外部取水的量。根据《工业企业产品取水定额编制通则》(GB/T 18220—2002)，工业生产的取水量，包括取自地表水（以净水厂供水计量）、地下水、城镇供水工程，以及企业从市场购得的其他水（如其他企业回用水量）或水的产品（如蒸汽、热水、地热水等），不包括企业自取的海水和苦咸水等以及企业为外供给市场的水的产品（如蒸汽、热水、地热水等）而取用的水量。

工业生产活动用水包括主要工业生产用水、辅助生产（包括机修、运输、空压站等）用水和附属生产（包括厂内绿化、职工食堂、非营业的浴室及保健站、厕所等）用水；不包括①非工业生产单位的用水，如厂内居民家庭用水和企业附属幼儿园、学校、对外营业的浴室、游泳池等的用水；②生活用水单独计量且生活污水不与工业废水混排的水量。

【重复用水量】指报告期内企业生产用水中重复再利用的水量，包括循环使用、一水多用和串级使用的水量（含经处理后回用量）。

指企业内部对工业生产活动排放的废水直接利用或经过处理后回收再利用的水量，不包括从城市污水处理厂回用的水量。每重复利用一次，则计算一次重复用水量。但锅炉、循环冷却系统等封闭式系统内的循环水不能计算重复用水量。重复用水量的计算原则：

(1) 开放原则。即水的循环在开放系统进行，循环一次计算一次，但锅炉、循环冷却系统等封闭式系统内的循环水不能计算重复用水量。

(2) "源头"计算原则。对循环水来说，使用后的水，又回流到系统的取水源头，流经源头一次，计算一次。循环系统中的中间环节用水不得计算重复用水量。

(3) 异地原则。对于非循环系统，根据不同工艺对不同水质的要求，在一个地方（工艺）使用过的水，在另外一个地方（工艺）中又进行使用，使用一次，计算一次。在同一地方（容器）多次使用的水，不得计算重复用水量。

(4) 经过净化处理后的水重复再用，在任何情况下都按照重复用水计算。

【煤炭消耗量】指报告期内企业所用煤炭的总消耗量。

【燃料煤消耗量】指报告期内企业厂区内用作燃料的煤炭消耗量（实物量），包括企业厂区内生产、生活用燃料煤，也包括砖瓦、石灰等产品生产用的内燃煤，不包括在生产工艺中用作原料并能转换成新的产品实体的煤炭消耗量。如转换为水泥、焦炭、煤气、碳素、活性炭、氮肥的煤炭。

【燃料油消耗量（不含车船用）】指报告期内企业用作燃料的原油、汽油、柴油、煤油等各种油料总消耗量，不包括车船交通用油量。

【焦炭消耗量】指报告期内企业消耗的焦炭总量。

【天然气消耗量】指报告期内企业用作燃料的天然气消耗量。

【其他燃料消耗量】指报告期内企业除了煤炭、燃油、天然气等以外，用作燃料的其他燃料消耗量。其他燃料应根据当地的折标系数折算为标准煤后统一填报。

各类能源的参考折标系数表

能源种类		折标系数	能源种类	折标系数
原煤		0.714 3	煤焦油	1.142 9
洗精煤		0.900 0	粗苯	1.428 6
其他洗煤	洗中煤	0.285 7	原油	1.428 6
	煤泥	0.285 7~0.428 6	汽油	1.471 4
型煤		0.5~0.7	煤油	1.471 4
焦炭		0.971 4	柴油	1.457 1
焦炉煤气		0.571 4~0.614 3 千克标准煤/米³	燃料油	1.428 6
高炉煤气		0.128 6 千克标准煤/米³	热力	0.034 12 千克标准煤/百万千焦 0.142 86 千克标准煤/1 000 千卡
天然气		1.330 0 千克标准煤/米³	电力	0.122 9 千克标准煤/千瓦·时
液化天然气		1.757 2	生物质能	大豆秆、棉花秆 0.543
液化石油气		1.714 3		稻秆 0.429
炼厂干气		1.571 4		麦秆 0.500
其他煤气	发生炉煤气	0.178 6 千克标准煤/米³		玉米秆 0.529
	重油催化裂解煤气	0.657 1 千克标准煤/米³		杂草 0.471
	重油热裂解煤气	1.214 3 千克标准煤/米³		树叶 0.500
	焦炭制气	0.557 1 千克标准煤/米³		薪柴 0.571
	压力气化煤气	0.514 3 千克标准煤/米³		沼气 0.714 千克标准煤/米³
	水煤气	0.357 1 千克标准煤/米³		—

注：除表中标注单位的能源外，其余能源折标系数单位均为：千克标准煤/千克。

各地的能源折标系数由当地环保部门协调统计部门提供。调查对象也可根据燃料品质分析报告，自行折标填报。

【用电量】指报告期内企业的用电量，包括动力用电和照明用电。

【工业锅炉数】指报告期内企业厂区内用于生产和生活的大于1蒸吨（含1蒸吨）的蒸汽锅炉、热水锅炉总台数和总蒸吨数，包括燃煤、燃油、燃气和燃电的锅炉，不包括茶炉。

【其中35蒸吨及以上的】指报告期内企业厂区内用于生产和生活的大于35蒸吨（含35蒸吨）的蒸汽锅炉、热水锅炉总台数和总蒸吨数。

【其中20（含）~35蒸吨的】指报告期内企业厂区内用于生产和生活的大于20蒸吨（含20蒸吨）小于35蒸吨的蒸汽锅炉、热水锅炉总台数和总蒸吨数。

【其中10（含）~20蒸吨的】指报告期内企业厂区内用于生产和生活的大于10蒸吨（含10蒸吨）小于20蒸吨的蒸汽锅炉、热水锅炉总台数和总蒸吨数。

【其中10蒸吨以下的】指报告期内企业厂区内用于生产和生活的小于10蒸吨的蒸汽锅炉、热水锅炉总台数和总蒸吨数。

【工业炉窑数】指报告期内企业生产用的炉窑总数，如炼铁高炉、炼钢炉、冲天炉、烘干炉窑、锻造加热炉、水泥窑、石灰窑等。

【废水治理设施数】指报告期内企业用于防治水污染和经处理后综合利用水资源的实有设施（包括构筑物）数，以一个废水治理系统为单位统计。附属于设施内的水治理设备和配套设备不单独计算。备用的、报告期内未运行的、已经报废的设施不统计在内。

只填报企业内部的废水治理设施，工业废水排入的城镇污水处理厂、集中工业废水处理厂不能算作企业的废水治理设施。企业内的废水治理设施包括一级、二级和三级处理的设施，如企业有2个排污口，1个排污口为一级处理（隔油池、化粪池、沉淀池等），另1个排污口为二级处理（如生化处理），则该企业有2套废水治理设施；若该企业只有1个排污口，经由该排污口的废水先经过一级处理，再经二级（甚至三级）处理后外排，则该企业视为1套废水治理设施。即针对同一股废水的所有水治理设备均视为1套治理设施，针对不同废水的水治理设备可视为多套治理设施。

【废水治理设施处理能力】指报告期内企业内部的所有废水治理设施实际具有的废水处理能力。

【废水治理设施运行费用】指报告期内企业维持废水治理设施运行所发生的费用。包括能源消耗、设备维修、人员工资、管理费、药剂费及与设施运行有关的其他费用等。

【工业废水处理量】指经各种水治理设施（含城镇污水处理厂、工业废水处理厂）实际处理的工业废水量，包括处理后外排的和处理后回用的工业废水量。虽经处理但未达到国家或地方排放标准的废水量也应计算在内。计算时，如遇有车间和厂排放口均有治理设施，并对同一废水分级处理时，不应重复计算工业废水处理量。

【工业废水排放量】指报告期内经过企业厂区所有排放口排到企业外部的工业废水量。包括生产废水、外排的直接冷却水、废气治理设施废水、超标排放的矿井地下水和与工业废水混排的厂区生活污水，不包括独立外排的间接冷却水（清浊不分流的间接冷却水应计算在内）。

直接冷却水：在生产过程中，为满足工艺过程需要，使产品或半成品冷却所用与之直接接触的冷却水（包括调温、调湿使用的直流喷雾水）。

间接冷却水：在工业生产过程中，为保证生产设备能在正常温度下工作，用来吸收或转移生产设备的多余热量，所使用的冷却水（此冷却用水与被冷却介质之间由热交换器壁或设备隔开）。

【直接排入环境的】指废水经过工厂的排污口或经过下水道直接排入环境中，包括排入海、河流、湖泊、水库、蒸发地、渗坑以及农田等。对应的排水去向代码为A、B、C、D、F、G、K。

【排入污水处理厂的】指企业产生的废水直接或间接经市政管网排入污水处理厂的废水量，包括排入城镇污水处理厂、集中工业废水处理厂以及其他单位的污水处理设施的废水量。对应的排水去向代码为E、L、H。

【工业废水中污染物产生量】指报告期调查对象生产过程中产生的未经过处理的废水中所含的化学需氧量、氨氮、石油类、挥发酚、氰化物等污染物和砷、铅、汞、镉、六价铬、总铬等重金属本身的纯质量。它可采用产排污系数根据生产的产品产量或原辅料用量计算求得，也可以通过工业废水产生量和其中污染物的浓度相乘求得，计算公式是：

$$污染物产生量（纯质量）＝工业废水产生量×废水处理设施入口污染物的平均浓度（无处理设施可使用排口浓度）$$

计算砷、铅、汞、镉、六价铬、总铬等重金属污染物时，上述计算公式中"工业废水产生量"为产生重金属废水的车间年实际产生的废水量，"废水处理设施入口污染物的平均浓度"为该车间废水处理设施入口的年实际加权平均浓度，如没有设施则为车间排口的年实际加权平均浓度。

【工业废水中污染物排放量】指报告期内企业排放的工业废水中所含化学需氧量、氨氮、石油类、挥发酚、氰化物等污染物和砷、铅、汞、镉、六价铬等重金属本身的纯质量。它可采用产排污系数根据生产的产品产量或原辅料用量计算求得，也可以通过工业废水排放量和其中污染物的浓度相乘求得，计算公式是：

$$污染物排放量（纯质量）＝工业废水排放量×排放口污染物的平均浓度$$

（1）如企业排出的工业废水经城镇污水处理厂或工业废水处理厂集中处理的，计算化学需氧量、氨氮、石油类、挥发酚、氰化物等污染物时，上述计算公式中"排放口污染物的平均浓度"即为污水处理厂排放口的年实际加权平均浓度。如果厂界排放浓度低于污水处理厂的排放浓度，以污水处理厂的排放浓度为准。

（2）计算砷、铅、汞、镉、六价铬等重金属污染物时，上述计算公式中"工业废水排放量"为车间排放口的年实际废水量，"排放口污染物的平均浓度"为车间排放口的年实际加权平均浓度。

【工业废气排放量】 指报告期内企业厂区内燃料燃烧和生产工艺过程中产生的各种排入空气中含有污染物的气体的总量，以标准状态（273K，10 132 5Pa）计。

工业废气排放总量＝燃料燃烧过程中废气排放量+生产工艺过程中废气排放量

【废气治理设施数】 指报告期末企业用于减少在燃料燃烧过程与生产工艺过程中排向大气的污染物或对污染物加以回收利用的废气治理设施总数，以一个废气治理系统为单位统计。包括除尘、脱硫、脱硝及其他的污染物的烟气治理设施。备用的、报告期内未运行的、已报废的设施不统计在内。锅炉中的除尘装置属于"三同时"设备，应统计在内。

【除尘设施数】 指专门设计、建设的去除废气烟（粉）尘的设施。

【脱硫设施数】 脱硫设施指专门设计、建设的去除废气二氧化硫的设施，具有兼性脱硫效果的设施，如湿法除尘等治理设施等其他可能具有脱硫效果的废气治理设施不计入脱硫设施。具有脱硫效果的生产装置，如制酸、水泥生产等不作为脱硫设施。

【脱硝设施数】 指在治理设施中采用选择性催化还原技术（SCR）、选择性非催化还原技术（SNCR）及其联合技术或采用活性炭吸附进行烟气脱硝的设施。具有脱硝效果的生产装置，如水泥生产等不作为脱硝设施。

【废气治理设施处理能力】 指报告期末企业实有的废气治理设施的实际废气处理能力。

【废气脱硫设施处理能力】 指报告期末企业实有的废气脱硫设施的实际废气处理能力。

【废气脱硝设施处理能力】 指报告期末企业实有的废气脱硝设施的实际废气处理能力。

【废气除尘设施处理能力】 指报告期末企业实有的废气除尘设施的实际废气处理能力。

【废气治理设施运行费用】 指报告期内维持废气治理设施运行所发生的费用。包括能源消耗、设备折旧、设备维修、人员工资、管理费、药剂费及与设施运行有关的其他费用等。

【二氧化硫产生量】 指当年全年调查对象生产过程中产生的未经过处理的废气中所含的二氧化硫总质量。

【二氧化硫排放量】 指报告期内企业在燃料燃烧和生产工艺过程中排入大气的二氧化硫总质量。工业中二氧化硫主要来源于化石燃料（煤、石油等）的燃烧，还包括含硫矿石的冶炼或含硫酸、磷肥等生产的工业废气排放。

【氮氧化物产生量】 指当年全年调查对象生产过程中产生的未经过处理的废气中所含的氮氧化物总质量。

【氮氧化物排放量】 指报告期内企业在燃料燃烧和生产工艺过程中排入大气的氮氧化物总质量。

【烟（粉）尘产生量】 烟尘是指通过燃烧煤、石煤、柴油、木柴、天然气等产生的烟气中的尘粒。通过有组织排放的，俗称烟道尘。工业粉尘指在生产工艺过程中排放的能在空气中悬浮一定时间的固体颗粒。如钢铁企业耐火材料粉尘、焦化企业的筛焦系统粉尘、烧结机的粉尘、石灰窑的粉尘、建材企业的水泥粉尘等。烟（粉）尘产生量指当年全年调查对象生产过程中产生的未经过处理的废气中所含的烟尘及工业粉尘的总质量之和。

【烟（粉）尘排放量】 指报告期内企业在燃料燃烧和生产工艺过程中排入大气的烟尘及工业粉尘的总质量之和。烟尘或工业粉尘排放量可以通过除尘系统的排风量和除尘设备出口烟尘浓度相乘求得。

【重金属产生量】 指报告期调查对象生产过程中产生的未经过处理的废气中分别所含的砷、铅、汞、镉、铬及其化合物的总质量（以元素计）。

【重金属排放量】指报告期内企业在燃料燃烧和生产工艺过程中分别排入大气的砷、铅、汞、镉、铬及其化合物的总质量（以元素计）。

【一般工业固体废物产生量】是指未被列入《国家危险废物名录》或者根据国家规定的危险废物鉴别标准（GB 5085）、固体废物浸出毒性浸出方法（GB 5086）及固体废物浸出毒性测定方法（GB/T 15555）鉴别方法判定不具有危险特性的工业固体废物。根据其性质分为两种：

（1）第 类一般工业固体废物：按照 GB 5086 规定方法进行浸出试验而获得的浸出液中，任何一种污染物的浓度均未超过 GB 8978 最高允放排放浓度，且 pH 值在 6~9 范围之内的一般工业固体废物；

（2）第 类一般工业固体废物：按照 GB 5086 规定方法进行浸出试验而获得的浸出液中，有一种或一种以上的污染物浓度超过 GB 8978 最高允许排放浓度，或者是 pH 值在 6~9 范围之外的一般工业固体废物。

主要包括：

代码	名称	代码	名称
SW01	冶炼废渣	SW07	污泥
SW02	粉煤灰	SW08	放射性废物
SW03	炉渣	SW09	赤泥
SW04	煤矸石	SW10	磷石膏
SW05	尾矿	SW99	其他废物
SW06	脱硫石膏		

不包括矿山开采的剥离废石和掘进废石（煤矸石和呈酸性或碱性的废石除外）。酸性或碱性废石是指采掘的废石其流经水、雨淋水的 pH 值小于 4 或 pH 值大于 10.5 者。

冶炼废渣：指在冶炼生产中产生的高炉渣、钢渣、铁合金渣等，不包括列入《国家危险废物名录》中的金属冶炼废物。

粉煤灰：指从燃煤过程产生烟气中收捕下来的细微固体颗粒物，不包括从燃煤设施炉膛排出的灰渣。主要来自电力、热力的生产和供应行业和其他使用燃煤设施的行业，又称飞灰或烟道灰。主要从烟道气体收集而得，应与其烟尘去除量基本相等。

炉渣：指企业燃烧设备从炉膛排出的灰渣，不包括燃料燃烧过程中产生的烟尘。

煤矸石：指与煤层伴生的一种含碳量低、比煤坚硬的黑灰色岩石，包括巷道掘进过程中的掘进矸石、采掘过程中从顶板、底板及夹层里采出的矸石以及洗煤过程中挑出的洗矸石。主要来自煤炭开采和洗选行业。

尾矿：指矿山选矿过程中产生的有用成分含量低、在当前的技术经济条件下不宜进一步分选的固体废物，包括各种金属和非金属矿石的选矿。主要来自采矿业。

脱硫石膏：指废气脱硫的湿式石灰石/石膏法工艺中，吸收剂与烟气中 SO_2 等反应后生成的副产物。

污泥：指污水处理厂污水处理中排出的、以干泥量计的固体沉淀物。

放射性废物：指含有天然放射性核素，并其比活度大于 2×10^4 Bq/kg 的尾矿砂、废矿石及其他放射性固体废物（指放射性浓度或活度或污染水平超过规定下限的固体废物）。

赤泥：指从铝土矿中提炼氧化铝后排出的污染性废渣，一般还氧化铁量大，外观与赤色泥土相似。

磷石膏：指在磷酸生产中用硫酸分解磷矿时产生的二水硫酸钙、酸不溶物，未分解磷矿及其他杂质的混合物。主要来自磷肥制造业。

其他废物：指除上述 10 类一般工业固体废物以外的未列入《国家危险废物名录》中的固体废物，如机械工业切削碎屑、研磨碎屑、废砂型等；食品工业的活性炭渣；硅酸盐工业和建材工业的砖、瓦、碎砾、混凝土碎块等。

一般工业固体废物产生量＝（一般工业固体废物综合利用量－其中：综合利用往年贮存量）一般工业固体废物贮存量（一般工业固体废物处置量－其中：处置往年贮存量）一般工业固体废物倾倒丢弃量

【一般工业固体废物综合利用量】 指报告期内企业通过回收、加工、循环、交换等方式，从固体废物中提取或者使其转化为可以利用的资源、能源和其他原材料的固体废物量（包括当年利用的往年工业固体废物累计贮存量）。如用作农业肥料、生产建筑材料、筑路等。综合利用量由原产生固体废物的单位统计。

工业固体废物综合利用的主要方式有：

序号	综合利用方式	序号	综合利用方式
1	铺路	9	再循环/再利用不是用作溶剂的有机物
2	建筑材料	10	再循环/再利用金属和金属化合物
3	农肥或土壤改良剂	11	再循环/再利用其他无机物
4	矿渣棉	12	再生酸或碱
5	铸石	13	回收污染减除剂的组分
6	其他	14	回收催化剂组分
7	作为燃料（直接燃烧除外）或以其他方式产生能量	15	废油再提炼或其他废油的再利用
8	溶剂回收/再生（如蒸馏、萃取等）	16	其他有效成分回收

【一般工业固体废物贮存量】 指报告期内企业以综合利用或处置为目的，将固体废物暂时贮存或堆存在专设的贮存设施或专设的集中堆存场所内的量。专设的固体废物贮存场所或贮存设施必须有防扩散、防流失、防渗漏、防止污染大气、水体的措施。

粉煤灰、钢渣、煤矸石、尾矿等的贮存量是指排入灰场、渣场、矸石场、尾矿库等贮存的量。

专设的固体废物贮存场所或贮存设施指符合环保要求的贮存场，即选址、设计、建设符合《一般工业固体废物贮存、填埋场污染控制标准》（GB 18599—2001）等相关环保法律法规要求，具有防扩散、防流失、防渗漏、防止污染大气和水体措施的场所和设施。

工业固体废物的贮存方式主要有：

序号	贮存方式
1	灰场堆放
2	渣场堆放
3	尾矿库堆放
4	其他贮存（不包括永久性贮存）

【一般工业固体废物处置量】 指报告期内企业将工业固体废物焚烧和用其他改变工业固体废物的物理、化学、生物特性的方法，达到减少或者消除其危险成分的活动，或者将工业固体废物最终置于符合环境保护规定要求的填埋场的活动中，所消纳固体废物的量。

处置方式如：填埋、焚烧、专业贮存场（库）封场处理、深层灌注、回填矿井及海洋处置（经海洋管理部门同意投海处置）等。

处置量包括本单位处置或委托给外单位处置的量。还包括当年处置的往年工业固体废物贮存量。

工业固体废物的主要处置方式有：

处置方式
围隔堆存（属永久性处置）
填埋
置放于地下或地上（如填埋、填坑、填浜）
特别设计填埋
海洋处置
经海洋管理部门同意的投海处置
埋入海床
焚化
陆上焚化
海上焚化
水泥窑共处置（指在水泥生产工艺中使用工业固体废物或液态废物作为替代燃料或原料，消纳处理工业固体或液态废物的方式）
固化
其他处置（属于未在上面5种指明的处置作业方式外的处置）
废矿井永久性堆存（包括将容器置于矿井）
土地处理（属于生物降解，适合于液态固废或污泥固废）
地表存放（将液态固废或污泥固废放入坑、氧化塘、池中）
生物处理
物理化学处理
经环保管理部门同意的排入海洋之外的水体（或水域）
其他处理方法

【一般工业固体废物倾倒丢弃量】 指报告期内企业将所产生的固体废物倾倒或者丢弃到固体废物污染防治设施、场所以外的量。倾倒丢弃方式如：

（1）向水体排放的固体废物；

（2）在江河、湖泊、运河、渠道、海洋的滩场和岸坡倾倒、堆放和存贮废物；

（3）利用渗井、渗坑、渗裂隙和溶洞倾倒废物；

（4）向路边、荒地、荒滩倾倒废物；

（5）未经环保部门同意作填坑、填河和土地填埋固体废物；

（6）混入生活垃圾进行堆置的废物；

（7）未经海洋管理部门批准同意，向海洋倾倒废物；

（8）其他去向不明的废物；

（9）深层灌注。

一般工业固体废物倾倒丢弃量计算公式是：

一般工业固体废物倾倒丢弃量＝一般工业固体废物产生量－一般工业固体废物贮存量－（一般工

业固体废物综合利用量-其中：综合利用往年贮存量）-（一般工业固体废物处置量-其中：处置往年贮存量）

【危险废物产生量】指当年全年调查对象实际产生的危险废物的量。危险废物指列入国家危险废物名录或者根据国家规定的危险废物鉴别标准和鉴别方法认定的，具有爆炸性、易燃性、易氧化性、毒性、腐蚀性、易传染性疾病等危险特性之一的废物。按《国家危险废物名录》（环境保护部、国家发展和改革委员会 2008 部令第 1 号）填报。

【危险废物综合利用量】指当年全年调查对象从危险废物中提取物质作为原材料或者燃料的活动中消纳危险废物的量。包括本单位利用或委托、提供给外单位利用的量。

【危险废物贮存量】指将危险废物以一定包装方式暂时存放在专设的贮存设施内的量。

专设的贮存设施指对危险废物的包装、选址、设计、安全防护、监测和关闭等符合《危险废物贮存污染控制标准》（GB 18597—2001）等相关环保法律法规要求，具有防扩散、防流失、防渗漏、防止污染大气和水体措施的设施。

【危险废物处置量】指报告期内企业将危险废物焚烧和用其他改变工业固体废物的物理、化学、生物特性的方法，达到减少或者消除其危险成分的活动，或者将危险废物最终置于符合环境保护规定要求的填埋场的活动中，所消纳危险废物的量。处置量包括处置本单位或委托给外单位处置的量。

【危险废物倾倒丢弃量】指报告期内企业将所产生的危险废物未按规定要求处理处置的量。

2．工业企业污染防治投资情况

【污染治理项目名称】指以治理老污染源的污染、"三废"综合利用为主要目的的工程项目名称，或本年完成建设项目"三同时"环境保护竣工验收的项目名称。

【项目类型】按照不同的项目性质，污染治理项目分为 3 类，并给予不同的代码。

1——老工业污染源治理在建项目；2——老工业污染源治理本年竣工项目；

3——建设项目"三同时"环境保护竣工验收本年完成项目。

【治理类型】按照不同的企业污染治理对象，污染治理项目分为 14 类：

1——工业废水治理；2——工业废气脱硫治理；3——工业废气脱硝治理；4——其他废气治理；5——一般工业固体废物治理；6——危险废物治理（企业自建设施）；7——噪声治理（含振动）；8——电磁辐射治理；9——放射性治理；10——工业企业土壤污染治理；11——矿山土壤污染治理；12——污染物自动在线监测仪器购置；13——污染治理搬迁；14——其他治理（含综合防治）。

【本年完成投资及资金来源】指在报告期内，企业实际用于环境治理工程的投资额。投资额中的资金来源，是指投资单位在本年内收到的用于污染治理项目投资的各种货币资金，包括排污费补助、政府其他补助、企业自筹。各种来源的资金均为报告期投入的资金，不包括以往历年的投资。

本年污染治理资金合计=排污费补助+政府其他补助+企业自筹

【竣工项目设计或新增处理能力】设计能力是指设计中规定的主体工程（或主体设备）及相应的配套的辅助工程（或配套设备）在正常情况下能够达到的处理能力。报告期内竣工的污染治理项目，属新建项目的填写设计文件规定的处理、利用"三废"能力；属改扩建、技术改造项目的填写经改造后新增加的处理利用能力，不包括改扩建之前原有的处理能力；只更新设备或重建构筑物，处理利用"三废"能力没有改变的则不填。

工业废水设计处理能力的计量单位为 t/d；工业废气设计处理能力的计量单位为（标态）m^3/h；工业固体废物设计处理能力的计量单位为 t/d；噪声治理（含振动）设计处理能力以降低分贝数表示；电磁辐射治理设计处理能力以降低电磁辐射强度表示（电磁辐射计量单位有电场强度单位：V/m、磁场强度单位：A/m、功率密度单位：W/m^2）。放射性治理设计处理能力以降低放射性浓度表示，废水计量单位为 Bq/L，固体废物计量单位为 Bq/kg。

3. 农业源污染排放及处理利用情况

【治污设施累计完成投资】养殖场/小区当前可用于养殖污染治理设施的累计投资总额。

【新增固定资产】指报告期内交付使用的固定资产价值。对于新建治污设施，本年新增固定资产投资等于总投资；对于改、扩建治污设施，本年新增固定资产投资仅指报告期内交付使用的改、扩建部分的固定资产投资，属于累计完成投资的一部分。

4. 城镇生活污染排放及处理情况

调查范围：

城镇生活包括"住宿业与餐饮业、居民服务和其他服务业、医院和独立燃烧设施以及城镇生活污染源"。

根据《关于统计上划分城乡的暂行规定》（国统字[2006]60 号），城镇是指在我国市镇建制和行政区划的基础上，经以下规定划定的区域。城镇包括城区和镇区。

（1）城区是指在市辖区和不设区的市中，包括：①街道办事处所辖的居民委员会地域；②城市公共设施、居住设施等连接到的其他居民委员会地域和村民委员会地域。

（2）镇区是指在城区以外的镇和其他区域中，包括：①镇所辖的居民委员会地域；②镇的公共设施、居住设施等连接到的村民委员会地域；③常住人口在 3 000 人以上独立的工矿区、开发区、科研单位、大专院校、农场、林场等特殊区域。

城镇生活源的基本调查单位为地（市、州、盟），其所属的区、县城以及镇区数据包含在所在地（市、州、盟）数据中。

【城镇人口】是指居住在城镇范围内的全部常住人口。相关数据以统计部门公布数据为准。

【生活煤炭消费量】指报告期内调查区域用作城镇生活的煤炭总量，包括批发和零售贸易业、餐饮业、居民生活以及生活供热三个部分，以统计部门的能源平衡表数据为准。其中，批发和零售贸易业、餐饮业和居民生活煤炭消费量从能源平衡表直接获取；生活供热煤炭消费量需由能源平衡表中的供热总煤耗扣减工业供热煤耗得到。生活煤炭消费量计算公式为：

生活煤炭消费量 = 批发和零售贸易业、餐饮业煤炭消费量＋居民生活煤炭消费量＋
生活供热煤炭消费量

生活供热煤炭消费量 = 供热总煤耗－工业供热煤耗（环境统计工业调查中 4 430 行业
煤炭消费总量）

【生活用水总量】指报告期内调查区域住宿业与餐饮业、居民服务和其他服务业、医院以及城镇生活等的用水总量（含自来水和自备水），包括居民家庭用水和公共服务用水的总量两个部分（包括自来水和自备水），以城市供水管理部门的统计数据为准。

【居民家庭用水总量】指城镇范围内所有居民家庭的日常生活用水。包括城镇居民、农民家庭、公共供水站用水，以城市供水管理部门的统计数据为准。

【公共服务用水总量】指为城镇社会公共服务的用水。包括行政事业单位、部队营区和公共设施服务、社会服务业、批发零售贸易业、旅馆饮食业等单位的用水。

【城镇生活污水污染物产生量】指报告期内各类生活源从贮存场所排入市政管道、排污沟渠和周边环境的量。

【城镇生活污水污染物排放量】指报告期内最终排入外环境生活污水污染物的量，即生活污水污染物产生量扣减经集中污水处理设施去除的生活污水污染物量。

【城镇生活污水排放量】用人均系数法测算。

如果辖区内的城镇污水处理厂未安装再生水回用系统，无再生水利用量，则

城镇生活污水排放量＝城镇生活污水排放系数×城镇人口数×365

反之，辖区内的城镇污水处理厂配备再生水回用系统，其再生水利用量已经污染减排核查确认，则

城镇生活污水排放量＝城镇生活污水排放系数×城镇人口数×

365－城镇污水处理厂再生水利用量

5．机动车

调查范围：

以直辖市、地区（市、州、盟）为单位调查机动车保有量，以及其中分车型、分年度登记数量；机动车调查对象包括汽车和摩托车。

机动车保有量分类登记数量信息由直辖市、地区（市、州、盟）环保部门协调同级公安交通管理部门负责提供，调查表由各级环保部门填写。

大型载客汽车：车长大于等于 6 米或者乘坐人数大于等于 20 人。乘坐人数可变的，以上限确定。乘坐人数包括驾驶员；

中型载客汽车：车长小于 6 米，乘坐人数大于 9 人且小于 20 人；

小型载客汽车：车长小于 6 米，乘坐人数小于等于 9 人；

微型载客汽车：车长小于等于 3.5 米，发动机汽缸总排量小于等于 1 升；

重型载货汽车：车长大于等于 6 米，总质量大于等于 12 000 公斤；

中型载货汽车：车长大于等于 6 米，总质量大于等于 4 500 公斤且小于 12 000 公斤；

轻型载货汽车：车长小于 6 米，总质量小于 4 500 公斤；

微型载货汽车：车长小于等于 3.5 米，载货质量小于等于 1 800 公斤；

三轮汽车（原三轮农用运输车）：以柴油机为动力，最高设计车速小于等于 50 公里/小时，最大设计总质量不大于 2 000 公斤，长小于等于 4.6 米，宽小于等于 1.6 米，高小于等于 2 米，具有三个车轮的货车；采用方向盘转向，由操纵杆传递动力、有驾驶室且驾驶员车厢后有物品存放的长小于等于 5.2 米，宽小于等于 2.8 米，高小于等于 2.2 米。

低速载货汽车（原四轮农用运输车）：以柴油机为动力，最高设计车速小于 70 公里/小时，最大设计总质量小于等于 4 500 公斤，长小于等于 6 米，宽小于等于 2 米，高小于等于 2.5 米，具有四个车轮的货车；

普通摩托车：最大设计时速大于 50 公里/小时或者发动机汽缸总排量大于 50 毫升；

轻便摩托车：最大设计时速小于等于 50 公里/小时，发动机汽缸总排量小于等于 50 毫升。

6．污水处理厂

填报范围：

城镇污水处理厂及其他社会化运营的集中污水处理单位，包括专业化的工业废水集中处理设施或单位填报本套表。

污水处理厂包括城镇污水处理厂、工业废（污）水集中处理设施和其他污水处理设施。

城镇污水处理厂：指城镇污水（生活污水、工业废水）通过排水管道集中于一个或几个处所，并利用由各种处理单元组成的污水处理系统进行净化处理，最终使处理后的污水和污泥达到规定要求后排放或再利用的设施。

工业废（污）水集中处理设施：指提供社会化有偿服务、专门从事为工业园区、联片工业企业或

周边企业处理工业废水（包括一并处理周边地区生活污水）的集中设施或独立运营的单位。不包括企业内部的污水处理设施。

其他污水处理设施：指对不能纳入城市污水收集系统的居民区、风景旅游区、度假村、疗养院、机场、铁路车站以及其他人群聚集地排放的污水进行就地集中处理的设施。

污水处理厂调查中，不包括氧化塘、渗水井、化粪池、改良化粪池、无动力地埋式污水处理装置和土地处理系统处理工艺。

【污水处理厂累计完成投资】指至当年末，调查对象建设实际完成的累计投资额，不包括运行费用。

【新增固定资产】指报告期内交付使用的固定资产价值。对于新建污水处理厂，本年新增固定资产投资等于总投资；对于改、扩建污水处理厂，本年新增固定资产投资仅指报告期内交付使用的改、扩建部分的固定资产投资，属于累计完成投资的一部分。

【运行费用】指报告期内维持污水处理厂（或处理设施）正常运行所发生的费用。包括能源消耗、设备维修、人员工资、管理费、药剂费及与污水处理厂（或处理设施）运行有关的其他费用等，不包括设备折旧费。

【污水设计处理能力】指截至当年末调查对象设计建设的设施正常运行时每天能处理的污水量。

【污水实际处理量】指调查对象报告期内实际处理的污水总量。

【再生水利用量】指调查对象报告期内处理后的污水中再回收利用的水量。该水量仅指总量减排核定 COD 或氨氮削减量的用于工业冷却、洗涤、冲渣和景观用水、生活杂用的水量。

【工业用水量】指调查对象报告期内污水再生水利用量中用于工业冷却用水等工业方面的水量。

【市政用水（杂用水）】指调查对象报告期内污水再生水利用量中用于消防、城市绿化等市政方面的水量。

【景观用水量】指调查对象报告期内污水再生水利用量中用于营造城市景观水体和各种水景构筑物的水量。

【污泥产生量】指调查对象报告期内在整个污水处理过程中最终产生污泥的质量。折合为 80%含水率的湿泥量填报。污泥指污水处理厂（或处理设施）在进行污水处理过程中分离出来的固体。

【污泥处置量】指报告期内采用土地利用、填埋、建筑材料利用和焚烧等方法对污泥最终消纳处置的质量。

【土地利用量】指报告期内将处理后的污泥作为肥料或土壤改良材料，用于园林、绿化或农业等场合的处置方式处置的污泥质量。

【填埋处置量】指报告期内采取工程措施将处理后的污泥集中堆、填、埋于场地内的安全处置方式处置的污泥质量。

【建筑材料利用量】指报告期内将处理后的污泥作为制作建筑材料的部分原料的处置方式处置的污泥质量。

【焚烧处置量】指报告期内利用焚烧炉使污泥完全矿化为少量灰烬的处置方式处置的污泥质量。

【污泥倾倒丢弃量】指报告期内不作处理、处置而将污泥任意倾倒弃置到划定的污泥堆放场所以外的任何区域的质量。

7. 生活垃圾处理厂（场）

填报范围：

所有垃圾处理厂（场）填报本表。

垃圾处理厂（场）包括垃圾填埋厂（场）、堆肥厂（场）、焚烧厂（场）和其他方式处理垃圾的处理厂（场）。其中，垃圾焚烧厂（场）不包括垃圾焚烧发电厂，垃圾焚烧发电厂纳入工业源调查。

【生活垃圾处理厂（场）累计完成投资】指至当年末调查对象建设实际完成的累计投资额，不包括运行费用。

【本年新增固定资产】指报告期内交付使用的固定资产价值。对于新建垃圾处理厂（场），本年新增固定资产投资等于总投资；对于改、扩建垃圾处理厂（场），本年新增固定资产投资仅指报告期内交付使用的改、扩建部分的固定资产投资，属于累计完成投资的一部分。

【本年运行费用】指报告期内维持垃圾处理厂（场）正常运行所发生的费用。包括能源消耗、设备维修、人员工资、管理费及与垃圾处理厂（场）运行有关的其他费用等，不包括设备折旧费。

【本年实际处理量】指报告期内对垃圾采取焚烧、填埋、堆肥或其他方式处理的垃圾总质量。

【渗滤液污染物排放量】指报告期内排放的渗滤液中所含的化学需氧量、氨氮、石油类、总磷、挥发酚、氰化物、砷和汞、镉、铅、铬等重金属污染物本身的纯质量。按年排放量填报。

表中所列出的污染物，调查对象要根据监测结果或产排污系数如实填报。

【焚烧废气污染物排放量】指报告期内垃圾焚烧过程中排放到大气中的废气（包括处理过的、未经过处理的）中所含的二氧化硫、氮氧化物、烟尘和汞、镉、铅等重金属及其化合物（以重金属元素计）的固态、气态污染物的纯质量。按年排放量填报。

表中所列出的污染物，调查对象要根据监测结果或产排污系数如实填报。

8. 危险废物（医疗废物）集中处理（置）厂

填报范围：

所有危险废物（医疗废物）集中处（理）置厂填报本表。集中式危险废物处（理）置厂指统筹规划建设并服务于一定区域专营或兼营危险废物处（理）置且有危险废物经营许可证的危险废物处置厂。

【危险废物（医疗废物）集中处（理）置厂累计完成投资】指至当年末，调查对象建设实际完成的累计投资额，不包括运行费用。

【本年新增固定资产】指报告期内交付使用的固定资产价值。对于新建危险废物（医疗废物）处置厂，本年新增固定资产投资等于总投资；对于改、扩建危险废物（医疗废物）处置厂，本年新增固定资产投资仅指报告期内交付使用的改、扩建部分的固定资产投资，属于累计完成投资的一部分。

【本年运行费用】指报告期内维持危险废物处置厂正常运行所发生的费用。包括能源消耗、设备维修、人员工资、管理费及与危险废物处置厂运行有关的其他费用等，不包括设备折旧费。

【危险废物设计处置能力】指调查对象设计建设的每天可能处置危险废物的量。

【本年实际处置危险废物量】指报告期内调查对象将危险废物焚烧和用其他改变危险废物的物理、化学、生物特性的方法，达到减少已产生的危险废物数量、缩小危险废物体积、减少或者消除其危险成分的活动，或者将危险废物最终置于符合环境保护规定要求的填埋场的活动中，所消纳危险废物的量。

【处置工业危险废物量】指调查对象报告期内采用各种方式处置的工业危险废物的总量。医疗废物集中处置厂不得填写该项指标。

【处置医疗废物量】指调查对象报告期内采用各种方式处置的医疗废物的总量。

【处置其他危险废物量】指调查对象报告期内采用各种方式处置的除工业危险废物和医疗废物以外其他危险废物的总质量，如教学科研单位实验室、机械电器维修、胶卷冲洗、居民生活等产生的危险废物。医疗废物集中处置厂不得填写该项指标。

【本年危险废物综合利用量】指报告期内调查对象从危险废物中提取物质作为原材料或者燃料的活动中消纳危险废物的量。

【渗滤液污染物排放量】指报告期内排放的渗滤液中所含的汞、镉、铅、总铬等重金属和砷、氰化物、挥发酚、石油类、化学需氧量、氨氮、总磷等污染物本身的纯质量。按年排放量填报。

表中所列出的污染物，调查对象要根据监测结果或产排污系数核算结果如实填报。

【焚烧废气污染物排放量】指报告期内危险废物焚烧过程中排放到大气中的废气（包括处理过的、未经过处理的）中所含的二氧化硫、氮氧化物、烟尘和汞、镉、铅等等重金属及其化合物（以重金属元素计）的固态、气态污染物的纯质量。按年排放量填报。

9．环境管理

【环保系统机构数/人数】指环保系统行政主管部门及其所属事业单位、社会团体设置情况，包括机构总数和在编人员总数。

【环保行政主管部门机构数/人数】指各级人民政府环境保护行政主管部门设置情况，不包括各类开发区等非行政区环保主管部门。

【环境监测站机构/人数】统计独立设置的核与辐射环境监测机构以外的环境监测机构设置情况。在机构数中应注明加挂了核与辐射环境监测站牌子的机构数，以及未加挂牌子但承担了核与辐射环境监测职能的机构数；在总人数中注明承担核与辐射环境监测职能的人数。

【环境监察机构数/人数】指独立设置的环境执法监察机构设置情况。

【核与辐射环境监测站机构数/人数】统计独立设置的核与辐射环境监测站机构设置情况。

【科研机构数/人数】指独立设置的环境科研机构设置情况。

【宣教机构数/人数】指独立设置的环境宣传教育机构设置情况。

【信息机构数/人数】指独立设置的环境信息机构设置情况。

【环境应急（救援）机构数/人数】指环保部门建立或与其他部门共同建立的专兼职环境应急救援机构数、人数。

【当年受理行政复议案件数】指调查年度内环保部门受理的所有行政复议案件数（含非环保案件），包括已受理但未办结的案件；但不包括非本统计年受理而在本统计年内办理或办结的案件。

【本级行政处罚案件数】

【电话/网络投诉数；电话/网络投诉办结数】开通12369环保举报热线电话和网络的地区，按受理情况填报；未开通地区，可根据当地环保部门的公开举报电话和网络信箱的投诉情况填报。只填报本级受理的投诉数，不填报上级转到下级处理的投诉数。

【来访总数】指各级环保部门接待上访人员的数量。同时统计批次、人次。

【来信、来访已办结数量】

【承办的人大建议】指国家、省、市、县环保部门承办的本年度本级人大代表建议数总和。

【承办的政协提案数】指国家、省、市、县环保部门承办的本年度本级政协提案数总和。

【本级环保能力建设资金使用总额】是指本级使用的、当年已完成的用于提高环境保护监督管理能力和环境科技发展能力的基本建设和购置固定资产的投资。具体包括：各级环境保护行政主管部门、各类环境保护事业单位、各类环境监测站点等的基本建设投资、购置固定资产的投资及监测、监察等环境监管运行保障经费（含办公业务用房和科研用实验室的建设费用；不含行政运行费用、生活福利设施的建设费用）。

环境保护事业单位包括：环境监测、环境监察、环境应急、固体废物管理、环境信息、环境宣传、核与辐射、环境科研以及环保部门驻外和派出机构等。

其中：

（1）监测能力建设资金使用总额是指各级环境监测机构、各类环境监测站点等当年已完成的基本建设投资和购置固定资产的投资总额。

(2) 监察能力建设资金使用总额是指各级环境监察机构、污染源监控中心、污染源自动监控设施等当年已完成的基本建设投资和购置固定资产的投资总额。

(3) 核与辐射安全监管能力建设资金使用总额是指各级核与辐射安全监管机构、核与辐射自动监测站点等当年已完成的基本建设投资和购置固定资产的投资总额。

(4) 固体废物管理能力建设资金使用总额是指各级固体废物管理机构当年已完成的基本建设投资和购置固定资产的投资总额。

(5) 环境应急能力建设资金使用总额是指各级环境应急管理机构当年已完成的基本建设投资和购置固定资产的投资总额。

(6) 环境信息能力建设资金使用总额是指各级环境信息机构、环境信息网络当年已完成的基本建设投资和购置固定资产的投资总额。

(7) 环境宣教能力建设资金使用总额是指各级环境宣教机构、培训基地等当年已完成的基本建设投资和购置固定资产的投资总额。

(8) 环境监管运行保障资金使用总额是指各级监测、监察、核与辐射安全、宣教等机构开展污染源与总量减排监管、环境监测与评估、环境信息等业务发生的环境监管运行保障经费。

【清洁生产审核当年完成企业数】是指本地区在调查年度完成清洁生产审核并通过评估或验收的企业数量。各地区在清洁生产审核的管理一般有两种情况：评估和验收分开实施，则统计该地区当年完成评估的企业数量；评估和验收同时进行，则统计该地区当年完成验收的企业数量。

【强制性审核当年完成数】是指本地区在调查年度完成强制性清洁生产审核并通过评估或验收的企业数量。各地区在强制性清洁生产审核的管理一般有两种情况：评估和验收分开实施，则统计该地区当年完成评估的企业数量；评估和验收同时进行，则统计该地区当年完成验收的企业数量。

【应开展监测的重金属污染防控重点企业数】指按照《重金属污染综合防治"十二五"规划》要求开展监测的重金属污染防控重点企业家数。重金属污染防控重点企业指《重金属污染综合防治"十二五"规划》中的4 452家重金属排放企业，主要考核重点企业废气、废水中重点重金属污染物达标排放情况，若企业关闭则不纳入统计基数。《重金属污染综合防治"十二五"规划》要求各地对4 452家重点企业每两个月开展一次监督性监测，重点企业应实现稳定达标排放。

【重金属排放达标的重点企业数】指重点重金属污染物达标排放的重点企业数。由各地环保部门按照《重金属污染综合方式"十二五"规划》要求对重点企业实施每两个月一次的监督性监测，涉及企业实际排放的铅、汞、镉、铬、砷等5种重点重金属中的一种或多种按排放标准评价其达标排放情况。若一次监测不达标则视为该企业不达标；未开展重金属监督性监测视为不达标。

【已发放危险废物经营许可证数】截至报告期末，由各级环境保护行政主管部门依法审批发放并处于有效期内的危险废物经营许可证数量，包括综合经营许可证和收集经营许可证。

【具有医疗废物经营范围的许可证数】指在"已发放危险废物经营许可证数"指标的统计范围内，核准经营危险废物类别包括HW01医疗废物的危险废物经营许可证数。

【监测用房面积】指开展环境监测工作所需的实验室用房、监测业务用房、大气、水质自动监测用房等的面积。

【监测业务经费】指完成常规监测、专项监测、应急监测、质量保证、报告编写、信息统计等工作所需经费，不含自动监测、信息系统运行费和仪器设备购置费等。

【监测仪器设备台套数及原值总值】指基本仪器设备、应急环境监测仪器设备和专项监测仪器设备等的数量。监测仪器设备的原值总值，是指通过政府采购、公开招投标或其他采购方式购买的各类监测仪器设备的购置金额。

【空气监测点位数】指位于本辖区、为监测所代表地区的空气质量而设置的常年运行的例行监测点位的数量。

【其中：国控监测点位数】指位于本辖区、由国家批准纳入国家城市环境空气质量监测网络的空气监测点位数。

【酸雨监测点位数】指位于本辖区、为了掌握当地、区域或全国酸雨状况，了解酸雨变化趋势而设立的例行采集降水样品的点位数。

【沙尘天气影响空气质量监测点位数】指位于本辖区、为反映监测沙尘天气对我国城市环境空气质量的影响，由环保部专门设立的环境空气质量监测点位数。

【地表水水质监测断面数】指位于本辖区、为反映地表水水质状况而设置的监测点位数。一般包括背景断面、对照断面、控制断面、削减断面等。

【其中：国控断面数】指位于本辖区、由国家组织实施监测的、为反映水体水质状况而设置的监测点位数。

【地表水集中式饮用水水源地监测点位数】指位于本辖区、为反映地表水集中式饮用水水源地水质状况而设置的监测点位数。

【近岸海域环境功能区点位数】指位于本辖区、为反映近岸海域功能区环境质量而布设的监测点位数。

【近岸海域环境质量点位数】指位于本辖区、为反映近岸海域环境质量而布设的环境监测点位数。

【开展污染源监督性监测的重点企业数】指按照相关要求开展污染源监督性监测的重点企业家数，以省、市、县本级实施监督性监测的企业数为准，委托监测的企业由受托方统计。

重点企业：由各级政府部门监控的占辖区内主要污染物排放负荷达到一定比例以上的，以及其他根据环境保护规划或者专项环境污染防治需要而列入监控范围的排污单位，包括国控、省控、市控企业和其他企业数。

污染源监督性监测：环境保护行政主管部门所属环境监测机构对辖区内的污染源实施的不定期监测，包括对污染源污染物排放的抽查监测和对自动监测设备的比对监测。

【已实施自动监控国家重点监控企业数】根据污染源自动监控工作进展情况，至本报告期末在环保部门污染源监控中心已经实现自动监控的国家重点监控企业数。

【已实施自动监控国家重点监控企业中水排放口数】已实施自动监控的国家重点监控企业中，环保部门污染源监控中心能够实施自动监控的水排放口数。

【已实施自动监控国家重点监控企业中气排放口数】已实施自动监控的国家重点监控企业中，环保部门污染源监控中心能够实施自动监控的气排放口数。

【COD 监控设备与环保部门稳定联网数】已实施自动监控的国家重点监控企业中，其 COD 自动监控设备正常运行、自动监控数据（浓度和排放量）能通过数据采集与传输设备与环保部门污染源监控中心稳定联网报送的企业数。

【NH$_3$-N 监控设备与环保部门稳定联网数】已实施自动监控的国家重点监控企业中，其 NH$_3$-N 自动监控设备正常运行、自动监控数据（浓度和排放量）能通过数据采集与传输设备与环保部门污染源监控中心稳定联网报送的企业数。

【SO$_2$ 监控设备与环保部门稳定联网数】已实施自动监控的国家重点监控企业中，其 SO$_2$ 自动监控设备正常运行、自动监控数据（浓度和排放量）能通过数据采集与传输设备与环保部门污染源监控中心稳定联网报送的企业数。

【NO$_x$ 监控设备与环保部门稳定联网数】已实施自动监控的国家重点监控企业中，其 NO$_x$ 自动监控设备正常运行、自动监控数据（浓度和排放量）能通过数据采集与传输设备与环保部门污染源监控中心稳定联网报送的企业数。

【排污费解缴入库户数】指调查年度经对账、实际解缴国库的排污费所对应的户数，同一排污者分期分批计征或解缴排污费的不重复计算户数。

【排污费解缴入库户金额】指调查年度经对账、实际解缴国库的排污费所累计金额。

【自然保护区个数】指全国（不含香港特别行政区、澳门特别行政区和台湾地区）建立的各种类型、不同级别的自然保护区的总数相加之和，应包括国家级、省级和市（县）级自然保护区。

【自然保护区个数国家级】至报告期末，经国务院批准建立的自然保护区的个数。

【自然保护区个数省级】至报告期末，经省（自治区、直辖市）人民政府批准建立的自然保护区的个数。

【自然保护区面积】指全国（不含香港特别行政区、澳门特别行政区和台湾地区）建立的各种类型、不同级别的自然保护区的总面积相加之和，应包括国家级、省级和市（县）级自然保护区。

【自然保护区面积国家级】至报告期末，国家级自然保护区的面积总和。

【自然保护区面积省级】至报告期末，省（自治区、直辖市）级自然保护区的面积总和。

【生态市、县建设个数】

【国家级生态市个数】至报告期末，环境保护部对外正式公告的国家级生态市个数。

【国家级生态县个数】至报告期末，环境保护部对外正式公告的国家级生态县个数。

【省级生态市个数】至报告期末，获得省政府或省环境保护主管部门命名的省级生态市个数。

【省级生态县个数】至报告期末，获得省政府或省环境保护主管部门命名的省级生态市个数。

【农村生态示范建设个数】

【国家级生态乡镇个数】指环境保护部公告命名的国家级生态乡镇数量。

【国家级生态村个数】指环境保护部公告命名的国家级生态村数量。

【国家有机食品生产基地数量】指符合《国家有机食品生产基地考核管理规定》相关要求的有机食品生产基地的数量。

【当年开工建设的建设项目数量】指一切基本建设项目和技术改造项目，包括饮食服务等三产项目（区域开发已列入规划环评管理，不再计入建设项目）。由省级环保部门根据统计部门数据填报。

【执行环境影响评价制度的建设项目数量】指调查年度开工并依法履行环境影响评价制度的建设项目数。由省级环保部门填报。

【当年审批的建设项目环境影响评价文件数量】指调查年度审批的建设项目数（包括新建项目、改扩建项目）。按审批权限统计，分别由国家、省级、地市级、县级环保部门对审批管理的项目进行统计。委托下级审批的项目，由受委托部门统计。

【当年审批的建设项目投资总额】指调查年度审批的建设项目工程总投资的汇总数额。

【当年审批的建设项目环保投资总额】指调查年度审批的建设项目环保投资的汇总数额。

【当年审查的规划环境影响评价文件数量】指调查年度审查的规划环境影响评价文件数量。按审查权限统计，分别由国家、省级、地市级环保部门对负责审查的规划环境影响评价文件进行统计。

【当年完成环保验收项目数】指调查年度完成环保验收的建设项目数。按审批权限统计，分别由国家、省级、地市级、县级环保部门对负责验收管理的项目进行统计。委托下级验收的项目，由受委托部门统计。

【环保验收一次合格项目数】指执行了环保"三同时"制度，经环保部门验收一次合格的建设项目数。

【经限期改正验收合格项目数】指未执行环保"三同时"制度，经限期改正后通过环保验收的建设项目数。

【**工业企业当年完成环保验收项目总投资**】指调查年度工业企业完成环保验收项目工程总投资的汇总数额。

【**工业企业当年完成环保验收项目环保投资**】指调查年度工业企业完成环保验收项目环保投资的汇总数额。

【**突发环境事件次数**】指调查年度发生突发性环境事件的次数。若突发环境事件涉及两个以上省（自治区、直辖市），由事发地省级环保部门负责上报。特别重大、重大、较大、一般环境事件的级别划分参考《突发环境事件信息报告办法》（环保部 17 号令）。

【**当年开展的社会环境宣传教育活动数/人数**】指以环保部、省级和地市级环保部门或环境宣教机构本级名义主办的面向社会的环境宣传教育活动，主要包括重要纪念日环保宣传纪念活动、公众参与环保宣传活动、环保主题大型文艺演出、环保影视作品公映等形式。人数为参加该活动的人数。

【**环境教育基地数**】环保部、省级和地市级环保部门或环境宣教机构命名的环境教育基地。